Handbook of Transdisciplinar

Gertrude Hirsch Hadorn · Holger Hoffmann-Riem ·
Susette Biber-Klemm · Walter Grossenbacher-Mansuy ·
Dominique Joye · Christian Pohl · Urs Wiesmann ·
Elisabeth Zemp
Editors

# Handbook of Transdisciplinary Research

Foreword by Jill Jäger

*Editors*
Gertrude Hirsch Hadorn
ETH Zurich
Zurich
Switzerland

Holger Hoffmann-Riem
Swiss Academies of Arts and Sciences
Berne
Switzerland

Susette Biber-Klemm
University of Basel
Basel
Switzerland

Walter Grossenbacher-Mansuy
Swiss Science and Technology Council
Berne
Switzerland

Dominique Joye
University of Lausanne
Lausanne
Switzerland

Christian Pohl
Swiss Academies of Arts and Sciences
Berne
Switzerland
*and*
ETH Zurich
Zurich
Switzerland

Urs Wiesmann
University of Berne
Berne
Switzerland

Elisabeth Zemp
University of Basel
Basel
Switzerland

ISBN: 978-1-4020-6698-6 (hb)
ISBN: 978-1-4020-6700-6 (pb)
e-ISBN: 978-1-4020-6699-3

Library of Congress Control Number: 2007936990



Printed on acid-free paper.

9 8 7 6 5 4 3 2 1

springer.com

The *Handbook of Transdisciplinary Research* results from a project of

Akademien der Wissenschaften Schweiz
Académies suisses des sciences
Accademie svizzere delle scienze
Academias svizras da las scienzas
Swiss Academies of Arts and Sciences

**td-net for transdisciplinary research,**
**a project of the swiss academies**

c/o SCNAT
Schwarztorstrasse 9
3007 Bern, Switzerland
tdnet@scnat.ch, www.transdisciplinarity.ch

The Editors gratefully acknowledge the financial support of

sc | nat

Swiss Academy of Sciences
Akademie der Naturwissenschaften
Accademia di scienze naturali
Académie des sciences naturelles

Department of
Environmental Sciences

**Swiss**
**National Centre**
**of Competence**
**in Research**
**North-South**

**National Research Programme 52**
**Childhood, Youth an Intergenerational Relationships in a Changing Society**

**NFP 48** Landschaften und Lebensräume der Alpen
**PNR 48** Paysages et habitats de l'arc alpin
**NRP 48** Landscapes and Habitats of the Alps

# Foreword

In a world characterised by rapid change, uncertainty and increasing interconnectedness there is a growing need for science to contribute to the solution of persistent, complex problems. These problems include not only some of the now broadly known environmental issues such as climate change and biodiversity loss, but also related issues such as poverty, security and governance. For all of these problems, progress in finding and implementing solutions has been very slow. The increase in availability of scientific knowledge has not been reflected in decisive action.

It is this mismatch between knowledge and action that lies behind the need for a Handbook of Transdisciplinary Research. As the editors point out in their introduction, a transdisciplinary orientation in research, education and institutions aims to overcome the disconnection between knowledge production, on the one hand, and the demand for knowledge to contribute to the solution of societal problems, on the other hand. This is achieved through transdisciplinary approaches in which researchers from a wide range of disciplines work together with stakeholders. Internationally the term 'transdisciplinary research' is defined in different ways, ranging from a diffuse conceptual term located above individual disciplines, to any research that involves stakeholders. The Handbook contributes to a clarification of both the concept and the term, and shows that the uniqueness of the approach lies in the partnership between members of different disciplines and stakeholders.

Given the progress in transdisciplinary approaches during the past two decades, the Handbook provides an opportunity to learn from projects that have already been carried out. The examples described show how knowledge requirements for problem solving have been met. The conceptual and methodological advances in transdisciplinary approaches, in a variety of problem areas, illustrate the broad applicability of the approach. The structured presentation of the examples will encourage a broad community of researchers, educators and users to explore ways in which the approach can be used and further developed. In particular, the Handbook clearly demonstrates to the research community the special requirements and opportunities of this line of research. The authors come from several continents and their projects deal with a wide range of issues. This is a clear indication that the interest in transdisciplinary approaches is widespread.

While the need for transdisciplinary research in a world facing complex problems of a persistent nature is very evident, we should not underestimate the barriers that will have to be overcome in order to turn this emerging form of research into a mainstream endeavour. There are barriers within the scientific community where many scientists prefer to continue their basic research and not confront issues and questions raised by non-scientists. While such basic research will remain important, tackling complex issues of concern to the public and the policy-makers will need input from scientists and non-scientists, resulting in a different type of research. There are also barriers within the area of research funding, where traditional funding agencies often struggle with the notion of participatory research and whether it is 'scientific' or not.

The examples in this book show that participatory, interdisciplinary research can contribute to the solution of complex persistent problems. Such research succeeds by building joint visions of the issue of concern, by finding a common language, by jointly discussing the trade-offs that result from particular choices, and above all through collaborative *learning*. The projects succeed through effective project management, which can deal with, and profit from, the different backgrounds of the participants. With the Handbook as a guide, the emerging community of peers in transdisciplinary research can be strengthened and expanded, so that the bridges between knowledge and action become stronger and progress can be made in tackling major issues faced by society.

Jill Jäger
Sustainable Europe Research Institute
Vienna, Austria

# Acknowledgements

At its first meeting in 2003, the advisory board of td-net decided to publish a 'Handbook of Transdisciplinary Research'. A key driver for this project was the strong belief that a handbook – assembling contributions on crucial topics in transdisciplinary research and written by scholars from all over the world – would be a big step forward in developing transdisciplinary practices. Although we realised that the task was challenging, we did not expect that it would take us more than three years to produce the result.

We started with a list of around ten core topics and sent out a call to about ninety colleagues, asking for comments and relevant published papers. We are very grateful to the more than thirty researchers who responded; they provided considerable assistance in shaping the ideas and structure of the Handbook. Their input was discussed at an advisory board workshop in March 2004, where we were joined by Wolfgang van den Daele from the Social Science Research Center Berlin (WZB). He was influential in steering us away from an abstract treatment of closely entangled topics towards the practical volume you have in your hands.

The editors – comprising members of the advisory board and collaborators from the td-net office – decided to base the Handbook on research projects in various problem-fields, structured according to the phases of a transdisciplinary research process. As a consequence, we searched for projects in a broad range of areas and developed a template for the contributions. About 20 colleagues reviewed proposals from project-teams and helped us select the ones that would best suit the Handbook. The structure of the Handbook developed further as draft contributions came in. It became clear to us that an additional section on cross-cutting issues in transdisciplinary research would help the analysis and sharing of research experiences. We therefore invited additional authors to join the project. In writing our introductory sections, we editors benefited considerably from Theo Koller's comments, and during the final editing from Gabriele Bammer's advice.

We are especially grateful to Paul Roos from Springer publishers for his enthusiastic support for the Handbook. He and the anonymous reviewers provided valuable advice on making the book more accessible to readers. Su Moore's English language editing was outstanding in improving the texts, most of which were written by non-native speakers. In the final stages of preparation of the chapters for publication, Betty van Herk from Springer was a reliable font of information and, in our team Manuela Gähwiler, with the support of Michiel Fehr, Urs Allenspach and Benji Sunarjo, edited and finalised the texts, figures and tables with never-ending patience.

In the process of shaping the many contributions into a coherent book, accessible to readers inside and outside the field of transdisciplinarity, we asked a lot of the authors. We are very grateful for their willingness to share the process of writing this Handbook of Transdisciplinary Research with us.

The Handbook was made possible by generous funding from a variety of sources. We are particularly grateful to the Swiss Academy of Sciences (SCNAT). Important financial and scientific support was received from the Swiss National Centre of Competence in Research (NCCR) North–South. Financial support was also received from the Swiss National Research Programme 'Landscapes and Habitats of the Alps' (NRP 48) and the Swiss National Research Programme 'Childhood Youth and Intergenerational Relationships in a Changing Society' (NRP 52), as well as from the Department of Environmental Sciences, ETH Zurich.

Gertrude Hirsch Hadorn
Holger Hoffmann-Riem
Susette Biber-Klemm
Walter Grossenbacher-Mansuy
Dominique Joye
Christian Pohl
Urs Wiesmann
Elisabeth Zemp

# Contents

# Contributors

**Peter Baccini** ETH Zurich, Zurich, Switzerland

**Gabriele Bammer** National Centre for Epidemiology and Population Health, ANU College of Medicine and Health Sciences, The Australian National University, Canberra, ACT, Australia; Hauser Center for Nonprofit Organizations, Harvard University, Cambridge, MA, USA; Integration and Implementation Sciences Network

**Mahamat Béchir** Centre de Support en Santé Internationale, N'Djaména, Chad

**Matthias Bergmann** Institute for Advanced Study, Berlin, Germany; Institute for Social-Ecological Research (ISOE), Frankfurt am Main, Germany

**Hans Peter Bernhard** University of Basel, Basel, Switzerland

**Susette Biber-Klemm** Program Sustainability Research, University of Basel, Basel, Switzerland

**Bassirou Bonfoh** Swiss Tropical Institute (STI), Basel, Switzerland; Centre Suisse de Recherches Scientifiques, Abidjan, Côte d'Ivoire

**Joseph Bonnemaire** Etablissement National d'Enseignement Supérieur Agronomique de Dijon, Dijon, France

**Patricia Burkhardt-Holm** Programme Man – Society – Environment (MGU), Department of Environmental Sciences, University of Basel, Basel, Switzerland

**Andrea Büchler** Faculty of Law, University of Zurich, Zurich, Switzerland

**Nuria Castells** United Nations Conference on Trade and Development, Geneva, Switzerland

**Carole Després** École d'Architecture, Université Laval, Québec, Canada

**Colette Diguimbaye** Laboratoire de Recherches Vétérinaires et Zootechniques de Farcha, N'Djaména, Chad

**Ottmar Edenhofer** Potsdam Institute for Climate Impact Research (PIK), Potsdam, Germany

**Aant Elzinga** Department of History of Ideas and Theory of Science, University of Göteborg, Göteborg, Sweden

**Andrée Fortin** Département de Sociologie, Université Laval, Québec, Canada

**Silvio Funtowicz** European Commission – Joint Research Centre, Ispra, Italy

**Elise Gatti** École d'Architecture, Université Laval, Québec, Canada

**Walter Grossenbacher-Mansuy** Centre for Technology Assessment (TA-SWISS), Swiss Science and Technology Council, Berne, Switzerland

**Ramon Guardans** National Reference Center for Persistent Organic Pollutants, Ministry of the Environment, Madrid, Spain

**Johannes Heeb** seecon gmbh, Wolhusen, Switzerland

**Hermann Held** Potsdam Institute for Climate Impact Research (PIK), Potsdam, Germany

**Karin E. Hindenlang** Swiss Federal Institute for Forest, Snow and Landscape Research (WSL), Birmensdorf, Switzerland; Grün Stadt Zürich, Zurich, Switzerland

**Gertrude Hirsch Hadorn** Department of Environmental Sciences, ETH Zurich, Zurich, Switzerland

**Holger Hoffmann-Riem** td-net, Swiss Academies of Arts and Sciences, Berne, Switzerland

**Kirsten Hollaender** Department of Sociology, University of Groningen, Groningen, The Netherlands

**Bernard Hubert** Institut National de la Recherche Agronomique (INRA), Paris, France; Ecole des Hautes Etudes en Sciences Sociales, Paris, France

**Thomas Jahn** Institute for Social-Ecological Research (ISOE), Frankfurt am Main, Germany

**Florent Joerin** École Supérieure d'Aménagement et de Développement, Université Laval, Québec, Canada

**Dominique Joye** Interdisciplinary Institute of Life Trajectories (ITB), University of Lausanne, Lausanne, Switzerland

**Ruth Kaufmann-Hayoz** Interdisciplinary Center for General Ecology (IKAÖ), University of Berne, Berne, Switzerland

**Boniface P. Kiteme** Centre for Training and Integrated Research in Arid and Semi-arid Lands Development (CETRAD), Nanyuki, Kenya

**Julie Thompson Klein** Department of Interdisciplinary Studies, Wayne State University, Detroit, MI, USA

**Maria S. Kopp** Institute of Behavioral Sciences, Semmelweis University, Budapest, Hungary

**Wolfgang Krohn** Institute for Science and Technology Studies, Bielefeld University, Bielefeld, Germany

**Marie Céline Loibl** Austrian Institute for Applied Ecology, Vienna, Austria

**Anne Luginbühl** Institute of Infectious Diseases and Institute of Geography, University of Berne, Berne, Switzerland; Federal Veterinary Office (FVO), Berne, Switzerland

**Bruno Messerli** Institute of Geography, University of Berne, Berne, Switzerland

**Paul Messerli** Institute of Geography, University of Berne, Berne, Switzerland

**Michel Meuret** Institut National de la Recherche Agronomique, Avignon, France

**GianPiero Moretti** École d'Architecture, Université Laval, Québec, Canada

**Franz Oswald** ETH Zurich, Zurich, Switzerland

**Moustapha Ould Taleb** Institut National de Recherches en Santé Publique, Nouakchott, Mauritania; Swiss Tropical Institute (STI), Basel, Switzerland

**Marianne Penker** Department of Economics and Social Sciences, University of Natural Resources and Applied Life Sciences, Vienna, Austria

**Pasqualina Perrig-Chiello** Faculty of Human Sciences, University of Berne, Berne, Switzerland

**Bettina F. Piko** Department of Psychiatry, Behavioral Science Group, University of Szeged, Szeged, Hungary

**Christian Pohl** td-net, Swiss Academies of Arts and Sciences, Berne, Switzerland; Department of Environmental Sciences, ETH Zurich, Zurich, Switzerland

**Jerome Ravetz** James Martin Institute for Science & Civilization, University of Oxford, Oxford, UK

**Arie Rip** University of Twente, Enschede, The Netherlands

**Michel Roux** Swiss Federal Institute for Forest, Snow and Landscape Research (WSL), Birmensdorf, Switzerland; SVIAL, Zollikofen, Switzerland

**Esther Schelling** Swiss Tropical Institute (STI), Basel, Switzerland; International Livestock Research Institute, Nairobi, Kenya

**Roland W. Scholz** Institute for Environmental Decisions (IED), Natural and Social Science Interface, ETH Zurich, Zurich, Switzerland

**Markus Schwaninger** Institute of Management, University of St. Gallen, St. Gallen, Switzerland

**Rainer J. Schweizer** Faculty of Law, University of St. Gallen, St.Gallen, Switzerland

**Heidi Simoni** Institute Marie Meierhofer, Zurich, Switzerland

**Marcel Tanner** Swiss Tropical Institute (STI), Basel, Switzerland

**Silvia Ulli-Beer** Research Department General Energy, Paul Scherrer Institute (PSI), Villigen, Switzerland; Interdisciplinary Center for General Ecology (IKAÖ), University of Berne, Berne, Switzerland

**Geneviève Vachon** École d'Architecture, Université Laval, Québec, Canada

**Lorrae van Kerkhoff** National Centre for Epidemiology and Population Health, ANU College of Medicine and Health Science, The Australian National University, Canberra, ACT, Australia; Integration and Implementation Sciences Network

**Alexander I. Walter** Institute for Environmental Decisions (IED), Natural and Social Science Interface, ETH Zurich, Zurich, Switzerland

**Arnim Wiek** Institute for Environmental Decisions (IED), Natural and Social Science Interface, ETH Zurich, Zurich, Switzerland

**Urs Wiesmann** Centre for Development and Environment (CDE), Institute of Geography, University of Berne, Berne, Switzerland

**Arnold Wilts** Department of Public Administration & Organization, Free University, Amsterdam, The Netherlands

**Kaspar Wyss** Swiss Tropical Institute (STI), Basel, Switzerland

**Hans Karl Wytrzens** Department of Economics and Social Sciences, University of Natural Resources and Applied Life Sciences, Vienna, Austria

**Elisabeth Zemp** Institute for Social and Preventive Medicine, University of Basel, Basel, Switzerland

**Jakob Zinsstag** Swiss Tropical Institute (STI), Basel, Switzerland

# Part I
# Introduction

# Chapter 1
# Idea of the Handbook

**Holger Hoffmann-Riem, Susette Biber-Klemm, Walter Grossenbacher-Mansuy, Gertrude Hirsch Hadorn, Dominique Joye, Christian Pohl, Urs Wiesmann and Elisabeth Zemp**

**Abstract** Transdisciplinary orientations in research, education and institutions try to overcome the mismatch between knowledge production in academia, and knowledge requests for solving societal problems. Addressing societal knowledge demands by designing research processes in a transdisciplinary way has several major implications. It becomes necessary to transgress boundaries between different academic cultures, such as between the humanities and the natural sciences. Furthermore, researchers have to step into problem fields and engage in mutual learning with people in the life-world. In doing so, disciplinary standards of knowledge production are sacrificed. Therefore, it is necessary to develop a state of the art for transdisciplinary forms of research. This is best done by learning from experiences. The Handbook is intended to enable learning from exemplary experiences in research and to provide a more systematic account of some cross-cutting issues. This chapter describes the idea behind the Handbook and the contents of the Handbook.

**Keywords:** Networks · Research programmes · Case studies · Cross-cutting Issues · Paradigm

## 1.1 Transdisciplinary Research

Both in the North and in the South societies are developing towards knowledge societies. Research is becoming an integral component of innovation and problem solving strategies in the life-world, affecting not only the private sector, public agencies, and civil society, but also personal life. Scientific knowledge production in these societal contexts involves a wide range of different disciplines, ranging from natural sciences and engineering to social sciences, economics and humanities.

The large-scale application of scientific knowledge in the life-world has had both beneficial and harmful consequences. One of the reasons for this is the

✉ H. Hoffmann-Riem
td-net, Swiss Academies of Arts and Sciences, Berne, Switzerland
e-mail: holger@hoffmann-riem.ch

fragmentation of scientific knowledge. As Brewer (1999) stated: '*The world has problems, but universities have departments*.' The pursuit of research within university departments has given rise to the ongoing specialisation of disciplines and thematic fields with fuzzy, somewhat arbitrary, shifting boundaries. The high degree of compartmentalisation of scientific knowledge is due to two interacting factors. Institutional structures and incentives in academia result in an 'ethnocentrism of disciplines' (Campbell, 1969). At the same time the concepts, theories and methods used in basic research are becoming ever more sophisticated.

Knowledge is requested by actors in the life-world to address serious problems, such as poverty, sickness, crime and environmental degradation, from the local to the global scale. These requests have led to calls for a 'new social contract for science' (Lubchenco, 1998) or a 'new commitment of science' (Cetto, 2000) to tackle the problems of the 21st century. To solve or at least mitigate these problems, sufficient research capacities are required. In addition to that, research practices, institutions, education and the underlying conception of science need to be transformed.

The 'ideal' of scientific knowledge underlying basic research is concerned with universal concepts, theories, models and methods. Therefore, basic research is built on an idealisation of the multitude of phenomena and relationships. This is also the case for basic research in the humanities, for instance in ethics. It is unclear to what extent idealised theories and models are able to adequately describe, explain, understand, reflect and handle the reality of concrete problem situations. Nevertheless, the use of scientific knowledge in the private sector, in public agencies and the life-world has been seen as a one-way transfer of allegedly reliable instrumental knowledge from experts to 'ignorant' users. Serious harm has been caused by ignoring the uncertainty of scientific knowledge, by neglecting the users' knowledge, and by failing to consider contextual conditions of applications. In view of these risks, which were often unknown until damage occurred, the model of progress in modern civilisation has been questioned by social movements, intellectuals from the humanities and by concerned scientists.

It is in this context that ideas about transdisciplinarity emerged in the 1970s. Transdisciplinary orientiations in research, education and institutions try to overcome the mismatch between knowledge production in academia, on the one hand, and knowledge requests for solving societal problems, on the other. Transdisciplinary research, therefore, aims at identifying, structuring, analysing and handling issues in problem fields with the aspiration '*(a) to grasp the relevant complexity of a problem (b) to take into account the diversity of life-world and scientific perceptions of problems, (c) to link abstract and case-specific knowledge, and (d) develop knowledge and practices that promote what is perceived to be the common good*' (Pohl and Hirsch Hadorn, 2007).

Knowledge requests from the life-world are only partly met by describing, analysing and interpreting empirical processes that influence problems and by giving projections into the future. This so-called systems knowledge (ProClim, 1997) has to be linked with answers to further, and equally important, questions of a different type: What are the subjects of concern? Where are changes needed in public institutions, in the private sector and in individual ways of acting? What policies will be effective for improving, mitigating and solving these problems in the private sector and civil society? This so-called target knowledge is about the needs, interests and reasons of various practitioners and stakeholders who may be affected directly or indirectly. Their needs have to be taken into account to develop better societal practices. Furthermore, so-called transformation knowledge is needed to improve existing practices. It addresses questions about technical, social, legal, cultural and other possible means of acting that aim to transform existing practices and introduce desired ones. In the case of transformation knowledge, established technologies, regulations, practices and power relations must be taken into account.

Attention has to be paid to the mutual interrelations between these three forms of knowledge. The investigation of systemic processes has to be related to the societal purposes and practices on which they depend and which they influence. If the needs, interests and reasons of practitioners and stakeholders are ignored or if systemic processes are not taken into account in developing transformation knowledge, major unexpected obstacles and unintended effects may result.

Addressing the threefold societal knowledge demands by designing research processes in a transdisciplinary way has several major implications. It becomes necessary to transgress boundaries between disciplines and especially between different academic cultures, such as between the humanities and the natural sciences. Furthermore, the doors of laboratories and libraries must be opened and researchers have to step into problem fields and engage in mutual learning with people in the life-world. In doing so, academic standards of knowledge production and quality control criteria are sacrificed. Therefore, it is necessary to develop a state of the art and to define quality criteria that are appropriate for transdisciplinary forms of research. This is best done by learning from experiences.

## 1.2 The Background of the Handbook

In the case of transdisciplinary research, learning from experience requires special effort because of the heterogeneity of the fields and the participants. There is a lack of strong institutional structures for transdisciplinary research that are necessary for establishing scientific communities in which a state of the art can be developed. Transdisciplinary research groups enter into a variety of problem fields such as health and environmental issues, technology development, and social conflicts. In addition, cooperation and mutual learning between participants from various institutional contexts and heterogeneous backgrounds take place on a project related basis. Frequently, they do not outlast the temporary context of the research project or programme. As a consequence of professional mobility and lack of systematisation, the lessons learned on the job are seldom passed on to others for capacity building. Furthermore, the diversity of research problems makes it necessary to consider where and how systematisation can be recommended.

In view of this situation, transdisciplinarity-net (td-net) was initiated by the Swiss Academies of Arts and Sciences as a forum for transdisciplinary research in order to facilitate capacity building and advancement in transdisciplinary research (www.transdisciplinarity.ch). Transdisciplinary efforts in Switzerland go back to the 1970s when a Man and Biosphere Programme was initiated in the Swiss Alps, based on an integrated systems approach and cooperation among scientists, inhabitants and decision-makers (Chapter 3). In 1979 a National Research Programme Man and Biosphere was initiated, funded by the Swiss National Science Foundation. During the 1990s, the Swiss Priority Programme Environment (SPPE) – also funded by the Swiss National Science Foundation – prepared the ground for a deepening of transdisciplinary research in Switzerland (www.sppe.ch). About 250 research projects covered subjects such as climate, biodiversity, soil, waste, social and economic issues and the North–South dimension. It was within the framework of SPPE that a discussion forum on transdisciplinarity was initiated. International workshops and cooperations among German, Austrian and Swiss researchers took place. Towards the end of the programme, the International Transdisciplinarity Conference in 2000 was initiated (Klein et al., 2001). It was at this conference that the Swiss Academic Society for Environmental Research and Ecology (SAGUF) started sagufnet as a network for transdisciplinary research. Sagufnet was transformed into td-net by the Swiss Academies of Arts and Sciences in 2003. Td-net is currently deepening international cooperations by contributing to the programmes, 'Socio-Ecological Research' (http://www.bmbf.de/en/972.php) in Germany, and 'proVision' (http://www.bmbwk.gv.at/forschung/fps/provision.xml) in Austria, and by maintaining close ties to the Integration and Implementation Sciences network (http://www.anu.edu.au/iisn/) in Australia, to mention a few.

The advisory board of td-net decided to propose 'Principles for Designing Transdisciplinary Research' (Pohl, 2007) and to edit this 'Handbook of Transdisciplinary Research' as a core means for enhancing transdisciplinary research. The Principles highlight the challenges in transdisciplinary projects, and suggest means of designing and shaping the research process to meet these challenges. Thus, the Principles provide the conceptual basis and structure for the Handbook. The Handbook is intended to enable learning from exemplary experiences in research and to provide a more systematic account of some cross-cutting issues. Due to the context in which td-net has emerged as a Swiss institution, a considerable number of authors are based in Switzerland. But the fact that authors come from several continents and are dealing with a broad range of issues indicates an emerging international college of peers in transdisciplinary research.

## 1.3 The Structure of the Handbook

The Handbook consists of six parts. The introductory part describes the idea behind the Handbook and the developments leading to transdisciplinarity as a form of research. The following three parts assemble the research projects, which are grouped according to the three phases of transdisciplinary research: (1) problem identification and structuring, (2) problem analysis, and (3) bringing results to fruition. The next part deals with cross-cutting issues. Contributions describe overarching challenges in transdisciplinary research and draw upon both the general literature and upon the projects in the handbook. The final part provides a summary and outlook. It includes explanations of the core terms used in transdisciplinary research and a synthesis of the Handbook's contents in 15 propositions for enhancing transdisciplinary research.

While handbooks in well-established fields usually focus on methods, the core of this Handbook consists of a collection of research projects. Although scientific research is characterised by systematic procedures, transdisciplinary research cannot be learned using a collection of methods. Transdisciplinary research needs concrete paradigms to help researchers understand problems in context and to structure the research accordingly. This will eventually give rise to formalisation. As Don Rosenblum points out in his analysis of interdisciplinary conceptions in basic and applied research (Rosenblum, 1997) little has been written about research practices. An important reason for this, he argues, is the absence of paradigms as 'shared examples' that enable both problem recognition and problem solving efforts by grounding concepts, methods, tools and standards for research as well as the building a scientific community and its institutions.

The same applies to transdisciplinary research: there is a need for projects that have the potential to be approved by a college of peers. Shared experiences contribute to the systematisation and formalisation of concepts, methods and tools as well as standards. The Handbook is designed to overcome this obstacle: it collects transdisciplinary projects that have the potential to become paradigmatic research examples because they illustrate how the knowledge requirements for problem solving in the life-world are met. The chapters show how the challenges of problem identification, problem structuring, problem analysis and bringing results to fruition can be tackled. In this way, they contribute to a grounded systematisation and formalisation of research practices and to their in-depth analysis. The projects demonstrate the potential of transdisciplinary research in a broad range of problem fields such as environment and health, urban and landscape development, and new technologies. There is a variety of subjects, approaches, methods and tools used. Whether, and in what sense, there will one day emerge a general methodology of transdisciplinary research is not yet clear (Bammer, 2005).

The project based chapters are arranged on the basis of the project phase (problem identification and structuring, problem analysis, and bringing results to fruition) that we consider to be particularly interesting in the corresponding project. This does not imply that the other two phases were not part of the project. Each chapter is of interest in several different respects. Authors describe how they have addressed the genesis of a problem and possible further developments. They explain how the need for change was determined, how desired goals were selected, or how technical, social, legal, cultural and other possible means of acting were aimed at transforming existing approaches. At the end of each of these chapters, authors point out results and develop recommendations for further research relevant to other problem fields.

The chapters on cross-cutting issues deal with the overarching challenges that are relevant for most transdisciplinary projects: participation, values and uncertainties, learning from case studies, project management, education, and finally integration.

The authors describe important lines of thinking that have shaped research on these issues. They structure recent developments and directions and comment on how some of the projects described in the Handbook deal with the issues.

As a guide to the reader the following figure (Fig. 1.1) matches the chapters on cross-cutting issues and project chapters. The columns of the table represent the cross-cutting issues. The issue of integration has been split into two columns to provide more specific information. The rows of the table contain the individual project chapters. Grey cells indicate that an issue is discussed in detail in a particular chapter.

The summary and outlook section encompasses a list of core terms used in transdisciplinary research and a synthesis of the Handbook's contents in 15 propositions for enhancing transdisciplinary research. The description of the core terms provides a conceptual synthesis of transdisciplinary research, while the propositions summarise strategic elements.

## 1.4 The Contents of the Handbook

In this section, we briefly describe the chapters in the following three parts of the Handbook: research projects, cross-cutting issues, and finally a 'summary and outlook'.

We consider eight projects to be of special interest with regard to the way they address problem identification and problem structuring: two large-scale programmes on rural development and natural resources, two projects on urban development, two projects on ecological (biodiversity) issues, and two projects on emerging sciences and technologies. These projects address the challenge of defining the proper object for research and of establishing an adequate conceptional framework from different angels.

In Chapter 3, Bruno Messerli and Paul Messerli describe UNESCO's worldwide Man and Biosphere programme, which was launched in the 1970s. This project can be seen as a pioneer project within transdisciplinary research, since it was based on transdisciplinary thinking long before the term 'transdisciplinarity' became popular. The Swiss Man and Biosphere programme stimulated research that integrated natural and social sciences in order to understand and shape the socio-economic development in mountainous regions, while taking into account the ecological carrying capacity. The Swiss programme was based on a recursive approach to problem identification and problem structuring. It combined a careful analysis of the system dynamics with target knowledge about land use, desired aesthetic and recreational qualities and transformation knowledge about key factors that could be utilised to develop management options. In the 1980s, a participatory process was initiated to develop a long-term strategy. Since then, similar concepts such as the Syndrome Mitigation Approach have been applied in developing countries and in mountain research programmes, and since the 1990s, with a particular emphasis on global change.

| Projects | | 22 Participation | 23 Values and Uncertainties | 24 Learning from Case Studies | 25 Management | 26 Education | 27 Integration: Concepts | 27 Integration: System Analysis |
|---|---|---|---|---|---|---|---|---|
| **Problem Identification and Problem Structuring** | | | | | | | | |
| 3 | From Local Projects to Global Change | ■ | ■ | | | | ■ | ■ |
| 4 | Sustainable River Basin Management | ■ | ■ | | | ■ | | ■ |
| 5 | Designing the Urban | | ■ | | | | ■ | ■ |
| 6 | CITY:*mobil*: | ■ | | | ■ | | ■ | |
| 7 | Shepherds, Sheep and Forest Fires | ■ | ■ | ■ | | | ■ | |
| 8 | Fischnetz | ■ | | ■ | ■ | | | ■ |
| 9 | Nanoscience and -Technologies | ■ | ■ | ■ | | | | ■ |
| 10 | Chimeras | | ■ | ■ | | | ■ | |
| **Problem Analysis** | | | | | | | | |
| 11 | Multilateral Environmental Agreements | ■ | ■ | | | | | ■ |
| 12 | Climate Protection vs. Economic Growth | | ■ | | | | ■ | ■ |
| 13 | Policy Analysis and Design | ■ | | | | | | ■ |
| 14 | Regional Development Strategies | ■ | ■ | | ■ | ■ | | ■ |
| 15 | Evaluating Landscape Governance | | ■ | ■ | | | ■ | |
| 16 | Children and Divorce | | | ■ | | | ■ | |
| **Bringing Results to Fruition** | | | | | | | | |
| 17 | Health Services for Nomadic Pastoralists | ■ | | | ■ | | ■ | |
| 18 | Water Associated Infection Risks | ■ | | ■ | | | ■ | |
| 19 | Behavioural Sciences in the Health Field | | ■ | | | ■ | ■ | |
| 20 | Coexistence of Ungulates and Trees | ■ | ■ | | ■ | | | ■ |
| 21 | Retrofitting Postwar Suburbs | ■ | ■ | | ■ | | ■ | |

Cross-cutting Issues

**Fig. 1.1** Cross-cutting issues (chapters 22–27) in projects (chapters 3–21)

The programme 'Sustainable River Basin Management in Kenya: Balancing Needs and Requirements' (Chapter 4) by Boniface P. Kiteme and Urs Wiesmann, focuses on integrated water resources management in the upper Ewaso N'giro catchment of Mount Kenya in Kenya. In this region the socio-economic dynamics and ecological processes over the past decades have resulted in a water crisis that poses a major sustainability problem in the catchment. The authors describe the impact of the societal and political context on the restructuring of the research programme: starting with 'Research for District and Project Planning'; then 'Broader topical and geographical scope to respond to sustainability requirements'; and further implementation and consolidation of knowledge by the combination of scientific knowledge systems and local knowledge system. These steps in recursive problem identification and structuring broadened the understanding of the main challenge of sustainable river basin management and provided useful insights into designing a multistakeholder, multilevel strategy. The approach is validated by the project being extended from the test area to other regions.

The projects on urban development propose frameworks for integrating perspectives of design, engineering and planning with analytical perspectives from the natural and social sciences. 'Designing the Urban: Linking Physiology and Morphology' (Chapter 5) by Peter Baccini and Franz Oswald focuses on a conception of urban development in the Lowlands of Switzerland that meets the requirements of sustainable development. Two types of perceptions were combined when formulating the questions with regard to the fabric of new urban systems: the morphological approach, based on urban planning experience; and the physiological approach, based on natural science, engineering and economics. The authors describe the process of interdisciplinary learning to arrive at a common system approach 'Netzstadt'. The urban system is defined by a limited set of elements, by four basic activities and by five essential system qualities. Methods and design tools for the reconstruction of urban systems aim at reducing the complexity of the mutual relationships between activities, territories and resources. The quality goals for a concrete urban project have to be determined by a participatory political process that ensures support and commitment. This procedure is systematised as the Synoikos method.

'CITY:*mobil*: A Model for Integration in Sustainability Research' (Chapter 6) by Matthias Bergmann and Thomas Jahn was realised in an interdisciplinary cooperation between engineers, planners, sociologists, and economists and was supported by the administration of the two German model cities of Freiburg and Schwerin (CITY:*mobil*). To disentangle the strong connection between 'mobility' and 'automobility' a three dimensional concept of mobility was developed. This was based on the idea of mobility/motion as a fundamental societal relation to nature mobility; of 'spatial mobility' in the sense that transportation is a technical realisation of transport; of 'socio-spatial mobility' in the sense of the movement of people between locations and the social purposes pursued therein; and of 'socio-cultural mobility' in the sense that the distinctive meanings of mobility and means of transportation have consequences for social positioning. Within this framework the research process was guided by the differentiation of interdisciplinary cooperations, by the integration of knowledge from research and practice into concepts of agency; and by the intervention, using the integrated results, into the practical and scientific discourse of urban transportation policy and planning.

The ecological projects have developed procedures for problem framing and structuring and focus on agents and agencies. 'Shepherds, Sheep and Forest Fires: A Reconception of Grazingland Management' (Chapter 7) by Bernard Hubert, Michel Meuret and Joseph Bonnemaire deals with grazing in the Mediterranean rangelands of southern France, which are subject to scrub encroachment. The authors describe five steps for identifying and structuring the research objects in such a way that biological processes are coupled with the farmer's management strategies. Firstly, it is necessary to identify an ecological problem. Secondly, the problem is reformulated with regard to the relevant agents and their practices. Thirdly, the research problem is focused. Fourthly, the problem-solving instrument is designed, and finally (innovative) research questions are addressed that investigate the processes related to the proposed problem-solving instruments. The chapter gives an example of how research on issues of practical relevance develops along the margins of disciplines through interdisciplinary cooperation and confrontation, causing disciplinary frameworks to evolve by inducing a reflexive process in the research activity.

'Fischnetz: Involving Anglers, Authorities, Scientists and the Chemical Industry to Understand Declining Fish Yields' (Chapter 8), described by Patricia Burkhardt-Holm, documents the health status of brown trout and the decline in fish numbers, the identification of causes and suggested measures for correction. The network of participants was central to the integration of already existing data and know-how and a prerequisite for jointly identifying knowledge gaps. The chapter describes the role of a network in formulating hypotheses and research questions and in initiating research projects. Collaboration throughout the project ensured an efficient exchange of results, ideas and conclusions leading to the setting of new priorities and to an agreement on further procedures and proposed measures.

Two projects focus on problem identification and structuring in the field of emerging sciences and technologies. In Chapter 9, Arie Rip describes the application of Constructive Technology Assessment to nanoscience and nanotechnologies. To enable reflexive co-evolution of science, technology and society, it is crucial to bridge the gap between actors in the life-world and nanoscientists and technologists. Interactive reflexive learning is required to develop better technologies in a better society. These activities encompass the mapping and analysis of the ongoing dynamics as well as an articulation of socio-technical scenarios about further developments, impacts and real-time experiments.

In Chapter 10, 'Chimeras and other Human–animal Mixtures in Relation to the Swiss Constitution: A Case for Regulatory Action', Hans-Peter Bernhard and Rainer J. Schweizer analyse a problem in the making: the artificial creation of human–animal mixtures for research and therapeutical purposes. Not surprisingly, normative elements are of crucial importance in this debate, but knowledge of biomedical research agendas is equally essential. To achieve a mutual understanding of the underlying biological facts and the arising normative issues, researchers followed an iterative process. They clarified definitions to achieve a commonly accepted terminology and to reach an agreement on the pertinent problems. Such an understanding is a prerequisite for timely legislation.

We consider six projects of special interest with regard to problem investigation. Four of these describe the use of some kind of systems analysis to produce target knowledge, which reflects the systemic relations of problem development or the systemic relations that play a role in transformation. These chapters demonstrate how feedback about implementation and recurring adaptations of problem structuring can improve the results. The remaining two projects deal with issues of global change and regulation. 'The Development of Multilateral Environmental Agreements on Toxic Chemicals: Integrating the Work of Scientists and Policy Makers' (Chapter 11) by Nuria Castells and Ramon Guardans, is based on examples of recent Multilateral Environmental Agreements concerning environment and health. The authors argue that the procedures established in the framework of these agreements provide a solid international base for stable and effective scientific, industrial and political cooperation. Emphasis is put on involving relevant stakeholders and on institutional innovations from the national to the international and global scale. When developing and using Integrated Assessment Models in the design of policy scenarios, the cost of implementing commitments should be estimated *ex ante*. Transdisciplinary cooperation is a prerequisite for handling complex environmental problems and for designing effective abatement scenarios. *Ex post*, in the implementing phase, cooperation among all stakeholders (scientists, policy makers and civil society) is required.

In 'Climate Protection vs. Economic Growth as a False Trade off: Restructuring Global Warming Mitigation' (Chapter 12), Hermann Held and Ottmar Edenhofer describe methods for Integrated Assessment of climate change mitigation. The project addresses the intertemporally optimal mix of investments into energy efficiency, transformation to renewable energy sources and carbon capture and sequestration. A social optimum is determined, consistent with economic growth theory, in such a way that certain boundary conditions related to the evolution of the climate system are observed. The project integrates paradigms from natural science, economic growth theory and engineering, which shape the public debate by a delicate entanglement of target knowledge and systems knowledge arguments. Therefore, the clarification not only of the validity, but also of the category of those arguments, is addressed. The project identifies stylised climate policies that comply with both systems knowledge and presently competing target knowledge bases of various interest groups and thereby catalyses a societal consensus on climate policy.

Two projects use systems analysis in problem investigation to deal with regional issues of sustainable development. Chapter 13, 'Policy Analysis and Design in Local Public Management: A System Dynamics Approach' by Markus Schwaninger, Silvia Ulli-Beer and Ruth Kaufmann-Hayoz, portrays the strengths and limitations of the computer assisted theory building approach of System Dynamics and Group Model Building. A tested system dynamics simulation model helps to address solid waste management issues. The model structures dynamic interaction between public policies and environmentally relevant behaviour, as well as public management problems, which are important for the design of effective policies. The study generated knowledge about system structure and transformation processes, and a computer based learning environment and a communication tool for political decisions. It is pointed out that transdisciplinary collaboration in action research has proven crucial for the societal relevance of problem oriented research and for societal learning processes.

The chapter, 'Constructing Regional Development Strategies: A Case Study Approach for Integrated Planning and Synthesis' (Chapter 14) by Alexander I. Walter, Arnim Wiek and Roland W. Scholz, presents a transdisciplinary integrated planning and synthesis (TIPS) approach. This is illustrated by a case study on sustainable regional development in a typical central European rural landscape, struggling with problems of structural change and migration. Numerous officials, representatives and inhabitants of the region, scientists and graduate students have been involved throughout the research process. TIPS puts emphasis on an integrated overall project architecture, which starts from a division (faceting) of the case, and then uses system analysis and variant construction for the problem investigation procedure. For implementation, which is closely linked with investigation, the preferences of the stakeholders are evaluated through multicriteria procedures. The results from the different facets are integrated to formulate possible development strategies for the region.

Other approaches for integrating the diversity of relevant perspectives and grasping the complexity of the problem focus on integrating different scientific cultures to investigate empirical effects of policies. Two projects, one on landscape management, the other on divorced parents and their children, address the empirical effects of implemented legal regulations. In Chapter 15, 'Evaluating Landscape Governance: A Tool for Legal-Ecological Assessments', Marianne Penker and Hans Karl Wytrzens describe an interdisciplinary approach for the systematic identification and evaluation of the ecological effects of legal regulations. The concept draws on the fact that legislation has no direct effect on the landscape, but has the ability to influence human activity. The project is therefore based on (legal) socio-economic factors, e.g. awareness of the law among those subject to it, and on (landscape) ecological factors, e.g. the effects of behavioural change on flows of material and energy. The applicability and feasibility has been demonstrated by three case studies. The impact mechanisms that were identified could help to improve the instruments' effectiveness. An *ex ante* evaluation of policies and legal regulations may even help to involve diverse stakeholder views in the process of policy design. This could minimise the unintended destruction of both socio-economic and ecological development options.

In Chapter 16, 'Children and Divorce: Investigating Current Legal Practices and their Impact on Family Transitions', Heidi Simoni, Pasqualina Perrig-Chiello and Andrea Büchler investigate the implementation of the revised Swiss Divorce Law in current judiciary practice and its effects on the course of life of affected children and their parents after the divorce. Data were collected simulataneously at three levels: firstly through analysis of court files and interviews with judges; secondly from written interviews with divorced parents; and thirdly from in-depth interviews with children and parents. The problem investigation had to ensure that all objections and perspectives of jurisprudence, psychology and sociology were incorporated into all parts of the project study and into the respective instruments for data collection. Through the discussion of the legal framework, its implementation and the everyday life of all parties involved, adequate knowledge will be gained about the best interests of the children, and children's rights. There is a close relationship between normative perceptions and persuasions on the one hand, and the acting and experiencing, on the other hand.

We consider five projects of special interest with regard to bringing results to fruition. Three projects deal with the iterative design and implementation of health services and education – two of them in Africa and one in Eastern Europe. The project, 'Towards Integrated and Adapted Health Services for Nomadic Pastoralists and their Animals: A North–South Partnership (Chapter 17) by Esther Schelling, Kaspar Wyss, Colette Diguimbaye, Mahamat Béchir, Moustapha Ould Taleb, Bassirou Bonfoh, Marcel Tanner and Jakob Zinsstag, was undertaken in Chad and starts from the concept of 'one medicine' – promoting a joint approach to human and veterinary medicine in order to develop adapted health and education systems for pastoralists. The strategy is one with strong action research components and comprises five stages: (1) the programme was launched and trust was built up by establishing institutional partnerships, interdisciplinary collaboration and stakeholder dialogue on the basis of accounted principles for North–South collaborations, (2) the inter- and transdisciplinary research into epidemiological pathways to disease and the identification of control strategies that are appropriate to the nomadic way of life, (3) the implementation began with four national workshops to formulate health service priorities and options, to readjust ongoing interventions and to discuss policy issues and strategies for ownership building, (4) joint vaccination campaigns for people and livestock were implemented with the help of local, trained staff with a strong commitment. Information campaigns were followed by an evaluation of the feasibility and costs, and (5) ensured subsequent regional activities and continued research.

Chapter 18, 'Sustainable Prevention of Water Associated Infection Risks: An Awareness Campaign Using Visual Media' by Anne Luginbühl, deals with infections that are associated with water–skin contact. These diseases are prevalent in tropical and subtropical environments. Within the project, environmental, socio-cultural and behavioural factors were examined to develop an effective awareness campaign based on local communication patterns. Qualitative and quantitative methodologies were used to identify the local populations' systems knowledge about infectious diseases. Based on these findings, a visual awareness campaign was developed to target illiterate population groups. To ensure that the information could be understood by the target group, the local population assisted in the design of the health education material.

In Chapter 19, 'Behavioural Sciences in the Health Field: Integrating Natural and Social Sciences', Bettina F. Piko and Maria S. Kopp describe the application of behavioural health sciences to increase understanding of health issues, and aid both prevention and therapy. The project is based on a framework called the 'biopsychosocial' model of human processes, which considers external and internal, genetic and environmental, somatic and psychosocial factors to be equally important in determining health and in inducing disease. The chapter outlines three elements of an integrated strategy, each building on the other with multiple feedback systems: first, to create an integrated course under the name 'Behavioural Science' in which students in the field of medicine and health sciences can be provided with a set of social and behavioural sciences, as applied to medicine; second, to develop a health status monitoring system by means of two surveys (Hungarostudy, which collects data on the health status of the Hungarian adult population; and South Plain Youth Study, which gathers data on the health status of the adolescent population); and third, to apply theoretical knowledge and empirical research results in the field of practice, in this case, practical prevention programmes and skills development training.

Two projects focus on models for developing a common understanding of the problem and for consensus building among stakeholders about effective measures. Chapter 20, 'Sustainable Coexistence of Ungulates and Trees: A Stakeholder Platform for Resource Use Negotiations' by Karin E. Hindenlang, Johannes Heeb and Michel Roux, addresses browsing by ungulates as a major problem for tree regeneration in Alpine forests. This chapter describes a 'platform for resource use negotiation' for solving such a regional forest–wildlife conflict in a mountainous environment. Collaborative learning was used to develop a common understanding of the systemic structure of the problem and an agreement on the management aims. Preconditions for successful solution include the recognition of a conflict situation combined with strong interest from the concerned parties – foresters, hunters, farmers and conservationists – to solve conflicts. Their representatives must be willing to participate according to the agreed principles of communication. Process related success factors include the creation of mutual trust; sufficient time for building a broad common knowledge base about the problem; and the development of a common understanding of the systemic structure with the help of mental modelling. A transparent procedure that motivates participants to cooperate further was also essential.

In Chapter 21, 'Retrofitting Postwar Suburbs: A Collaborative Design Process', Carole Després, Andrée Fortin, Florent Joerin, Geneviève Vachon, Elise Gatti and GianPiero Moretti tell of the five year research and planning process that focused on the future of Quebec City's aging postwar suburbs. The project was initiated by the Interdisciplinary Research Group on Suburbs (GIRBa). GIRBa's ultimate aim was to generate knowledge that could be applied to urban design, planning, management and policy making. The transdisciplinary and collaborative research process was adopted to establish a consensual diagnosis, define planning objectives and to design a master plan for Quebec City's first ring suburbs. The chapter outlines the procedure and evaluates the success of the transdisciplinary process on reaching, and sharing, a better understanding of postwar suburbs, on the planning and policy orientations of the involved institutions, as well as on generating ongoing collaborative work between participants.

The 'cross-cutting issues' section of the Handbook focuses on the overarching challenges of transdisciplinary projects: participation, values and uncertainties, learning from case studies, project management, education, and integration. Chapter 22, 'Participation', by Aant Elzinga, begins with a historical perspective on the development of transdisciplinarity and participation. Based on the general literature and of the Handbook's project chapters, this chapter addresses the role of participiption in problem identification and structuring, learning and analysis, and in the implementation phase of projects, and illustrates how participation can fulfill a range of purposes. It also addresses the need to problematise the concept of participation and ends with the question of who is empowered by participation.

In Chapter 23, Silvio Funtowicz and Jerome Ravetz examine the issue of values and uncertainties in policy related research. They distinguish between three different forms of inquiry: 'applied science', 'professional consultancy', and 'post-normal science', which is characterised by a high level of uncertainty and undeniable value-loading. In such a context, even the best professional expertise becomes insufficient. There is an urgent need for dialogue with the 'extended peer community' and this requires a new set of methodologies.

Since many research projects were based on case studies, the question of what can be learned from case studies in order to develop valid scientific knowledge arises. This question is addressed in Chapter 24 by Wolfgang Krohn. It begins with the distinction between ideographic and nomothetic knowledge, which was first discussed in the 19th century. An analysis of the project chapters in the Handbook suggests that transdisciplinary research often combines these two perspectives. While some projects focus on local management strategies, others place more emphasis on generalisable results. The different projects can be grouped into four research categories: causal analysis, recursive planning, technological implementation, and counselling. These differ with regard to the importance of situational factors and the intensity of research. Each is characterised by different questions and approaches.

Successful transdisciplinary projects require effective project management. The management of transdisciplinary projects is the topic of Chapter 25 by Kirsten Holländer, Marie Céline Loibl and Arnold Wilts. Conflicts may arise among team members from different disciplinary backgrounds. At the same time conflicts are fundamental to transdisciplinary research. Project management also has to take into account that team members come from different social systems and have different interests. Hence it becomes important to develop shared goals among researchers and between researchers and non-academic stakeholders. At the same time, the flow of information has to be managed. This requires some form of 'controlled confrontation' to gain the full benefit from the heterogeneity of the transdisciplinary teams.

In Chapter 26, Julie Thompson Klein deals with transdisciplinary education. Klein begins by explaining early frameworks of transdisciplinarity and early curriculum. Several examples are given of the way transdisciplinarity has been integrated into university degrees. Further, there is a need for stakeholder education and for the education of professionals at their workplace. Especially in the health field there are numerous courses that bring together specialists with a lay community. Different reports emphasise the need for the cultivation of transdisciplinary skills. The chapter ends with a number of lessons.

Finally, Chapter 27, by Christian Pohl, Lorrae van Kerkhoff, Gertrude Hirsch Hadorn and Gabriele Bammer, views integration as a core feature of transdisciplinary research. It begins with lessons from history and an overview of the current state of transdisciplinary research. The chapter then addresses conceptual and practical challenges of integration in transdisciplinary research on the basis of some of the projects that are described in the Handbook. Four different forms of collaboration and four different approaches to integration are discussed. Funding, the capacity to undertake transdisciplinary research, and demonstrated success of transdisciplinary projects are seen as the prerequisites for further progress in transdisciplinary research.

The final part provides a summary and outlook. Chapter 28, by Christian Pohl and Gertrude Hirsch Hadorn, describes core terms in transdisciplinary research as a guide for the reader. The explanations are taken from the 'Principles for Designing Transdisciplinary Research' (Pohl and Hirsch Hadorn, 2007) and from Chapter 2 of this Handbook. Authors of the Handbook were provided with a preliminary and shorter version of the term descriptions, but they were free to use the terms in their own way. In Chapter 29 the editors propose a synthesis of the Handbook's contents in 15 propositions for enhancing transdisciplinary research. These comprise the definition, scope and process of transdisciplinary research, stumbling blocks in transdisciplinary practice and cornerstones for enhancing transdisciplinary research.

## References

Bammer, G.: 2005, 'Integration and Implementation Sciences: Building a New Specialization', *Ecol Soc* 10, 6, Retrieved, 6 December 2006 from http://www.ecologyandsociety.org/vol10/iss2/art6/.

Brewer, G.D.: 1999, 'The Challenges Of Interdisciplinarity', *Pol Sci* 32, 327–337.

Campbell, D.T.: 1969, 'Ethnocentrism of Disciplines and the Fish-scale Model of Omniscience'. In: Sherif, M. and Sherif, C.W. (eds), *Interdisciplinary Relationships in the Social Sciences*, Aldine, Chicago, pp. 328–348.

Cetto, A.M. (ed.): 2000, *World Conference on Science, Science for the Twenty-first Century: A New Commitment*, UNESCO, Paris, 544pp.

Klein, J.T., Grossenbacher-Mansuy, W., Häberli, R., Bill, A., Scholz, R.W., and Welti, M.: 2001, *Transdisciplinarity: Joint Problem Solving among Science. An Effective Way for Managing Complexity*, Birkhäuser Verlag, Basel, 332pp.

Lubchenco, J.: 1998, 'Entering the Century of the Environment: A New Social Contract for Science', *Science* 279, 491–497.

Pohl, C. and Hirsch Hadorn, G.: 2007, *Principles for Designing Transdisciplinary Research, Proposed by the Swiss Academies of Arts and Sciences*, oekom, München, 124pp.

ProClim: 1997, Research on Sustainability and Global Change – Visions in Science Policy by Swiss Researchers, CASS/SANW, Bern, Retrieved 3 December 2006 from http://www.proclim.ch/Reports/Visions97/Visions_E.html.

Rosenblum, D.: 1997, 'In the Absence of a Paradigm: The Construcion of Interdisciplinary Research', *Issues in Integrative Stud* 15, 113–123.

# Chapter 2
# The Emergence of Transdisciplinarity as a Form of Research

**Gertrude Hirsch Hadorn, Susette Biber-Klemm, Walter Grossenbacher-Mansuy, Holger Hoffmann-Riem, Dominique Joye, Christian Pohl, Urs Wiesmann and Elisabeth Zemp**

**Abstract** The birth of science is based on a strict dissociation of scientific knowledge from the various aspects of practical knowledge. The ideal of scientific knowledge as it was shaped in antiquity is still influential today, although the conception of science and the relationship between science and the life-world has undergone major changes. The emergence of transdisciplinary orientations in the knowledge society at the end of the 20th century is the most recent step. The Handbook focuses on transdisciplinarity as a form of research that is driven by the need to solve problems of the life-world. Differences between basic, applied and transdisciplinary research, as specific forms of research, stem from whether and how different scientific disciplines, and actors in the life-world, are involved in problem identification and problem structuring, thus determining how research questions relate to problem fields in the life-world. However, by transgressing disciplinary paradigms and surpassing the practical problems of single actors, transdisciplinary research is challenged by the following requirements: to grasp the complexity of the problems, to take into account the diversity of scientific and societal views of the problems, to link abstract and case specific knowledge, and to constitute knowledge with a focus on problem-solving for what is perceived to be the common good. Transdisciplinary research relates to three types of knowledge: systems knowledge, target knowledge and transformation knowledge, and reflects their mutual dependencies in the research process. One way to meet the transdisciplinary requirements in dealing with research problems is to design the phases of the research process in a recurrent order. Research that addresses problems in the life-world comprises the phase of problem identification and problem structuring, the phase of problem investigation and the phase of bringing results to fruition. In transdisciplinary research, the order of the phases and the amount of resources dedicated to each phase depend on the kind of problem under investigation and on the state of knowledge.

✉ G. Hirsch Hadorn
Department of Environmental Sciences, ETH Zurich, Zurich, Switzerland
e-mail: hirsch@env.ethz.ch

G. Hirsch Hadorn et al. (eds.), *Handbook of Transdisciplinary Research*, 

**Keywords:** Knowledge society · Problem fields in the life-world · Conception of science · Research process · Knowledge forms

## 2.1 Science and Life-world: From Dissociation to Transdisciplinary Orientations in the Knowledge Society

At the cradle of science in Greek antiquity the idea evolved that science is basically a cognitive faculty for explaining the development of natural things, including humans. Scientific explanations must be based on principles inherent to natural things, which Aristotle (384–322 BCE) saw as their universal unchangeable form. Aristotle claimed that humans are capable of capturing these evident first principles in 'contemplation', which is the meaning of the Greek term 'theoria'. In antiquity, 'theory' meant the knowledge about self-evident principles on which scientific demonstration is based. This kind of scientific knowledge (*epistême*) is of no use for day to day living. To lead their life, humans need skills to act (*praxis*) and to produce (*poiêsis*), and they need prudence (*phronêsis*) to deliberate about things that allow choice. So, the birth of science is based on a strict dissociation of scientific knowledge from the various aspects of practical knowledge (Aristotle, 2003).

The distinction between scientific and practical knowledge gives rise to the ideal that scientific knowledge is universal, explanatory, demonstrated to be true by a standard method, teachable and learnable. As a consequence, science has to be detached from practical life or the life-world. The term 'life-world' is used for what the phenomenologist Edmund Husserl (1859–1938) called 'Lebenswelt' – meaning the ongoing lived experiences, activities and contacts that make up the world of an individual or corporate life. Alfred Schütz (1899–1959) introduced the term into sociology to describe the structural properties of social reality as grasped by the agent – the agent's life-world (Schütz and Luckmann, 1973). Jürgen Mittelstraß uses the term in defining 'transdisciplinarity' as a form of research that transcends disciplinary boundaries to address and solve problems related to the life-world (Mittelstraß, 1992).

The ideal of scientific knowledge as it was shaped in antiquity is still influential today, although the conception of science and the relationship between science and the life-world has undergone major changes. Important transformations have taken place. The enlightenment started with the dissociation of the natural sciences from philosophy, followed in the 19th century by the establishment of the humanities and the social sciences as separate specialised disciplines in universities. The emergence of transdisciplinary orientations in the knowledge society at the end of the 20th century is the most recent step in reshaping the conception of science and the distinctions between science and the life-world.

The conception of science in the modern period is shaped by the dissociation of the natural sciences from philosophy. The foundations of theory oriented experimental interventions into nature were in place by the end of the 16th century. Modern natural science retains the idea that scientific knowledge is about general principles that explain processes in nature. However, these principles are conceived as causal laws in the sense of abstract idealised models that relate events in time and space as causes and effects, and explain the variation of events due to those causal influences. These laws are tested by experimental research, by intervening in nature under standardised laboratory settings. Consequently, while theories are still elaborated by deductive ordering, statistical methods are now used to prove that a theory holds true in a general way for processes of a certain kind, replacing antiquity's demonstration by deductive reasoning from self-evident first principles. Newtonian mechanics, which reduces the plurality of phenomena in nature to some basic laws, is the leading example of the modern conception of science.

Interestingly, a paradigm for conceiving the complexity in science emerges as early as the 18th century. Johann Heinrich Lambert (1728–1777) developed a systems approach to structure complexity as a set of interrelated elements. Lambert described various types of systems such as systems of scientific knowledge, belief systems of cultures, religions and narratives, including systems that are constructed to realise desired states. The latter systems are formed by uniting objectives and means, the parts of which are correlated on the basis of natural causalities and voluntary decisions. Thus human agency becomes a subject of scientific knowledge. Lambert was a distinguished mathematician, philosopher and natural scientist in his own time: although his 'systematology' received little attention it became an early forerunner of present day systems thinking (Rescher, 1979).

Newtonian science is also closely related to practical issues such as the production of goods. Science in the modern period is concerned with empirical laws and is carried out as research by intervening into nature in technically equipped experimental settings. The close relations between modern science and technology open ways for science based technological innovation, which can be used in industry for the production of commercial goods. The benefit to society of progress in science and technology is the core argument of Francis Bacon (1561–1626) in his 'novum organum' for a new science in the early 17th century (Bacon, 2000). Bacon was convinced that collaboration among scientists is most important for scientific progress, which is for the sake of societal benefit. This idea was important in the founding of the *Royal Society* in 1662. With the rise of the liberal market economy in the 19th century the use of knowledge from natural sciences and technology in industrial production began to play a major role in welfare economics. The instrumental interest in scientific knowledge from economics and society became an external driving force for the investment of resources in the progress of modern science and its experimental, quantitative and mathematical perfection. The purpose of improving the standard of living by improved quality, and increasing quantities of goods has been uncontested in society for a long time. As a result, many see scientific activity as free from extra-scientific societal values. Such an understanding does not take into account the way modern science and technology is imbedded in economic activities, cultural orientations and political measures or how these shape and legitimate scientific development as external drivers. Awareness of the way science is embedded in extra-scientific values and institutions has grown with the various steps in the debate about the modern conception of science and the role of science in the life-world.

In the 19th century, the conception of modern science was criticised as a model for all of science. When the humanities and history dissociated from philosophy the model of science that explained events by universal causal laws based on experimental testing for their fields was reflected. Wilhelm Dilthey (1833–1911) advocated a hermeneutic paradigm to achieve an understanding of cultural ideals and historical configurations, which constituted the identity of a cultural epoch in the history of mankind. He conceived of the humanities as hermeneutic sciences that rely on a method of understanding the meaning of life by interpreting its expressions in texts and other symbols. Wilhelm Windelband (1848–1915) also saw the methods and subjects of history as distinct from those of the natural sciences (Chapter 24). In his famous inaugural address as rector of Strassburg University in 1894 he argued that the natural sciences explain general aspects of empirical events by universal laws, while history investigates the individuality of empirical phenomena, giving them values to aid the understanding of their meaning and importance.

The methodological division within the sciences continued with the development of the social sciences. The emergence of the social sciences during the 19th and first half of the 20th century was influenced by the serious problems experienced by the country workers and the industrial working class due to major economic, social and political transformations. The destructive influence of colonialism in the south got little attention at that time. However, the social risks of industrialisation and migration in the north attracted attention and stimulated innovative developments in academia, for example, in the 1920s, the Chicago School of Sociology in the United States, and also shaped Human Ecology (Groß, 2004). In Europe developments began earlier with Karl Marx (1818–1883), Max Weber (1864–1920) and Emile Durkheim (1858–1917), whose thoughts were taken up by Talcott Parsons (1902–1979) in his seminal theoretical work 'The Structure of Social Action' (Parsons, 1968). Max Weber related research in social sciences with knowledge demands for societal problem-solving. For him, the stimuli behind the posing of scientific problems were always practical problems, which thus coincided with specifically oriented motives and values (Weber, 1949). He started with empirical investigations of social problems in Germany and made major theoretical contributions to the rise and shaping of sociology as a science of societal agency with his conception of ideal-types for understanding societal institutions. Ideal-types structure institutions and agencies analytically, and thus organise their complexity as value oriented complex functional wholes. Weber was well aware of the individual complexity of concrete settings. He therefore called his general analytical schemes ideal-types. Ideal-types are theoretical idealisations in the sense that they are grounded rational constructs of societal institutions and agencies. They are useful in describing and analysing empirical phenomena to the degree that they approximate empirical cases. The degree by which an ideal-type diverges from empirical observations indicates whether another ideal-type needs to be developed for understanding these phenomena (Weber, 1949).

Since then, actions of individuals and institutions have been a prominent subject for investigation in the social sciences, giving rise, in the 20th century, to a long-lasting debate about the conception of the empirical sciences and their relation to societal values (Chapter 23). The beginning of the debate is labelled 'Werturteilsstreit'. Starting from a clear cut distinction between facts and values, proponents of a 'neutrality of science in societal value issues' like Weber, argue that empirical sciences are about what is either true or false, while the normative distinction in the sphere of values is that of right or wrong. Weber rejected the possibility of an ethical foundation for value judgments, and restricted the tasks of empirical science to the analysis of the rationality of societal forms of agency as a means to certain ends, while pointing out that negative side-effects of behaviour should be involved in judging its rationality (Weber, 1949).

According to this view the benefit of the social sciences for practical life is – analogous to the natural sciences – an instrumental one. In this case, this is beneficial for structuring and regulating the effectiveness and efficiency of human agency. Since then the controversial debate within the social sciences and philosophy is about whether scientific investigation is restricted to the instrumental rationality of knowledge. Subsequently, the position of Sir Karl Popper (1902–1994) and Hans Albert (born 1921) in the 1960s has been to restrict scientific investigation of extra-scientific values to the functional analysis of means to certain ends. Jürgen Habermas (born 1929) in his critique of positivism in 'Knowledge and Human Interest' argues for three types of scientific rationality related to specific standards in research (Habermas, 1968): (1) the instrumental rationality of the empirical sciences and their standards of quantification and experimental testing, (2) the rationality of the historical sciences, which concerns the role of knowledge in creating meaning for life and constituting personal identity in societal contexts, based on rules for hermeneutic interpretation, and (3) the sciences of action, which are about societal transformations (in his later works this is based on communicative rationality as communicative action). According to this conception participants engage in deliberation, following the regulative percept of an 'ideal speech-situation' (Habermas, 1984, 1987). In transdisciplinary research, Habermas' conception of communicative rationality is broadly referred to, providing foundations for models of dialogue and knowledge claims (Chapter 21). This typology of the sciences and their rationality replaces the strict distinction in antiquity of science as *epistême*, on the one hand, and the knowledge of the life-world as *poiêsis, praxis* and *phronêsis*, on the other hand, by relating different conceptions of science with different types of interests: production, action and deliberation.

Of major importance for transdisciplinary research is a further alternative to the positivist view and its ideal of a physicalistic unitary science, namely the development, beginning in the 1940s, of systems theory in a broad range of fields. Systems theory was proposed by Ludwig von Bertalanffy (1901–1972) in biology; and developed by Norbert Wiener (1894–1964) in cybernetics; by John von Neumann (1903–1957) in game theory; by Claude Elwood Shannon (1916–2001) in information theory; and by Niklas Luhmann (1927–1998) in sociology, to mention some eminent individuals. Systems theory studies the abstract organisation of phenomena, independent of their substance, type, or spatial or temporal scale of existence. It investigates both the principles and the mathematical models used to describe them. These developments give rise to the idea of an abstract structural unity of scientific knowledge against the background of the progressive fragmentation of the sciences into more and more specialised disciplines and thematic fields. The continuing differentiation in research and higher education, as well as in social institutions in general, becomes a major risk for modern civilisation, because specialisation prevents the recognition of possible negative side effects. Multidisciplinary research approaches an issue from the perceptions of a range of disciplines; but each discipline works in a self-contained manner with little cross-fertilisation among disciplines, or synergy in the outcomes. The growing awareness of these kinds of risks therefore stimulates integrative approaches labelled 'interdisciplinarity' or 'transdisciplinarity'. It is in this context that Erich Jantsch (1929–1980) an others argued for innovations in planning for society at large, in a government–industry–university triangle which included a far-reaching reorganisation of higher education into an education–innovation system. He proposed that knowledge should be organised into hierarchical goal oriented systems. Blueprints for such coordinated frameworks, for which he introduced the term transdisciplinarity, would be general systems theory and organisation theory; that is, the study of organisations by the means of systems theory. He distinguished four levels within such a system: purposive (meaning values), normative (social systems design), pragmatic (physical technology, natural ecology, social ecology) and empirical (physical inanimate world, physical animate world, human psychological world). Values were crucial to his transdisciplinary systems because this approach involved activities at all levels of the education–innovation system being coordinated towards a common purpose (Jantsch, 1972).

Joseph Kockelmans (born 1923) argues against restricting problem oriented research to theoretical frameworks. This is in the spirit of general systems theory or structuralism, proposed by Jean Piaget (1896–1980) to address the unity of the sciences against the background of fragmentation of knowledge. Kockelmans suggests the term 'crossdisciplinary work' for research which 'is primarily concerned with finding a reasonable solution for the problems that are so investigated, whereas transdisciplinary work is concerned primarily with the development of an overarching framework from which the selected problems and other similar problems should be approached' (Kockelmans, 1979). From this, it becomes clear that there are several quite different cognitive motives for transcending boundaries between disciplines, such as unity of knowledge in general, grasping the complexity of concrete issues, and innovation in basic research as, for instance, in the case of molecular biology. On the other hand, a variety of terms such as interdisciplinarity, crossdisciplinarity, transdisciplinarity and others have been coined to distinguish between the forms and functions of crossing disciplinary boundaries. Unfortunately these terms do not always have the same meaning, due to independent developments and different related motives.

While these emerging ideas about inter-, cross- and transdisciplinarity are widely discussed with comparatively little impact on research or on institutional structures in universities, systems analysis and modelling are advancing to become leading paradigms in the natural and social sciences. They are used for describing complexity and for analysing the risks that global change poses to life-support systems as a result of the manifold and poorly understood negative side-effects related to

the increasing use of natural resources, and to population growth (Forre
Meadows et al., 1972; Chapter 3). It is in this context, that in 1986 U
coined the term 'Risk Society' (Beck, 1992). In his bestseller he points at [illegible]
transformations in so much of everyday life in the industrial society, together with unintended and poorly understood damage to natural resources and life-support systems. As a consequence he sees the sciences becoming reflective, meaning that they will become increasingly busy understanding and handling the negative side-effects of the use of science based technological innovations in society. Beck insists that these effects have to be understood as hybrids that no longer match the separation of natural events and societal meaning. Modernisation itself induces hazards and insecurities, which call for precautionary and systematic ways of dealing with hazards as essentially political issues. Social sciences and humanities become involved in activities such as technology assessment, ethical committees on morally sensitive technologies as well as research into the ethical, legal and social implications (ELSI) of technologies (Chapter 9 and Chapter 10) such as ELSI research within the Human Genome Program. According to Beck, society and the sciences undergo intertwined transformations into a risk society in 'Science beyond Truth and Enlightenment' (Beck, 1992).

Beck is criticised for his rather vague statements about the transformations of science. Silvio Funtowicz and Jerome Ravetz clear some of these grounds with their conception of post-normal science (Funtowicz and Ravetz, 1993; Chapter 23), which has been the result of their analysis of the management of high uncertainty and decision stakes in policy related scientific inquiries. They find that the paradigm of normal science is inadequate to ensure the validity of knowledge about these kinds of issues. Therefore routine scientific expertise is inadequate and professional knowledge and judgment are insufficient to address these policy issues. They argue that in such cases science must engage in dialogue with all those who have a stake in the decision. Quality assurance of scientific inputs to the policy process is perceived as mutual learning, with stakeholders as an extended peer community. While science is becoming an agent in the policy process, ideas about reflectivity and democratisation of science attract broad attention, especially within the community of science and technology studies (STS). These ideas undergo various interpretations and adaptations. In 1994 Michael Gibbons and his colleagues published their 'New Production of Knowledge', in which they contrasted Mode 1 of knowledge production, the Newtonian model, with a Mode 2, emerging in the field of research and development, which has features such as transdisciplinarity, heterogeneity, reflectivity, social accountability, and context- and user-dependency (Gibbons et al., 1994).

Through scientists entering into dialogue and mutual learning with societal stakeholders, science becomes part of societal processes, contributing explicit and negotiable values and norms in society and science, and attributing meaning to knowledge for societal problem-solving. Within such hermeneutic frameworks problem-solving includes reflection, the transformation of attitudes, the development of personal competences and ownership, along with capacity building, institutional transformations and technology development. Mutual learning connects transdisciplinary orientations to action research, a conception aimed at mutual benefit to theory and practice. Action research is driven by three principles: (1) The location of the starting point in social reality, i.e. people's interpretation of reality. Action research, therefore, is related to interpretative approaches of social research, which extend back to the Chicago school of sociology of the 1920s. (2) Action in field research, aimed at learning about the consequences of different forms of social action. Research directed toward the solving of social problems was developed by Kurt Lewin (1890–1947). To achieve this, action, research and education must form an interlinked triangle (Lewin, 1951). (3) The so-called subject status of the research object or participation. Jakob L. Moreno (1889–1974), proposed that researchers and the people studied should both research and be researched, and both should participate in the situation and intervene to create change in accordance with their competences. Action research was adopted in studies about religious and racial prejudices and in projects concerning education and social work during the social upheaval of the 1960s and 1970s. Thanks to the work of Chris Argyris and Donald Schön on experiential and organisational learning, as well as on theories of action (Argyris and Schön, 1996), action research has found its way into transdisciplinary projects concerned with sustainable development of companies, landscape research and Local Agenda 21, to note a few examples.

During the past thirty years similar changes have been taking place in the design of research projects in development cooperation. Where, at the beginning, researchers defined the problems and the solution, now, the affected population's participation is supported in the research process. Experience shows that without participation, the resulting measures and outcomes are likely to be rejected or ignored by the local population. New approaches and methods, such as rapid rural appraisal (RRA) and participatory rural appraisal (PRA), are being developed. In addition to integrating the local population into the research process in an active way, the diversity and complexity of social, political, economic and environmental problems has to be adequately met (Chapter 3, Chapter 4 and Chapter 17). 'Diversity' means that empirical dimensions relevant to describing and analysing processes are heterogeneous in the sense that they belong to different disciplines or to the perceptions of different actors, and that there are plural values and norms that do not fit together in a systematic way. Diversity of dimensions or values is in contrast to homogeneity with respect to the disciplines and life-world perceptions involved. 'Complexity' is used for the interrelations among heterogeneous dimensions, or plural values and norms. Thus complexity is in contrast to simplicity.

Against the background of the perceived diversity and complexity of development problems and trends it becomes obvious that the formerly dominant set of theories – modernisation and dependency theories in the social sciences – have only limited explanatory power. The United Nations Conference on Environment and Development (UNCED) in Rio de Janeiro in 1992 had the commitment of the statesmen from most nations to sustainable development. It marked a paradigm shift in thinking about developmental issues. Sustainable development is a global socio-political model for changing practices and institutions in order to achieve more equitable opportunities within and between generations while taking into account the limitations imposed by the state of technology and social organisation on the

environment's ability to meet present and future needs (World Commissic
ronment and Development, 1990). Promoting sustainable development th
cessitates overcoming narrow preoccupations and compartementalised c
involving people from civil society, the private sector and public agencies as actors in participatory deliberation and decision making. Thus sustainable development is a way to conceive the common good as the basic principle of public legislation in a complex world. Agenda 21, a UN program, is a comprehensive blueprint for action to be taken globally, nationally and locally.

The Rio Conference and The World Summit on Sustainable Development held in Johannesburg in 2002 were highlighting the importance of science to sustainable development, and stressing the need to transform research by involving stakeholders and promoting mutual learning between science and the life-world (Hirsch Hadorn et al., 2006). When a group of authors in the 1970s first published 'Finalization in Science: The Social Orientation of Scientific Progress' (Böhme et al., 1983), in which they showed how the external orientation of scientific development was becoming more and more important for the development of the sciences, compared to putting forward endless internal scientific frontiers, the scientific world was shocked. The conception of science in the analysis of finalization is still that of scientific disciplines. These disciplines are now being responsive to problems in the life-world, which are external to their cognitive domain. Therefore, a problem oriented restructuring of knowledge was needed to produce knowledge that was valid for specialist problem-solving and translatable into technological innovation. In today's knowledge society, with sustainable development as its normative model, even forms of research and the role of science in societal change are altering. Science is not only a resource, but an 'agent of change' (Krohn and van den Daele, 1998): society is not only integrating scientific knowledge but adopting scientific research for societal problem-solving and innovation.

It is this bigger picture of ongoing and intertwined transformations in academia and the life-world during the modern period, especially in the last 30 years, which has to be kept in mind when looking at the challenges of, and opportunities for, transdisciplinary research. From this historical perspective it becomes clear why transdisciplinary research is a fuzzy and contested field, shaped by various lines of thinking, heterogeneous conceptions of science and approaches to research, with a variety of terminologies and definitions. The next section will comment on some of these and propose a structure as guidance for readers.

## 2.2 Transdisciplinarity as a Form of Research

Disciplines shape scientific research by forming the primary institutional and cognitive units in academia, on which the internal differentiation of science into specialised curricula, professions and research, is based (Stichweh, 1992). Members of a discipline are specialists who build a scientific community (Kuhn, 1963). Members communicate within their community, share basic assumptions and examples about meaningful problems, standards for reliable and valid methods, as well as what is considered a good solution to a problem. What modern science gains and preserves is based to a large extend on disciplinary structures. However, boundaries between disciplines are changing: by increasing specialisation through internal differentiation within the disciplines, and by the integration of disciplines.

One intellectual motive for transgressing disciplinary boundaries and integrating different disciplinary perspectives has been the search for innovation in fundamental scientific understanding of specific problems, often linked with innovation in investigative methods. Among the many examples are developments in the social sciences (Sherif and Sherif, 1969), in biology (Bechtel, 1986) and recently in the field of nanotechnology and nanoscience. Endeavours of this kind, which are motivated by factors internal to the scientific knowledge system, are often termed 'interdisciplinarity'. Migration and collaboration by researchers between disciplines, which take place in such interdisciplinary endeavours, change the landscape of disciplines by the transformation of existing disciplines, and the emergence of new ones.

A second motive stems from the knowledge demands of the knowledge society. Societal knowledge demands for a better understanding of, and solutions to, concrete issues in the life-world, function as an external driver for transgressing disciplinary boundaries and integrating different disciplinary perspectives. Although this is a different kind of endeavour, it is sometimes also called 'interdisciplinarity', as for instance in the following definition:

> Interdisciplinary research (IDR) is a mode of research by teams of individuals that integrates information, data, techniques, tools, perspectives, concepts, and/or theories from two or more disciplines or bodies of specialized knowledge to advance fundamental understanding or to solve problems whose solutions are beyond of the scope of a single discipline or area of research practice. (National Academies, 2005)

It is helpful to make a terminological distinction between these two different endeavours. One suggestion is to specifiy interdisciplinarity in the context of knowledge demands in the life-world with the help of additional terms such as 'interdisciplinary problem-solving' (Clark, 1999; Deppert, 1998) or 'goal oriented interdisciplinarity' (Hubert and Bonnemaire, 2000); or defining them as 'Mode 1 interdisciplinarity' and 'Mode 2 interdisciplinarity' (Bruce et al., 2004). It is, however, more useful to use a different term, such as the term 'transdisciplinarity' (Jantsch, 1972) as we do in the 'Handbook of Transdisciplinary Research'. Various more or less radical views of the transformation of science involving the transgression of disciplinary boundaries in addressing issues in the life-world in research, have emerged. These have resulted in the coining of specific terms such as 'post-normal science' (Funtowicz and Ravetz, 1993), 'Mode 2 of knowledge production' (Gibbons et al., 1994), and 'policy science' (Clark, 2002).

Jürgen Mittelstraß (1992) argues that transdisciplinarity is primarily a form of research for addressing and reflecting on issues in the life-world. Against the background of harm and serious risk posed by technologies and growth that does not fit within the disciplinary paradigms of academia, he calls for the transgression of disciplinary boundaries for identifying, structuring and analysing problems in research. Contrary to the more pragmatic approach of transdisciplinarity as a form of research, others argue for a further intellectual endeavour on a fundamental theoretical level. They conceive of transdisciplinarity as a theoretical unity of all of our knowledge, which they think is needed to respond adequately to knowledge demands for problem-solving in the life-world. (Nicolescu, 1996; Max-Neef, 2005). In coining the term 'transdisciplinarity' in 1972, Erich Jantsch envisioned a systems theory approach for the purpose oriented integration of knowledge to grasp the complexity of problems in the life-world:

> Transdisciplinarity: The co-ordination of all disciplines and interdisciplines in the education/innovation system on the basis of a generalized axiomatics (introduced from the purposive level) and an emerging epistemological ("synepistemic") pattern. (Jantsch, 1972)

Systems theory has been influential in shaping a range of transdisciplinary schools, among them 'human ecology' (Ehrlich et al., 1973), the 'Man and Biosphere' research concept (Chapter 3), 'ecological economics' (Costanza et al., 1997), 'sustainability science' (Kates et al., 2001) and 'socio-ecological research' (Becker and Jahn, 2006; Chapter 6).

Scholars in Social Studies of Science (STS) investigate the transformation of knowledge production in applied and policy oriented research, as opposed to basic research. Among other features, they stress the need for the participation of stakeholders in the research process (Chapter 22). Because of the high level of uncertainty of knowledge and the high decision stakes involved, Funtowicz and Ravetz in their conception of 'post-normal science' argue for mutual learning between scientists and stakeholders with stakeholders belonging to an 'extended peer community', and for quality control of knowledge in policy oriented research (Funtowicz and Ravetz, 1993; Chapter 23). In the context of application, Gibbons and colleagues in their 'Mode 2 of knowledge production' (Gibbons et al., 1994) insist on mutual learning for 'socially robust knowledge' (Nowotny, 1999). The collaboration of both researchers and actors in the life-world can be found in many definitions of 'transdisciplinarity'. This is highlighted by Julie Klein's definition:

> The core idea of transdiscipinarity is different academic disciplines working jointly with practitioners to solve a real-world problem. It can be applied in a great variety of fields. (Klein et al., 2001)

In summary, there are about four core concerns which show up in definitions of 'transdisciplinarity' or related terms: first the focus on life-world problems; second the transcending and integrating of disciplinary paradigms; third participatory research; and fourth the search for unity of knowledge beyond disciplines. While the first two concerns are widely shared, there is disagreement over whether, and to what extent participatory research is needed for taking into account societal views in investigation issues. There is even more disagreement about the importance of the search for unity of knowledge in addressing issues in the life-world.

The 'Handbook of Transdisciplinary Research' focuses on transdisciplinarity as a form of research that is driven by the need to solve problems of the life-world. To cover a broad range of research experiences and bring it to fruition in the 'Handbook of Transdisciplinary Research' a broad conception of 'transdisciplinary research' is used. This conception is developed and explained in more detail in td-net's 'Principles for Designing Transdisciplinary Research' (Pohl and Hirsch Hadorn, 2007). Based on a synthesis of what can be found in the literature the conception refers to cognitive features of the starting point, the requirements and the goals of a transdisciplinary research process:

> There is a need for TR when knowledge about a societally relevant problem field is uncertain, when the concrete nature of problems is disputed, and when there is a great deal at stake for those concerned by problems and involved in dealing with them. TR deals with problem fields in such a way that it can: a) grasp the complexity of problems, b) take into account the diversity of life-world and scientific perceptions of problems, c) link abstract and case specific knowledge, and d) constitute knowledge and practices that promote what is perceived to be the common good. (Pohl and Hirsch Hadorn, 2007)

Features of the starting point, namely high uncertainty of knowledge and high decision stakes are at the core of post-normal science (Funtowicz and Ravetz, 1993). Furthermore, grasping the complexity of the problems, taking into account the diversity of life-world and scientific perceptions of problems, as well as linking abstract and case specific knowledge, are widely shared concerns in transdisciplinary research processes. However, the goal, which is to constitute knowledge and practices that promote what is perceived to be the common good, is seldom explicitly stated in a definition, although it is sometimes implied, for instance by the term sustainability science (Kates et al., 2001). An exception is policy science (Clark, 2002), which explicitly refers to the common interest as the normative principle for assessing problem-solving measures.

Transdisciplinary research relates to three types of knowledge: systems knowledge, target knowledge and transformation knowledge. The terms are coined in 'Research on Sustainability and Global Change – Visions in Science Policy by Swiss Researchers' (ProClim, 1997). The definition of 'systems knowledge' as knowledge of the current status; of 'target knowledge' as knowledge about a target status; and 'transformation knowledge' as knowledge about how to make the transition from the current to the target status (ProClim, 1997); however, can be open to a technocratic misunderstanding. Therefore these forms of knowledge are described by the types of questions to be addressed by transdisciplinary research (Pohl and Hirsch Hadorn, 2007):

- Questions concerning the genesis, further development and interpretation of a problem in the life-world are answered by systems knowledge of empirical processes and interactions of factors – including the interpretations given to these in the life-world.
- Questions related to determining and explaining the need for change, desired goals and better practices that call for target knowledge.
- Questions about technical, social, legal, cultural and other possible means of acting that aim to transform existing practices and introduce desired ones, which have to be answered by transformation knowledge.

There are similar distinctions between forms of knowledge using different terminologies by other authors (Becker et al., 1999; Deppert, 1998; Costanza et al., 1997; Jantsch, 1972).

These three forms of knowledge remind us of Aristotle's forms of knowledge, namely: science (*episteme*); life-world action (*praxis*); production (*poiêsis*); and prudence (*phronêsis*) – now transformed as goals of transdisciplinary research. Being goals of transdisciplinary research means that the investigation of each of the three types of questions requires that explicit assumptions with regard to the other two types are made. Thus, instead of being investigated in isolation, research questions that refer to target knowledge should be examined on the basis of specific assumptions about systems relations and with a view to finding out about specific options for transforming existing practices. The opposite is also important: an empirical analysis of systemic relations must refer to the transformation of a specific social practice with a specific objective in mind; and studies of possible change options need to be based on specific knowledge about systems relations and goal oriented practices. Furthermore, instead of being conceived of in a sequential order as in the classical technical model of problem-solving, these three forms of knowledge form a triangle reflecting the mutual dependencies (Fig. 2.1).

This integrative perspective of the three forms of knowledge also displays an important difference to Habermas' conception of the different types of scientific rationality. This becomes clear when looking at the particular challenges that each form of knowledge has to face to produce valid knowledge. Systems knowledge needs strategies for dealing with uncertainties. On the one hand, these uncertainties are the result of transferring abstract insights from a laboratory, model or theory to a concrete case underlying specific conditions. On the other hand, depending on the

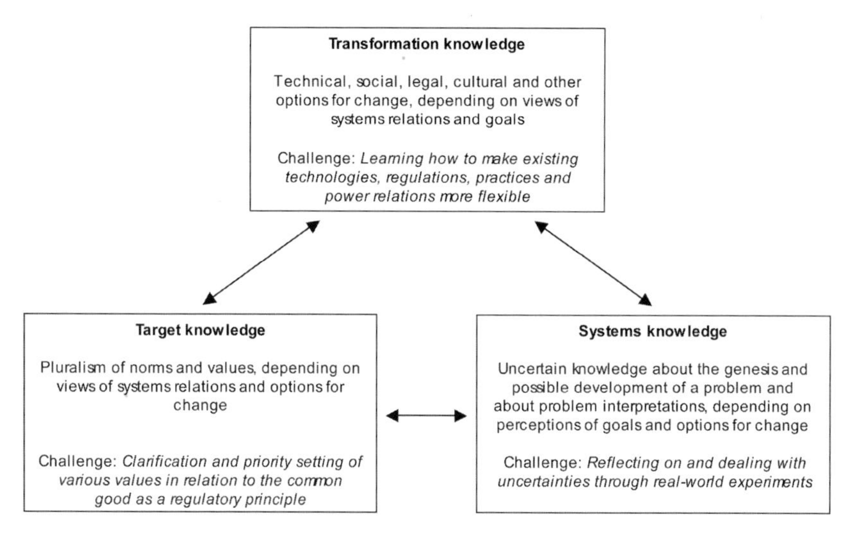


**Fig. 2.1** Interdependencies between systems, target and transformation knowledge and their particular challenges (Pohl and Hirsch Hadorn 2007, p. 38), adapted

approach to and interpretation of the problem, these uncertainties may be attributed different degrees of importance, leading to different assessments of targets and of transformation knowledge. In the case of target knowledge, against the background of multiple interests, needs and attitudes of stakeholders, a standard is needed to clarify the variety of positions, and to guide deliberation about their significance for problem-solving. This is the role of referring to and interpreting the common good, or more specifically, the sustainability model in addressing target and transformation knowledge. In the case of transformation knowledge, when designing technical, social, legal, cultural and other options, the challenge is to learn about the flexibility of existing infrastructure, current laws, and to a certain degree about current power relations, cultural opinions and people's capabilities for learning (Argyris and Schön, 1996). It is necessary to learn to make what is existing more flexible in order to have any chance of successfully navigating between 'the Scylla of political irrelevance and the Charybdis of technical inadequacy' (Guston and Sarewitz, 2002).

The starting point, requirements and goals of transdisciplinary research have major implications for the way problems and research questions must be identified and structured. To understand these challenges a comparison of transdisciplinary, applied and basic research is helpful. For this purpose, we conceive of basic, applied and transdisciplinary research as three ideal-types, thus doing justice to the fact that projects have characteristics of certain forms of research to a higher or lower degree: projects can combine features of different research forms. Differences between basic, applied and transdisciplinary research, as specific forms of research, stem from whether and how scientific disciplines, and actors in the life-world, are involved in problem identification and problem structuring, thus determining how research questions relate to problem fields in the life-world. In Fig. 2.2 each form of research is represented by a three column diagram to show the different ways of determining research questions and how those questions relate to problem fields. In the diagrams, scientific disciplines form the column to the left. The range of disciplines from sciences and humanities is illustrated by economics, ethics, molecular biology, and ecology. Disciplines evolve and change over time by means of internal differentiation as well as by integration. To keep the figure simple, this is not indicated in the diagrams. The right hand column shows actors in the life-world. In this third column the private sector, civil society, and public agencies, are each comprised of a range of institutions acting from local to global scales. To keep the diagrams simple, specific institutions and stakeholders are not included. The private sector, civil society, public agencies and disciplines of sciences and humanities interact as four policy cultures in dealing with policy issues in the knowledge society (Elzinga, 1996). The middle column lists a few examples of problem fields: poverty, land degradation, disease, and hunger. Borders between the problem fields are fuzzy as they build complex clusters with mutual influences and overlaps.

When identifying and structuring problems to determine research questions it is necessary to reduce the diversity and complexity of elements, structures and processes in problem fields by distinguishing between important and irrelevant aspects. Reducing diversity and complexity has to be based on knowledge. In Fig. 2.2, circles indicate the knowledge bases. The way knowledge bases in the sciences and in the

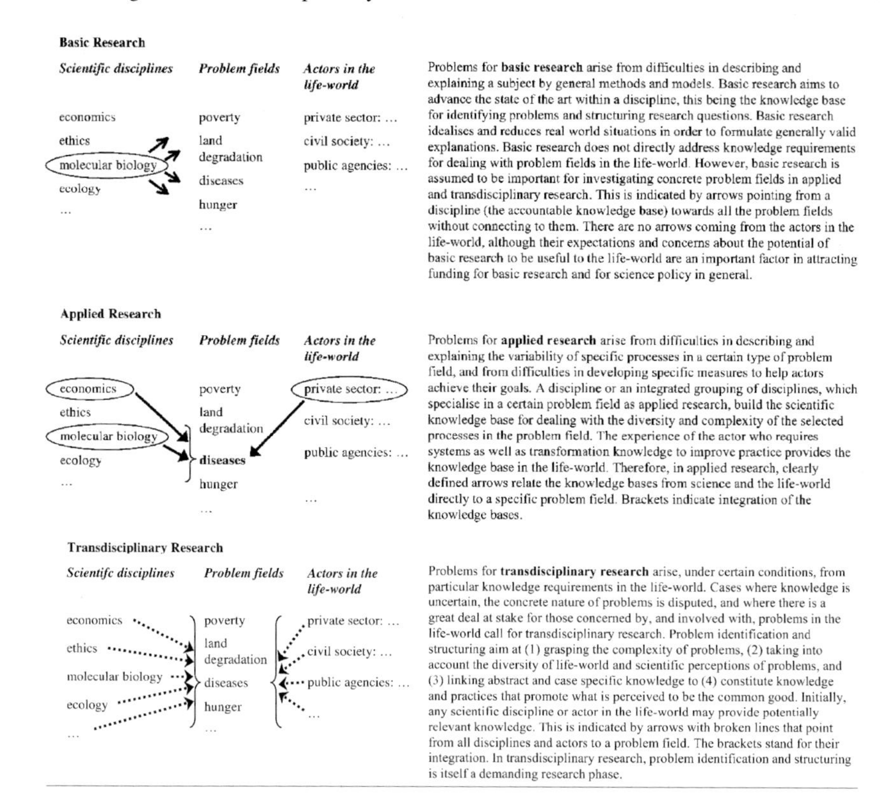


Each form of research is represented by a diagram, which is based on the same three columns. Scientific disciplines form the left hand column: economics, ethics, molecular biology, and ecology are examples. The right hand column shows actors in the life-world: the private sector, civil society, and public agencies, each comprising a range of institutions that act from local to global scales. The middle column shows examples of problem fields in the life-world: poverty, land degradation, diseases, and hunger being just a few. Circles indicate the knowledge base in science or the life-world used to identify problems and determine research questions by reducing the diversity and complexity of elements, structures and processes. The arrows point from the knowledge bases to the problem fields. They represent the specific ways that scientific disciplines and actors in the life-world are involved in identifying problems and determining research questions. The brackets indicate the means of integration. Basic, applied and transdisciplinary research are conceived of as three ideal-types, meaning that projects may combine features of each research form, or change their features during the research process, to a greater or lesser degree (adapted from Hirsch Hadorn et al., 2006).

**Fig. 2.2** Problem identification and structuring in basic, applied and transdisciplinary research (Hirsch Hadorn et al. 2006, p. 123–124) adapted

life-world are involved in reducing diversity and complexity for problem structuring and determining research questions in relation to problem fields, is different in each form of research. This is indicated by the different arrows that relate the knowledge base involved in determining research questions to the problem fields in each form of research.

Problems for basic research arise from difficulties in explaining and describing a subject by generally valid principles of a discipline. Members of a discipline share examples of good quality problems and solutions, concepts, methods, and standards for research using institutions such as journals, textbooks and educational programmes. These elements constitute the paradigm of a discipline (Kuhn, 1963). A paradigm ensures that research questions can be answered in a valid and reliable way. To arrive at theoretical explanations in basic research, problems must be modelled and investigated under standardised conditions: an idealisation of what happens in real world settings. Therefore, many factors that could contribute to the genesis and development of issues in a problem field have to be ignored. These factors could be a subject for other disciplines, again in an idealised way, but from a different perspective. For example, in the case of biological processes identified as a research problem for molecular biology with assumed relevance for some diseases, researchers will not concern themselves with cultural practices or economic conditions that also contribute to the causes of epidemics or disease. Such factors are a subject for investigation by other disciplines.

As a result, research questions in basic research aim at advancing the state of the art and do not directly address knowledge requirements for dealing with problem fields. It is, however, claimed that explanations of basic research, e.g. in molecular biology, are important for applied research investigations in concrete problem fields such as poverty, disease, land degradation and hunger. Therefore, in the Basic Research section of Fig. 2.2, the arrows that indicate the knowledge base for problem structuring start from one discipline as the only knowledge base. Arrows point towards all problem fields because knowledge of basic research is conceived of as universal knowledge. However, arrows do not connect with problem fields, because basic research does not address the way problems occur in problem fields. That is the task of applied research. For the same reasons, there are no arrows coming from actors in the life-world. Their knowledge, interests and concerns do not directly contribute to problem structuring in basic research, although their expectations and concerns about the potentials of basic research for the life-world have major implications for basic research funding and for science policy in general. The potential of knowledge gained in basic research to improve the approach to problem fields by actors in the life-world is still an issue that needs to be realised in applied and transdisciplinary research.

Applied research describes and explains the variable processes in a certain problem field and develops specific measures and support for the actors concerned. For this purpose the knowledge base in science is built by disciplines specialising in the applied research of a specific problem field. The diversity and complexity of factors in the problem field can trigger the integration of knowledge of further disciplines and development of an interdisciplinary conception of the problem, also called 'Mode 2 interdisciplinarity' (Bruce et al., 2004). Furthermore, the interests and knowledge of the actor, who requires systems knowledge as well as transformation knowledge to improve his practice, are involved in shaping research questions. This is indicated by arrows that relate the relevant knowledge bases from science and the life-world to the problem field. The example in Fig. 2.2 shows molecular biology and economics being applied to investigate a certain disease and develop a drug that can be produced by the pharmaceutical industry. Applied research is often funded by the private sector or by public agencies asking for knowledge in the search for innovation or to improve their dealing with issues in the life-world.

Transdisciplinary research is needed when knowledge about a societally relevant problem field is uncertain, when the concrete nature of problems is disputed, and when there is a great deal at stake for those concerned by problems and involved in dealing with them. However, by transgressing disciplinary paradigms and by surpassing the practical problems of single actors, transdisciplinary research can no longer build on clearly defined knowledge bases in science and the life-world. As a consequence, research problems that are soluble thanks to disciplinary paradigms of problem structuring and to the restriction of the range of interests taken into account, are turned into issues that scientists and agents can grapple with. The challenges mentioned give a reason for defining transdisciplinary research by the following requirements and goal: to grasp the complexity of the problems, to take into account the diversity of scientific and societal views of the problems, to link abstract and case specific knowledge, and to constitute knowledge with a focus on problem-solving for what is perceived to be the common good.

A way to meet these requirements is to design the phases of the research process in a recurrent order. Problem-solving research comprises the phase of problem identification and problem structuring, the phase of problem investigation and the phase of bringing results to fruition. Traditionally these phases follow a sequential order, with an emphasis on problem investigation. In transdisciplinary research, the order of the phases and the amount of resources dedicated to each phase depend on the kind of problem under investigation and on the state of knowledge (Fig. 2.3).

Problem identification and problem structuring is the phase of the transdisciplinary process in which researchers and actors in the life-world jointly work on identifying and understanding the nature of specific problems in a problem field. Participants are engaged in jointly framing and structuring the fuzzy issues in a problem field with regard to: the genesis and possible further developments of a problem; determining and explaining the need for change, desired goals and better ways of acting; and to technical, social, legal, cultural and other possible means of transforming existing practices. Thus the knowledge demands of systems, target and transformation knowledge are determined. In transdisciplinary research the phase of problem identification and structuring is very resource demanding because it cannot build on a specific knowledge base, as can basic and applied research. Instead, a

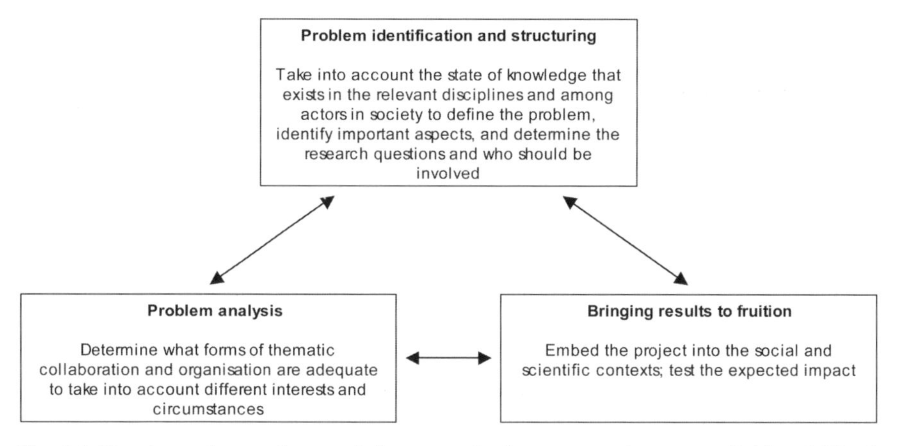


**Fig. 2.3** The three phases of research in a transdisplinary research process (Pohl and Hirsch Hadorn, 2007, p. 42)

broad range of participants and competences have to be involved to properly identify the relevant scientific disciplines and actors in the life-world, to evaluate the existing knowledge in academia and the life-world about the problems and to learn about the needs and interests at stake. This information can function as the knowledge base to determine the knowledge available for problem-solving, the questions that need to be addressed in research, and the competences that are required for the investigation and deliberation of results.

Of course one outcome of this phase could be that some knowledge is available and – instead of engaging in deeper investigation – allows the designing and implementing of measures to test the effectiveness of that knowledge to proceed. But this kind of real world experiment (Chapter 24) may lead to surprising results, which call for the investigation or even a restructuring of problems and research questions. Another possible outcome of problem structuring and investigation is that quite different competences and participants may be required in a project than was initially expected. Furthermore, problem identification and structuring, on the one hand, and problem analysis, on the other hand, can overlap. This makes an iterative treatment of phases a more rational approach for achieving valid results, than a sequential treatment.

The requirements and goals of transdisciplinary research also have implications for the phase of problem analysis. In order to grasp the relevant complexity of relations when research questions are structured into more specific ones for detailed analysis, an adequate understanding is needed of the way in which the diverse aspects and perspectives are integrated. In addition, quality assurance of knowledge has to take into account mutual influences between systems knowledge, target knowledge and transformation knowledge. From this it becomes clear that knowledge is related to conceptual, epistemological and methodological uncertainties (Chapter 23).

Furthermore, although research results are expected to be valid for problem-solving in concrete settings, some abstraction in cognitive conceptions is important, otherwise transdisciplinary projects would be restricted to counselling and lack a core element of the mission of research: to find out what can be learned for other problem situations. Instead of defining standard conditions for idealisation, generalisation of knowledge has to be achieved by transferring models and methods from the context in which they have been developed, to other contexts, while carefully validating the conceptions of each setting. Therefore, problem analysis and bringing results to fruition are best conceived of as iterative and integrated steps (Chapter 24). In such a situation researchers and actors in the life-world collaborate to achieve: (1) recurrent validation and adaptation of empirical models in concrete situations, (2) ongoing deliberation about goals, and (3) recurrent monitoring of experiments and effects in order to adapt conceptions and transformation strategies.

Bringing the results to fruition, as a phase of the transdisciplinary process, relies on the synthesis of knowledge and the translation of that knowledge. This takes into account the context of the actors in the life-world who are involved in transforming practices to promote what is perceived to be the common good. Transformations can comprise new insights, and as a consequence alter the perception of a problem, thus influencing policy making and individual behaviour. Because of the uncertain empirical knowledge, contested purposes and habits relating to existing practices, it is important that practitioners learn about the strengths and weaknesses of problem-solving strategies and develop competences for implementing and monitoring progress in order to be able to adapt strategies and purposes. This influences problem-solving strategies: from the implementation of definitive (technological) solutions, to social learning about problem-solving strategies including the design of technologies and institutional structures as well as changing attitudes.

In this chapter, transdisciplinary research has been characterised by its starting point, goals and requirements. It has been argued that transdisciplinary research is necessary when knowledge about a societally relevant problem field is uncertain, when the concrete nature of problems is disputed, and when there is a great deal at stake for those concerned by the problems and involved in investigating them. In such situations, the knowledge required as a starting point for research comprises the genesis and possible further development of problems. This includes: the interpretation in the life-world (systems knowledge); knowledge about the need for change, desired goals and better ways of acting (target knowledge); and knowledge about technical, social, legal, cultural and other means of redirecting the existing behaviour (transformation knowledge). Consequently, the challenge for researchers is to grasp the relevant complexity of the problems, to take into account the diversity of life-world and scientific perceptions of problems, to link abstract and case specific knowledge, and to develop knowledge and practices that promote what is perceived to be the common good. Against this background transdisciplinary research can be distinguished from both basic and applied research by the way scientific disciplines and actors in the life world are involved in problem identification and structuring. Concurrently building a knowledge base for relating research questions to problem fields and problem-solving in the life-world has design implications, for problem analysis and for bringing results to fruition. One of these implications is to design the research process as a recurrent ordering of phases.

## References


Argyris, C. and Schön, D.: 1996, *Organizational Learning II: Theory, Method, and Practice*, Addison-Wesley, Reading, 305pp.

Aristotle: 2003, *The Nicomachean Ethics* (H. Rackham, trans.), Harvard University Press, Cambridge, 650pp.

Bacon, F.: 2000, *The New Organon*, Cambridge University Press, Cambridge, 252pp.

Bechtel, W. (ed.): 1986, *Integrating Scientific Disciplines*, Martinus Nijhoff, Dordrecht, 354pp.

Beck, U.: 1992, *Risk Society. Towards a New Modernity* (M. Ritter, trans.), Sage Publications, London, 260pp.

Becker, E. and Jahn, Th. (ed.): 2006, *Soziale Oekologie. Grundzüge einer Wissenschaft von den gesellschaftlichen Naturverhältnissen*, Campus, Frankfurt, 521pp.

Becker, E., Jahn, Th., and Stiess, I.: 1999. Exploring Uncommon Ground: Sustainability and the Social Sciences. In: E. Becker and Th. Jahn (eds.), *Sustainability and the Social Sciences*, Zed Books, London, pp. 1–22.

Böhme, G., van den Daele, W., Hohlfeld, R., Krohn, W., and Schäfer, W.: 1983, *Finalization in Science: The Social Orientation of Scientific Progress* (P. Burgess, trans.), Reidel, Dordrecht, 315pp.
Bruce, A., Lyall, C., Tait, J., and Williams, R.: 2004, Interdisicplinary Integration in Europe: The Case of the Fifth Framework Programme, *Futures* 36, 45–470.
Clark, T.W.: 1999, Interdisciplinary Problem Solving: Next Steps in the Greater Yellowstone Ecosystem, *Pol Sci* 32, 393–414.
Clark, T.W.: 2002, *The Policy Process: A Practical Guide for Natural Resource Professionals*, Yale University Press, New Haven, 215pp.
Costanza, R., Cumberland, J., Daly, H., Goodland, R., and Norgaard, R.: 1997, *An Introduction to Ecological Economics*, St. Lucie Press, Boca Raton, 275pp.
Deppert, W.: 1998, Problemlösen durch Interdiszplinarität. In: W. Theobald (ed.), *Integrative Umweltbewertung. Theorie und Beispiele aus der Praxis*, Springer, Berlin, pp. 3–64.
Ehrlich, P.R., Ehrlich, A.H., and Holdren, J.P.: 1973, *Human Ecology: Problems and Solutions*, Freeman, San Francisco, 304pp.
Elzinga, A.: 1996, Shaping Worldwide Consensus: The Orchestration of Global Climate Change Research. In: A. Elzinga and C. Landström (eds.), *Internationalism and Science*, Taylor Graham Publishing, Cambridge, pp. 22–253.
Forrester, J.W.: 1961, *Industrial Dynamics*, M.I.T. Press Massachusetts Institute of Technology, Cambridge, 464pp.
Funtowicz, S.O. and Ravetz, J.R.: 1993, Science for the Post-Normal Age, *Futures* 25, 739–755.
Gibbons, M., Limoge, C., Nowotny, H., Schwartzman, S., Scott, P., and Trow P.: 1994, *The New Production of Knowledge. The Dynamics of Science and Research in Contemporary Societies*, Sage Publications, London, 179pp.
Groß, M.: 2004, Human Geography and Ecological Sociology: The Unfolding of a Human Ecology, 1890–1930 and Beyond, *Soc Sci Hist* 28, 57–605.
Guston, D.H. and Sarewitz, C.: 2002, Real-Time Technology Assessment, *Tech Soc* 24, 93–109.
Habermas, J.: 1968, Erkenntnis und Interesse. In: J. Habermas (ed.), *Technik und Wissenschaft als Ideologie*, Suhrkamp, Frankfurt am Main, pp. 146–168.
Habermas, J.: 1984, *The Theory of Communicative Action, Vol. 1, Reason and the Rationalization of Society*, Beacon Press, Boston, 465pp.
Habermas, J.: 1987, *The Theory of Communicative Action, Vol. 2, Lifeworld and System, A Critique of Functionalist Reason*, Beacon Press, Boston, 457pp.
Hirsch Hadorn, G., Bradley, D., Pohl, C., Rist, St., and Wiesmann, U.: 2006, Implications of Transdisciplinarity for Sustainability Research, *Ecol Econ* 60, 119–128.
Hubert, B. and Bonnemaire, J.: 2000, La construction des objets dans la recherche interdisciplinaire finalisée: de nouvelles exigences pour l'évaluation, *Nat Sci Soc* 8, 5–19.
Jantsch, E.: 1972, Towards Interdisciplinarity and Transdisciplinarity in Education and Innovation. In: L. Apostel et al. (eds.), *Problems of Teaching and Research in Universities*, Organisation for Economic Cooperation and Development (OECD) and Center for Educational Research and Innovation (CERI), Paris, pp. 97–121.
Kates, R.W., Clark, W.C., Corell, R., Hall. J.M., Jaeger, C.C., Lowe, I., McCarthy, J.J., Schellnhuber, H.J., Bolin, B., Dickson, N.M., Faucheux, S., Gallopin, G.C., Grübler, A., Huntley, B., Jäger, J., Jodha, N.S., Kasperson, R.E., Mabogunje, A., Matson, P., Mooney, H., Moore, B.I., O'Riordan, T., and Svedin, U.: 2001, Sustainability Science, *Science* 292, 641–642.
Klein, J.T., Grossenbacher-Mansuy, W., Häberli, R., Bill, A., Scholz, R.W., and Welti, M.: 2001, *Transdisciplinarity: Joint Problem Solving among Science. An Effective Way for Managing Complexity*, Birkhäuser Verlag, Basel, 332pp.
Kockelmans, J.J.: 1979, Why Interdisciplinarity? In: J.J. Kockelmans (ed.), *Interdisciplinarity and Higher Education*, The Pennsylvania State University Press, University Park and London, pp. 125–160.
Krohn. W. and van den Daele, W.: 1998, Science as an Agent of Change: Finalization and Experimental Implementation, *Soc Sci Inform* 37, 191–222.

Kuhn, Th.: 1963, *The Structure of Scientific Revolutions*, University of Chicago Press, Chicago, 172pp.
Lewin, K.: 1951, *Field Theory in Social Science*, Harper and Brothers Publishers, New York, 346pp.
Max-Neef, M.A.: 2005, Foundations of Transdisciplinarity, *Ecol Econ* 53, 5–16.
Meadows, D.L, Meadows, D.H., Randers, J., and Behrens, W.W., III, 1972: *The Limits to Growth*, Universe Books, New York, 205pp.
Mittelstraß, J.: 1992, Auf dem Weg zur Transdisziplinarität, *GAIA* 5, 250.
National Academies (eds): 2005, *Facilitating Interdisciplinary Research*, The National Academies Press, Washington, 306pp.
Nicolescu, B.: 1996, *La Transdisciplinarité – Manifeste, Éditions du Rocher Monaco*, Retrieved November 1, 2006, from www.nicol.club.fr/ciret/english/visionen.htm.
Nowotny, H.: 1999, The Need for Socially Robust Knowledge, *TA-Datenbank-Nachrichten* 8, 12–16.
Parsons, T.: 1968, *The Structure of Social Action*. Vols. I & II, The Free Press, New York, 775pp.
Pohl, C. and Hirsch Hadorn, G.: 2007, *Principles for Designing Transdisciplinary Research. Proposed by the Swiss Academies of Arts and Sciences*, oekom, München, 124pp.
ProClim: 1997, *Research on Sustainability and Global Change – Visions in Science Policy by Swiss Researchers*. CASS/SANW, Bern, Retrieved December 3, 2006, from http://www. proclim.ch/Reports/Visions97/Visions_E.html.
Rescher, N.: 1979, *Cognitive Systematization. A Systems-theoretic Approach to a Coherentist Theory of Knowledge*, Blackwell, Oxford, 211pp.
Schütz, A. and Luckmann, T.: 1973, *The Structures of the Life-World* (R.M. Zaner and T. Engelhardt, trans.), Northwestern University Press, Evanston and Heinemann, London, 335pp.
Sherif, M. and Sherif, C.D. (eds.): 1969, *Interdisciplinary Relationships in the Social Sciences*, Aldine, Chicago, 360pp.
Stichweh, R.: 1992, The Sociology of Scientifc Disciplines: On the Genesis and Stability of the Disciplinary Structure of Modern Science, *Sci Context* 5, 3–15.
Weber, M.: 1949, *The Methodology of the Social Sciences* (E.A. Shils and H.A. Finch, trans.), The Free Press, New York, 188pp.
World Commission on Environment and Development: 1990, *Our Common Future*, Oxford University Press, Oxford, 400pp.

# Part II
# Problem Identification and Problem Structuring

# Chapter 3
# From Local Projects in the Alps to Global Change Programmes in the Mountains of the World: Milestones in Transdisciplinary Research

**Bruno Messerli and Paul Messerli**

**Abstract** In the 1970s and 1980s UNESCO's worldwide Man and Biosphere Programme (MaB) provided a great stimulus to overcome the large gap between natural and social sciences. The global project number six, Man's Impact on Mountain Ecosystems, led to the Swiss National Research Programme 'Socio-Economic Development and Ecological Carrying Capacity in a Mountainous Region'. It had a profound effect on mountain research in general and on an interesting collaboration between different alpine countries in particular. Even though the expression transdisciplinarity was not yet known and defined, participating scientists from different disciplines were forced to develop methods and models for a true inter- and transdisciplinary cooperation, as discussed in Section 3.2.

In the 1980s and 1990s the discussion about transdisciplinarity began. In the developed world transdisciplinary research means striving for concrete problem solving in the social and political context through cooperation between science and society. In the developing world transdisciplinary research needs to concentrate on certain key processes and limiting factors in cooperation with the local population and political authorities. This concept was further developed as the so-called 'Syndrome Mitigation Research'. This approach proposes to compare clusters of problems linked to global change in mountain areas, in order to develop adequate mitigation strategies towards sustainability, as discussed in Section 3.3.

In the 1990s and 2000s fragile mountain ecosystems became significant on a global level as sensitive indicators for 'Global Environmental and Climate Change', and as treasures of natural and cultural resources (water, mining, forestry and agriculture, biological and cultural diversity, recreation and tourism). Therefore, the 'Mountain Research Initiative' (MRI), a project of the 'Global Change Programmes', was founded. In mountain areas natural and human processes are especially closely connected and this means that inter- and transdisciplinarity have a very special significance for mountain research and development, as discussed in Section 3.4.

✉ B. Messerli
Institute of Geography, University of Berne, Berne, Switzerland
e-mail: bmesserli@bluewin.ch

G. Hirsch Hadorn et al. (eds.), *Handbook of Transdisciplinary Research*,

The similarities between Fig. 3.2 as a first approach in 1978 and Fig. 3.5 as the current global approach in 2005 is an impressive testimony to the enduring significance of the MaB research concept that, from the beginning, called for an integration of natural and social sciences in a problem oriented perspective.

**Keywords:** Mountain research · UNESCO-MaB programme · Integrated approach natural-social sciences · Syndrome mitigation research · Global change programmes

## 3.1 Background

The second half of the 20th century was a period with rapidly growing human impacts on all different ecosystems of our planet. In this new century, we will probably experience a further acceleration in the pace of environmental change, resource use and vulnerability of societies and economies. It becomes more and more evident that major natural processes – from the local to the global scale – are influenced by human activities, creating a much higher degree of complexity through the interference of processes which are normally studied within the domain of natural or social sciences (Steffen et al., 2004). This implies a need to bridge the gap between the two cultures of science in order to advance our understanding of contemporary driving forces and their rapidly growing impacts on ecosystems, especially on the most sensitive and fragile mountain ecosystems.

In the late 1960s outstanding strategic thinkers in UNESCO realised the growing imbalance between human activities and the environment. As a consequence the rather discipline oriented International Biological Programme (IBP) was replaced in 1971 by the integrative Man and Biospshere Programme (MaB). From the 14 MaB projects, covering the main ecosystems of the world, number 6 had the title: Man's Impact on Mountain Ecosystems. The related research concept was created in a workshop at Salzburg in 1973, the national activities began in Switzerland in 1976 and the research process started in 1979 as a National Research Programme, funded by the Swiss National Science Foundation. In 1981, to mark the 10th anniversary of the MaB Programme, an international conference: 'Ecology in Practice – Establishing a Scientific Basis for Land Management' was organised in Paris by UNESCO. The mountain programme was presented and introduced with the following statement: 'Man is the most important factor in the rapidly changing mountain landscape of today. The use of land and resources is very different in the mountains of developing and developed countries. In both areas there is a need for basic ecological research. But ecological research is not sufficient; decisions and measures are required to bring about change in how land and resources ought to be used. In turn, measures and decisions are not in themselves sufficient. A systems approach is needed for promoting more fruitful cooperation among scientists, inhabitants and decision makers. But at the same time, the systems approach is not sufficient. We need a new thinking, and a new way of thinking needs time' (Messerli, 1984).

## 3.2 The Swiss MaB Mountain Programme – 1979–1987

More than 25 years ago, UNESCO's Man and Biosphere (MaB) Programme had a stimulating effect on research in the Alps and on the cooperation between different alpine countries. After two years of preparation, the Swiss MaB Programme 'Socio-Economic Development and Ecological Carrying Capacity in Mountain Regions' began in 1979, in close collaboration with the Austrian Obergurgel experiment. For the first time a system approach was used to explore the transformation of a mountain community (Obergurgel) by rapidly growing tourism (Moser, 1975). There were three main reasons for a Swiss contribution to the MaB mountain project, strongly supported by the federal administration:

- The European Council had launched, some years before, a campaign to create awareness for ecological problems related to the development of tourism.
- Switzerland, as a country which was experienced in how to tackle regional development in mountain areas, was challenged by the UNESCO initiative to make knowledge and experience available to other countries.
- The National Research Programme on 'Regional Disparities in Switzerland', was set up only one year before the MaB Mountain Programme, but the ecological dimension was neglected in the programme design.

Financed by the Swiss National Science Foundation, about 40 research projects from very different disciplines were initiated in four different and carefully selected test areas of the Swiss Alps: Grindelwald, Aletsch, Pays d'Enhaut, and Davos. Although these four regions were very different from each other in their natural, economic, cultural and political conditions, they were all located in the Alps – in the Swiss periphery – outside the powerful political and economic centres. Such mountain regions are often controlled by external forces and factors which claim the region's resources and impact on the internal political decision making processes. At a time in which the expression 'sustainable' was still unknown, the overall aim of the research programme was to identify the external and internal driving forces, to understand the transformation of these communities and the relationship between inhabitants and their environment, to define desirable and undesirable structures and processes, and finally to answer the question of how a long-term, well-balanced development could be achieved.

Before new methods and models in general and interdisciplinary approaches in particular are discussed, we should keep in mind the dramatic change in mountain communities from the second half of the 19th to the second half of the 20th century: without understanding the past we shall never understand the future. For Obergurgel, for instance, Moser (1975) was able to reconstruct the situation in this mountain village around 1860: Obergurgel could be represented as a closed system (Fig. 3.1).

Strong and direct links between inhabitants and the natural environment and resources were characteristic at this stage. Over generations the farmers had learned the meaning of 'sustainability', because the system was virtually closed and

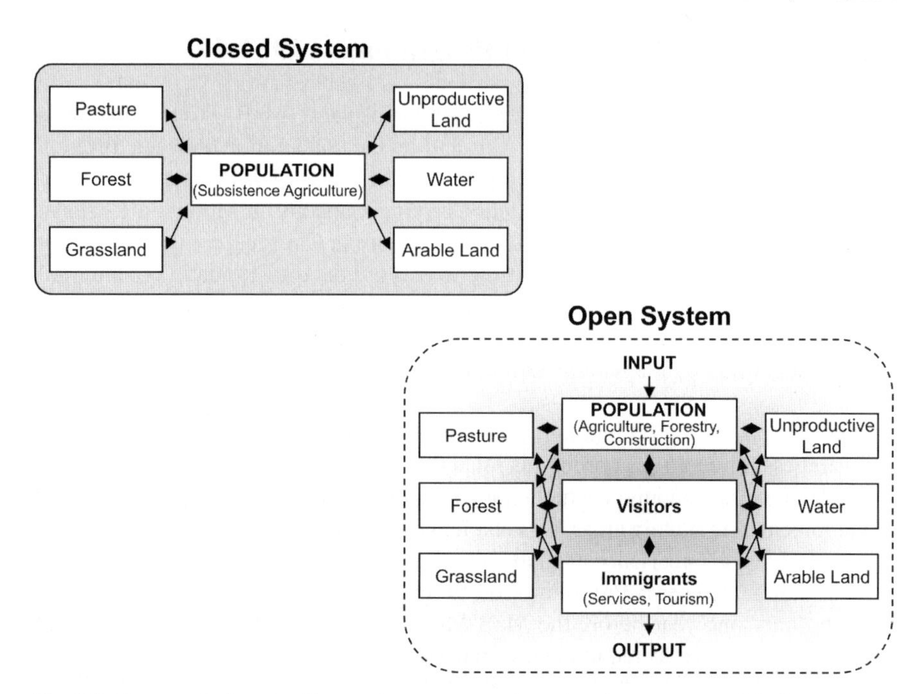


**Fig. 3.1** Obergurgel, Austria: Change from a more or less closed system around 1860 to an open system in the recent years (Explanations see text). Modified, especially the open system, after Moser (1975)

resources were limited. If the population increased to exceed the 'carrying capacity' of the area, some had to leave. In response, a strong social order was developed to maintain the population number in balance with the livelihood capability of the system. Under these circumstances a modelling approach would have been relatively simple: the grassland, forest, and arable land required to support a specific number of people could easily be calculated. On this basis it would be possible to develop a number of scenarios (for instance climatic change or population growth) and draw quite reasonable conclusions. However, these immediate 'challenge and response' relationships between a human population and its environment no longer exist, due to the opening up of the system, and the increasing influence of external flows, factors and forces. Figure 3.1 also represents a modified open system that can be applied to many mountain communities in the world (Messerli, 1983). There is now an input and output of people, capital, energy, technology, information, goods and services in many different forms. The number of relations not only increases but also becomes more complex due to the fact that different actors relate differently to the land use system. Even the areas that are perceived as unproductive under traditional rules become important for touristic infrastructure or nature conservation today. To model such a system without knowing whether some key elements are externally or internally controlled becomes much more difficult. The 'carrying capacity' of the community can therefore no longer be restricted to the matter and energy output of the system. In view of the challenge of how to conceptualise an open man–environment–system a rather pragmatic approach was adopted that should allow for an investigation of basic issues:

- Which socio-economic system processes have triggered, or could trigger, fundamental changes in land use patterns? These processes (including their internal and external control) should be traced.
- Which dimensions of the natural system and its components react particularly sensitively to changes in land use?
- Which consequences for the future utilisation of the alpine living-space result from the detected and potential changes in the natural system and its components?
- Which aesthetic and recreational qualities convey the image of the landscape to the various groups of actors; and which emotional ties to this living-space influence residents' decisions regarding land use and the behaviour and attitudes of tourists?

The MaB research concept (Fig. 3.2 with its four relations) was drawn up from a pure interaction logic between the natural and the socio-economic systems that is translated and shaped by the land use patterns. It can be used to indicate the fundamental action and reaction loops, including the influence of external factors (relation 1). The economic, political, demographic and cultural activities emanating from the socio-economic system do not only influence the type and intensity of land use but also the ecosystem's quality and services (relation 2). If the intensity of land

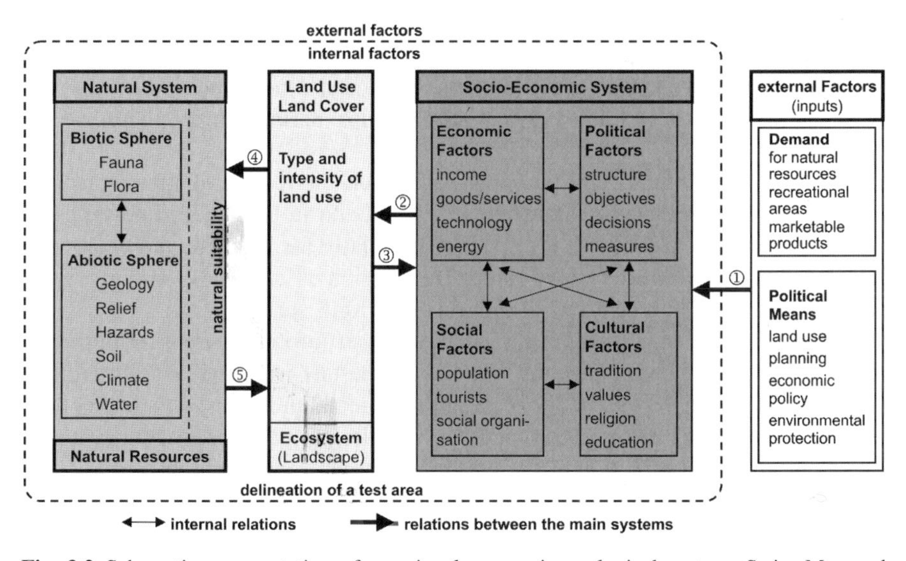


**Fig. 3.2** Schematic representation of a regional economic–ecological system: Swiss Man and Biosphere Programme (MaB). (Explanations see text; Messerli B. and Messerli P., 1978)

use can be absorbed by the natural system, there will be an expected feedback to the local economy and society in the form of harvestable goods, water supply and air quality (relation 3). If it cannot be absorbed, the natural system will be negatively affected (relation 4) and in response alter the land use system (relation 5) with possible consequences for the socio-economic system as well. This preliminary model was a first attempt to clarify which fundamental relationships could and should be investigated. It follows that this needs an interdisciplinary effort in which the land use subsystem makes the link between the demands of society and the capacity of nature (Messerli, 1986; Ives and Messerli, 1990).

The research process was launched from this starting point. The principal difficulty in defining related projects in the test areas proved to be the formulation of the research questions and mandates within the contributing disciplines. In no way did these research questions follow on from the basic issues presented above. The general character of the problem had to be met from the bottom up with research questions that were methodologically treatable by the involved disciplines. In addition, no-one had a theoretical background and knowledge broad enough to adequately offer a scientifically sound conceptualisation of the MaB scheme. Consequently, only a step-wise approach was feasible; taking into account first, the interactions between socio-economic development and land use changes and second, the effects of these changes on the natural system and its components. Corresponding data concepts were developed and, although today's GIS techniques were not yet available, this lead in two cases (Grindelwald and Davos) to spatial information systems in which socio-economic variables were coupled with those of the natural system. Moreover, the selection of disciplines and scientists followed a bottom-up and top-down procedure, ensuring that all relevant competences covering the main subsystems in the model were represented in the research teams. A good mixture of young innovative personnel and experienced senior scientists was an additional aspect that had to be considered.

A rough but not extreme simplification of the step-wise analysis of the man–environment system and its relevant relationships permits three phases to be described:

*The explorative phase*: This phase was characterised by the attempt to construct an extensive database and guaranteed both resolution in time (e.g. time-series of socio-economic development) and space (particularly for land use and ecological information) for the entire research area. The division of labour was strongly disciplinary and deliberately covered all aspects of the MaB scheme.

*Phase of progressive integration*: The amount of information concerning each test area increased rapidly once the first research results were available. Interdisciplinary cooperation started, not only because of mutual problems with data collection and with survey and analysis methods, but also in some cases spontaneously. The application of simulation techniques led to a more systematic investigation of existing interdependencies and to the first, admittedly incomplete, ideas of quantitative modelling.

*Synthesis phase*: Three processes were central to project synthesis in the test areas:

- the operational linkages following the MaB impact analysis (socio-economic development → land use consequences → effects in the natural system),
- the elaboration of key factors for scenario construction, and
- quantitative (simulation model) and qualitative (scenario technique) analysis of various development paths.

At the programme synthesis level, the research results from the test areas were newly challenged from a problem oriented viewpoint and concentrated into generalised statements and conclusions.

Our holistic research approach claimed to consider an extensive range of topics, so as not to restrict the research process from the start. The fundamental dilemma of this approach accompanied us throughout, and led to a balancing act between the holistic and reductional procedures. Since our knowledge must be illuminated from the viewpoint of the individual disciplines, sufficiently precise research questions are vital for actuating the research process. Once this basic knowledge is available and the participants are accepted experts within the research group, the process of linkage can begin and must be supported by formalised aids to communication (e.g. information systems, quantitative models, scenario techniques). It is dangerous to degrade the disciplines to mere suppliers of data from the start, for their specific knowledge builds on an implicit understanding of the system and should be skilfully used.

The lack of high-level theory concerning the essential relationships in a man–environment system is also an expression of the differing ways in which the problems under consideration are regarded from natural- and social-scientific viewpoints. Every attempt to devise a comprehensive model is immediately halted by this deficit and by an oversupply of theoretical concepts for the individual subsystems in the MaB model. Consistency cannot be obtained: this leaves only the pragmatic approach which concentrates on the operational linkage of the different subsystems. In the research process, this state of affairs is manifested in a continuous balancing act between theoretical and data-controlled approaches.

The linkage of the various subsystems primarily conflicts with the problem of aggregation. The relationships modelled in the economic and socio-cultural systems must produce outputs which are relevant to land use, so that land use alterations can be incorporated into the natural system. Conversely, the information supplied about the natural system, the landscape and the quality of life must be able to be absorbed from the socio-economic system as inputs relevant to action and behaviour. The complete operationalisation of this flow of information between the socio-economic and natural systems leads towards micromodelling, a process which we were not able to accomplish. Our pragmatic approach led to a two-stage problem resolution, through the reasoned combination of the socio-economic development scenarios in aggregated form with the spatial land use scenarios in a spatial resolution, which allowed for impact analyses.

Figure 3.3 shows the final MaB approach for formulating long-term strategies. The different levels represent different stages of knowledge and knowledge integration. Simulation models fulfilled a double function in this respect. During their

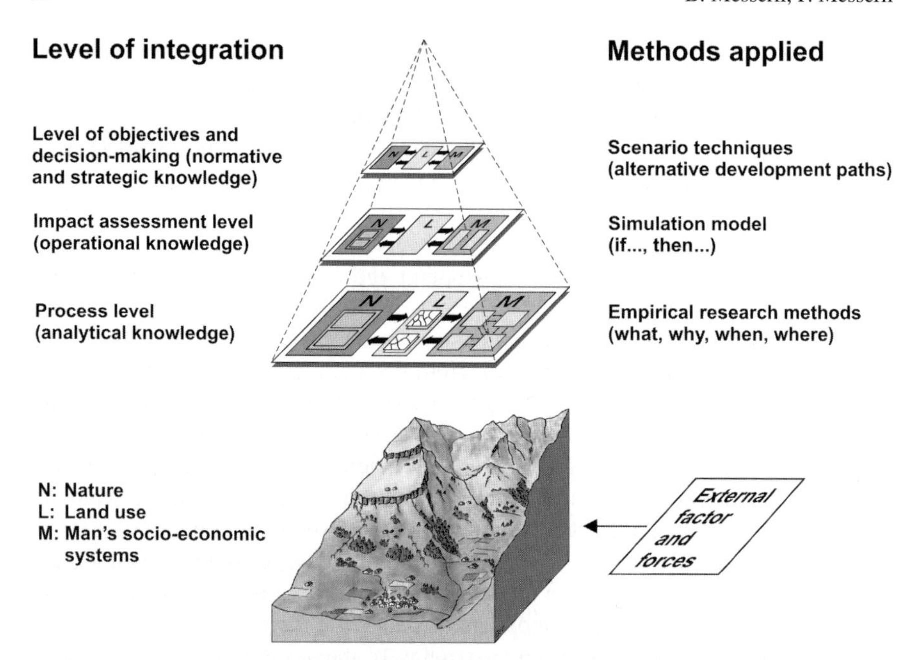


**Fig. 3.3** Further developoment of the MaB–model through different levels of contents and methods in order to reach a satisfactory base for scenarios, strategies and decision-making (Messerli P., 1986)

construction they stimulated the research process by provoking new research questions, by discovering data gaps, and by challenging linkage hypotheses. During the application stage, the models produced indispensable quantitative data structures for various development paths, including that of changing land use patterns, as basic information for the impact assessment. Scenario techniques were applied to analyse the pathways to alternative futures of the man–environment system. If each scenario gives a possible yet very strongly idealised picture of the future, there is an essential gain in strategic knowledge that allows for a choice between different paths and policies (Messerli, 1989).

In order to illustrate how the research process was continued in a transdisciplinary second step, we focus on the MaB project Grindelwald, from 1986 until the present, which is the most comprehensive study with a well defined implementation phase.

*Grindelwald* lies close to the highest summits of the Bernese Alps offering a spectacular landscape and winter sport opportunities initiated by British pioneers as early as the 19th century. The official inauguration of the railway to the Jungfraujoch (3,454 m above sea level) in 1912, may be a symbol of this first boom of tourism. Other astonishing projects to reach the mountain tops (e.g. Wetterhorn, Jungfrau) failed with the commencement of the First World War. Due to this long experience of tourism and economic development outside agriculture, and because the fast expansion of the winter sport infrastructure had major impacts on the natural systems and the landscape, the MaB research project was very welcome and was supported by the local authorities and population.

However, some scepticism from local stakeholders was also addressed by the project for two reasons: scepticism against scientific knowledge far removed from their own life-experience; and scepticism against any revelation of findings that could damage the image of the resort.

Due to this ambivalent attitude of the locals, the research team decided to limit the presentation of the project findings during the field research phase, because of its complex nature. Two articles in the local newspaper concerning tourists' critical appraisal of the transport infrastructure and the sceptical attitudes of the residents towards immigrants attracted attention and raised the level of discussion about the project. However, once the research process came to an end, an official invitation was received to present the main findings in the assembly hall of the commune. And this was the real beginning of a well concerted interaction between scientific experts, local authorities, and stakeholders. This interaction shows all the characteristics of a transdisciplinary process, i.e. the confrontation and integration of scientific conceptual knowledge with everyday contextual knowledge. An abbreviated summary of the main activities is presented next. Following Wiesmann (2001):

*1984–1987*: Based on the results of the MaB Programme, the community of Grindelwald initiated a participatory process in order to formulate certain policy guidelines. This process involved all inhabitants and more than 70 local institutions of the civil society. Members of the MaB research team acted as resource persons. This attempt resulted in policy guidelines, 'Grindelwald 2000', that explicitly formulated long-term goals covering the environmental, the socio-cultural and the economic dimension of sustainability. The goals were addressed to all relevant stakeholders. The resulting guidelines, 'Grindelwald 2000', can be seen as a first realisation of a 'Local Agenda 21' in the Alps.

*1986–2000*: Governmental and non-governmental organisations of Grindelwald developed and implemented measures compatible with the policy guidelines. A key concern thereby was to break the structural forces that endanger the positive balance between tourism, agriculture and the environment. Of utmost importance were measures in the following four fields: controlling investments in the transport sector, steering and channelling building investments without endangering the labour market, reviewing the village and landscape planning and diversifying supply components in tourism.

*1999–2001*: The community commissioned the former MaB research team to scientifically monitor the present situation according to the political guidelines. The community evaluated the achievements of the goals and their priorities. This assessment disclosed that the guidelines were generally successfully implemented, with a broad range of predominantly positive direct and indirect effects. However, the evaluation also led to the conclusion that the time had come to reconsider the goals set for the sustainable development of the community as some of the goals had become obsolete and a new generation had taken the political lead.

*2005*: The local government of Grindelwald decided to prove the adequacy of the political guidelines with respect to a balanced development path that meets ecological, economic and social interests and which change over time.

This short description of the continuing political–scientific reciprocal learning process for quantitative growth shows very clearly how innovative political authorities and an actively participating population can use scientific knowledge for the benefit of their future development. The Swiss MaB Programme has led to advances in the theory and practice of human–environment research in other mountain regions of Europe, Russia and Africa (Price, 1995) as will be stated in the following chapter. If the Man and Biosphere Programme encouraged many new initiatives for interdisciplinary research with the goal of acquiring action oriented knowledge for sustainable development, it inevitably promoted transdisciplinary research where local and regional stakeholders had to be addressed. The involvement of non-experts became a prerequisite in order to link two sorts of knowledge: conceptual scientific knowledge and contextual knowledge from real-life experience.

## 3.3 The Need for Transdisciplinarity in the Developing World Context – Since 1986

Fieldwork on highly specialised topics began in the mountains of Africa in the late 1960s, but integrated projects with natural- and social science components only started in the 1980s and reached an initial culmination with the first conference on African mountains and the foundation of the African Mountain Association in Addis Ababa 1986 (Messerli and Hurni, 1990). Research projects in the African context, for example on the critical problem of soil erosion in the mountains and highlands of East and North Africa, clearly revealed that pure natural science approaches were of little help when striving for counter measures that could be implemented locally. Without understanding the demographic, social and economic conditions of the local population, the process knowledge about soil erosion and derived technical solutions remain outside the harsh realities of everyday life. Moreover, fieldwork in the mountains of tropical and subtropical Africa made it clear that in a context of difficult and unstable political conditions, where people struggle for survival, the MaB model approach, which had been developed in the socio-economic context of well developed alpine communities, had to be adapted to the developing mountain world with its extreme problems of land degradation, water scarcity, population growth, poverty and even hunger. FAO (2002) estimates the total number of mountain people at 718 million in the year 2000. Of these, 625 million live in developing countries. Based on information currently available FAO estimates that more than half of the mountain population in the developing countries, in the range of 250–370 million people, are vulnerable to food insecurity. This again can increase the pressure on the natural resources and the highly vulnerable life support system. Therefore, the researchers are forced to set priorities and to start emergency actions in partnership with the local population, as soon as the key processes and the limiting factors are clearly identified. Moreover, the gained knowledge must be introduced into the decision making processes at all levels from political authorities to the individual farmer (Messerli et al., 1988). The important experience from a development context that implies immediate actions is reflected in the modified MaB model (Fig. 3.4), being the basic concept of man–environment research as it emphasises the need to restrict the systematic to a pragmatic approach and to involve the decision makers at all levels right from the beginning of the project.

For the developing world the focus on priority problems or key processes became a common basis for research on development. However, a next step to conceptualise the hypothesis that environmental problems appear in typical clusters or syndromes led to the so-called 'syndrome mitigation research'. The term syndrome of global change was borrowed from the 'German Advisory Council on Global Change', which first looked at combinations or clusters of problems of non-sustainable development. The Swiss 'National Centre of Competence in Research (NCCR) North-South' revised the German version, which was criticised for focusing solely on the negative aspects of development. Therefore, the priorities are: innovative approaches to mitigate syndromes, participatory and sustainability oriented research, and normative dimensions of sustainability. This new perspective in 'syndrome mitigation research' envisages three main goals (Hurni et al., 2004; Messerli et al., 2005):

- To promote disciplinary, interdiscplinary and transdisciplinary research focusing on sustainable development ('transdisciplinary' in this context is understood as an approach based on collaboration with local people that takes in account their rich traditional knowledge),

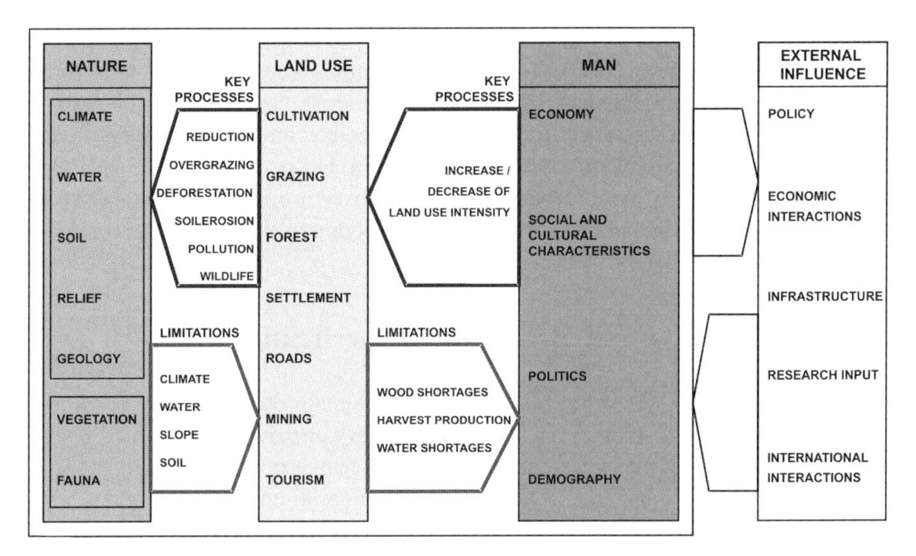


**Fig. 3.4** Adapation of the the MaB-model to the conditions of the mountains and highlands in Ethiopia. Key processes and limiting factors determine transdisciplinary and integrated approaches (Hurni in Messerli et al., 1988)

- To help develop institutions and train staff in these fields of research, in partner countries and in Switzerland,
- To support societies and institutions in partner countries in their autonomous efforts to address syndromes of global change over the long term (Hurni et al., 2004).

Today the syndrome mitigation programme is working worldwide in nine regions, seven of them being in developing countries: East Africa, the Horn of Africa, Central Asia, South Asia, the Mekong Region, Central America and the Andes. In order to find out whether or not core problems in all the mountain regions were of a similar kind and complexity, representatives of the local population, national institutions and foreign experts were invited to evaluate the situation. As a result, similar core problems could be identified in these different mountain regions. This means that a comparable approach focused on the mitigation of these worldwide common problems can make a real and true contribution to Global Change Research.

In retrospect it is remarkable to notice how the theory driven approach, represented in the original MaB model, shifted towards a more problem oriented and action driven approach. In the mountain context of Africa the shift was towards the global syndrome mitigation approach. The methodological rigor of the syndrome approach lies in its comparative component, partially realised already in the MaB mountain programme (Price, 1995; Funnell and Price, 2003) and now systematically included in the research design of each local project, allowing for the verification of the existence of certain types of syndromes in connection with global change.

According to our understanding, transdisciplinary research, as it was defined by Thompson Klein et al. (2001), has a much longer tradition than indicated by the use of the term, which only became popular in the 1990s. In the 1970s, when the concept for the different UNESCO – MaB Programmes were developed, the words 'transdisciplinarity' and 'sustainability' did not yet exist. Nevertheless, the participation of the local population, political authorities and technological experts was an important element in all the programmatic guidelines. However, since the late 1980s the differentiation between inter- and transdisciplinary projects began to evolve as it is discussed and ultimately defined in the introduction to this book.

## 3.4 Global Change and Mountain Research Since 1992

Although UNESCO's MaB Programme covered the main ecosystems on Earth and the growing impact of human activities on the environment, it was not a global programme in the present understanding of global change research. During the second half of the 1980s and the 1990s, the scientific community was forced to take new responsibilities by creating new structures and organisations to follow the rapid progress of economic globalisation, urbanisation and industrialisation and their consequences for climate and the environment, from the local to the global scale. The International Council of Scientific Unions (ICSU) decided in 1986, during a General

Assembly at the University of Bern, to start the natural science dominated International Geosphere – Biosphere Progamme (IGBP). This emphasis on the natural sciences created exactly the same difficulties as had occurred 10 years earlier in the MaB Programme. In a similar way the early preparation of the MaB Programme in the Swiss Alps was dominated by natural scientists and it was very difficult to motivate scientists from the social, economic, legal and cultural disciplines to cooperate. Without discussing the reasons for these different attitudes, the MaB committee had to overcome these difficulties in order to meet the challenge of a problem oriented research approach. A solution was possible on the national level under the defined setting of the Swiss National Science Foundation and the competent leadership of a multidisciplinary team of experts. On the global level, it took 10 years from the foundation of the IGBP (1986), before the International Social Science Council (ISSC), in cooperation with ICSU's IGBP, followed up with the foundation of the International Human Dimensions on Global Environmental Change Programme (IHDP) in 1996 – thanks to strong support from the German government and an efficient headquarters in Bonn. It was a pity that during those 10 years, mainly due to insufficient funding, an attempt to start such a human dimension programme failed in Switzerland.

The real breakthrough for the mountains as a global topic was the so-called 'Earth Summit' in Rio de Janeiro (1992), which included the chapter 'Managing Fragile Ecosystems – Mountain Sustainable Development' in the Agenda 21. Five years later at the special UN–General Assembly in New York (1997), supported by UNU and UNESCO, began a new period for mountain research and development (Messerli and Ives, 1997). The necessary awareness of the significance of mountains as sensitive indicators of natural and human induced global change and as treasures of natural and cultural resources for highland and lowland populations was not only sufficiently established and continuously growing, but culminated in three events or initiatives:

- The International Year of Mountains (2002), was announced in order to raise the awarenesss of the global public and formulate a global policy for the problems in the mountains of the world.
- The International Mountain Partnership (2002) with its secretariat at FAO in Rome, was founded in order to create a cooperation between different countries for concrete problem solving.
- The Mountain Research Initiative, with its secretariat in Switzerland and supported by the Swiss National Science Foundation, was founded in order to organise the scientific community for a transdisciplinarity global change research programme in the mountains of the world (Becker and Bugmann, 2001; Huber et al., 2005).

These three events represent by their connection a certain transdisciplinarity. Looking back over the time from the first publication of the MaB model in 1978 to the Global Change Programmes of today, it would be interesting to know if certain ideas remained the same or whether everything changed. With this question in mind, it was a great surprise to see the most recent science plan and implementation strategy of the new IGBP/IHDP Global Land Project (GLP, 2005), which is replacing the Land Use – Land Cover Project (LUCC, 1995). The main aim of this new research programme is to improve the coupled human–terrestrial environment system in the context of the Earth system functioning. Figure 3.5 is the leading diagram of GLP, showing the analytical structure of this new programme. It is astonishing to recognise a great similarity between this and the first MaB model in Fig. 3.2. The system structure and the system interactions are almost the same; however, there has been great progress in the modelling of those interactions and processes since the 1970s and 1980s. In addition, the studies of natural and human induced driving forces create a new understanding of interactions from the local to the global scale. All the same, it is interesting to see a similar structure for an integrated transdisciplinary research approach for the MaB Programme on the local level and 27 years later for the GLP programme on the global level.

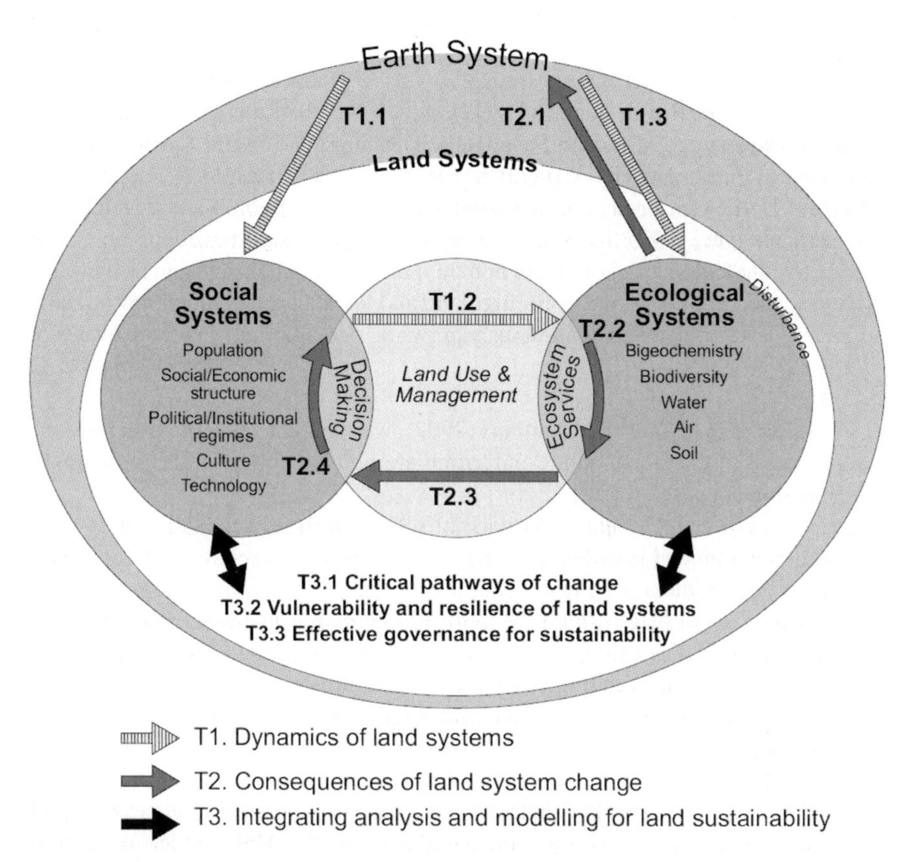


**Fig. 3.5** Global Land Project (GLP) of IGBP and IHDP. The leading figure for the science plan and the implementation strategy (GLP, 2005)

Not only do we have to consider the extension from the local to the global level, but also the development from inter- to transdisciplinarity in this time period. This looks very different. In almost all series of Global Change publications, interdisciplinarity is used not only for the cooperation between different disciplines, but also for the participation with actors outside academia in general or with experts and political authorities in particular. This simplified and generalised use of the term interdisciplinarity may show that the clear definition of inter- and transdisciplinarity is not of special interest in a global collaboration. It is much more important to focus on the cooperation between natural and human sciences and on disentangling natural and human driving forces. But it is evident, solutions are not possible without cooperation, without a common language and without integrated approaches. For the MaB Programme it was the same: interdisciplinarity was understood to be a cooperation between disciplines and with actors and participants outside science; because the study of a certain test area in order to develop an economic–ecological long term strategy needed the cooperation with the local population, experts and opinion leaders. The definitions for inter- and transdisciplinarity are today clearly formulated and accepted (Thompson Klein et al., 2001). Although the MaB Programme did not yet know and the Global Change Programmes did not care, both used transdisciplinary approaches and methods. For both programmes, different in time and different in scale, it was clear that integration was not the same as summation. Moreover, most decisions about the environment include human values and beliefs. The understanding or predicting of sustainability needs an identification of the characteristics of resilient systems and it needs dynamical relationships between knowledge production, policy formulation, and decision making (Wasson and Underdal, 2002).

## 3.5 The Development of Inter- and Transdisciplinarity in the Last 30 Years

In the executive summary of a recent IGBP synthesis with the title ‘Global Change and the Earth System, a Planet under Pressure’, Steffen et al. (2004: 2) wrote: ‘The interactions between environmental change and human societies have a long and complex history, spanning many millennia. They vary greatly through time and from place to place. Despite these spatial and temporal differences, in recent years a global perspective has begun to emerge that forms the framework for a growing body of research within the environmental sciences. Crucial to the emergence of this perspective has been the dawning awareness of two fundamental aspects of the nature of the planet. The first is that the Earth itself is a single system, within which the biosphere is an active, essential component. Second, human activities are now so pervasive and profound in their consequences that they affect the Earth at a global scale in complex, interactive and accelerating ways; humans now have the capacity to alter the Earth system in ways that threaten the very processes and components, both biotic and abiotic, upon which humans depend’. This statement by one of today's leading scientific communities confirms the challenge that the MaB Programme had taken up as early as the late 1970s. In retrospect the scientific community concerned with environmental change and development processes in mountain areas had already travelled a long way along the track towards a transdisciplinary approach, integrating conceptual knowledge from different disciplines and contextual knowledge from relevant stakeholders and decision makers.

*The Swiss MaB Programme* was typically theory driven, aiming to produce system knowledge about the interaction of socio-economic development and environmental change. The integration of analytical knowledge from the engaged disciplines was mainly done by modelling approaches on different levels of data aggregation. When the system dynamics came to be understood by the scientific body, the problem perception of the stakeholders and decision makers came into play in order to elaborate targets and action plans in a transdisciplinary process, followed by the implementation of a monitoring system.

*The African Mountain Research* was problem- and action driven right from the beginning. It started with the identification and analysis of key processes that accounted for the severe environmental degradation. The MaB concept served only as a reference at this stage of the process, looking for straightforward actions in order to stop the deteriorating land use practice (e.g. Food for Work Programme). However, we know that the effects of these short term actions were not sustainable because they neglected the institutional framework and the external constraints of the local stakeholders. This insight led to the development of a multilevel and multistakeholder approach in the follow-up period of the project. It has been learned that the behaviour logics and the internal and external constraints of all relevant stakeholders must be taken into account right from the beginning. Target knowledge and behavioural knowledge therefore became an integral part of the transdisciplinary approach as it is manifested in later concepts.

*Global Change Research in Mountain Areas* aims to extend the 'option space' (development alternatives) for local and regional stakeholders in response to the environmental and socio-economic challenges of global change. Again this approach is problem- and action driven, and based on the challenges and problems as they are perceived by the local population. Target and action knowledge (Fig. 3.6 and Fig. 3.7), and a clear participatory orientation, become as prominent as system knowledge in this final stage of the research scheme (CASS-ProClim, 1997).

If we synthesise our experience with the man–environment research in mountain areas from the early and local MaB projects to the recent and global Mountain Research Initiative (MRI) within the framework of interdisciplinary and transdisciplinary research explored in the CASS-ProClim (1997) document (Fig. 3.6), which became a standard in the field, then the step by step progress becomes quite visible and indicative. However, this does not mean that the methodological and methodical challenges of knowledge confrontation and integration are already solved satisfactorily. Nevertheless, the World Social Science Report emphasises: 'The convergences between natural and social sciences becomes greater to the degree one views both as dealing with complex systems . . .' (UNESCO, 1999). This can be seen as common ground to bridge the gap.

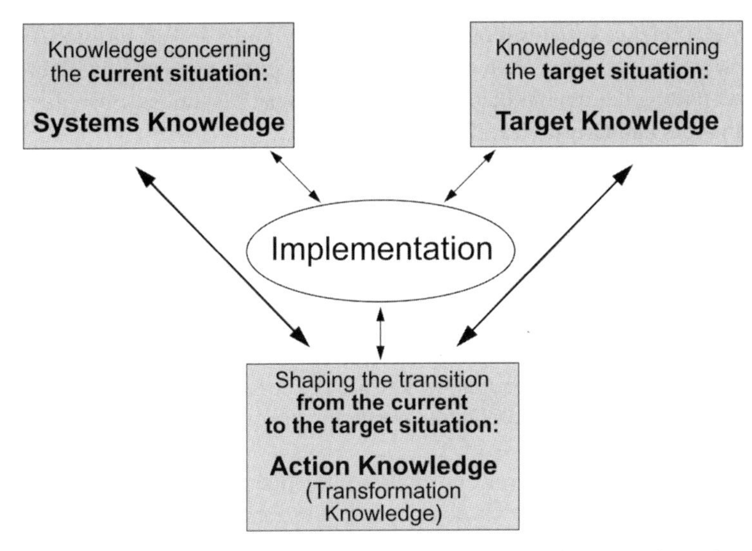


**Fig. 3.6** Types of knowledge in an transdisciplinary context (slightly modified: 'Action Knowledge' instead of 'Transformation Knowledge', CASS-ProClim, 1997)

The World Conference on Science in Budapest (1999) stated in the final Declaration on Science and the Use of Scientific Knowledge: 'Greater interdisciplinary efforts, involving both natural and social sciences, are a prerequisite for dealing with ethical, social, cultural, environmental, gender, economic and health issues' (UNESCO-ICSU, 2000). But it is a matter of fact that the scientific communities have frequently found it difficult to form a clear distinctive definition and understanding of successful collaborative research between different disciplines in relation to Global Environmental Change. In the search for 'best practices', new experiences should be more systematically evaluated and reflexively proved. The

| **Types of knowledge** | **How to produce knowledge** | | | **Experienced in** | | |
|---|---|---|---|---|---|---|
| | Discipl. | Interdiscipl. | Transdiscipl. | MaB-CH | African Mountains | Global Change in Mountains |
| Systems knowledge | X | X | (X) | X | X | X |
| Target knowledge | | | X | (X) | ? | X |
| Action knowledge | (X) | (X) | X | (X) | X | X |

**Fig. 3.7** Experienced knowledge production related to inter- and transdisciplinarity

international science organisations have the responsibility to call for more collaborative research because we are more and more frequently confronted with problems of a cross-cutting nature and the scientific community has the responsibility to find answers to this rapidly growing challenge.

## 3.6 Recommendations

- The distinction between theory, problem and action driven research, is justified in view of different kinds of knowledge production. Problem driven research includes target and transformation knowledge, whereas theory driven research aims at system knowledge by conceptualising how systems function and behave (Fig. 3.6 and Fig. 3.7).
- Knowledge integration across disciplinary boundaries requires a common conceptualisation of the research objects and questions in order to find a common language that structures the new field of discoveries. Interdisciplinary research creates an added value for all related disciplines.
- Transdisciplinary research requires an additional step in knowledge integration in order to make use of generalised conceptual knowledge in a specific societal and environmental context. Transformation and translation processes between researchers and stakeholders outside academia require intensive and time consuming interactions.
- Funding agencies are normally not prepared to cope with these new kinds of research proposals, which require a different pool of expertise and new evaluation procedures. Special funding is necessary to keep out competition from established disciplinary research. However, best practices need to be established in this field by learning on the job using the peer review procedures.
- The MaB Programme postulated and launched a new standard for environmental research, which has been, and which will be crucial for the ongoing changing environment and for the future management of ecosystems and limited natural resources. This new standard is broadly accepted in the scientific community and by the funding agencies. The breakthrough, however, depends on a new research generation which is able to establish a new research tradition and which is fully supported by the research agencies.
- In the framework of IHOPE (Integrated History and Future of People on Earth) and the Global Change Programmes it was said: 'It will be necessary to continuously incorporate the perspectives of non-disciplinary experts. These stakeholders include policy makers who must formulate and justify frameworks for future development, resource managers who must interpret and implement those frameworks and ultimately the communities who will either suffer or benefit from these policies and decisions' (Leemans and Costanza, 2005). This means, any comprehensive analysis of the causes and consequences of global change must go beyond scientific and economic considerations (O'Sullivan, 2004; Steffen et al., 2004).

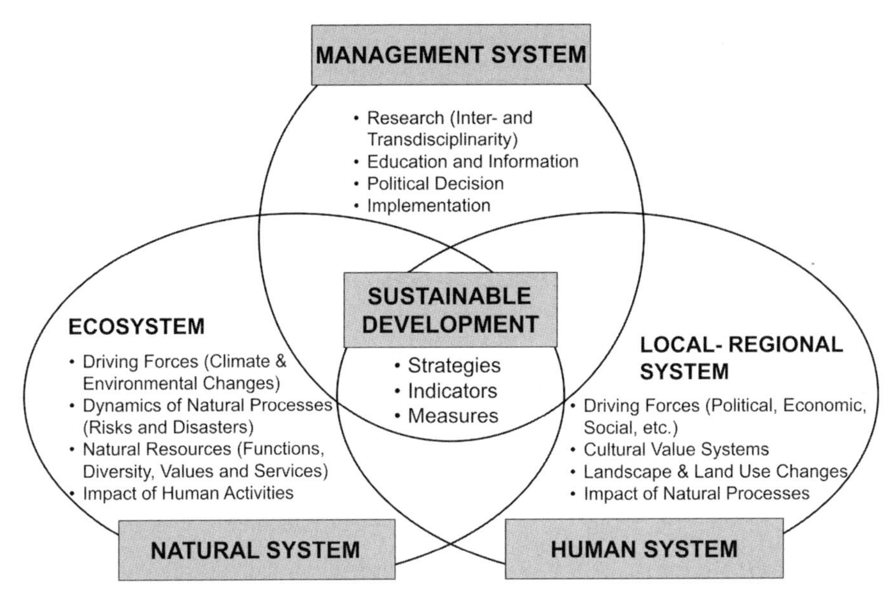


**Fig. 3.8** Key elements of disciplinary, interdisciplinary and transdisciplinary contributions to mountain research from the local to the global level

Finally Fig. 3.8 tries to illustrate and synthesise the development from the local MaB model to the environmental change approach, by integrating the natural and human system with their driving forces, specific resources and important values to produce better information, education, policy and management towards sustainability.

## References

Becker, A. and Bugmann, H. (eds): 2001, *Global Change and Mountain Regions. The Mountain Research Initiative*. IGBP Report 49, GTOS Report 28, IHDP Report 13, IGBP Secretariat, Stockholm, 86pp.

CASS-ProClim: 1997, *Research on Sustainability and Global Change – Visions in Science Policy by Swiss Researchers*, Swiss Academy of Science, Bern, 32pp.

FAO: 2002, Environment, Poverty and Food Insecurity: The Vulnerability of Mountain Environments and People, Special Feature. In: *The State of Food Insecurity in the World, 2002*, Rome, pp. 14–19.

Funnell, D.C. and Price, M.F.: 2003, Mountain Geography: A Review, *Geogr J* 3, 183–190.

GLP: 2005, *Global Land Project. Science Plan and Implementation Strategy*, IGBP Report 53, IHDP Report 19, IGBP Secretariat Stockholm, 64 pp (see also IGBP *Global Change Newsletter* 2005, 63, 16–17).

Huber, U., Bugmann, H. and Reasoner, M. (eds.): 2005, *Global Change and Mountain Regions. An Overview of Current Knowledge. Advances in Global Change Research*, Springer, Dordrecht, Netherlands, 650pp.

Hurni, H., Wiesmann, U. and Schertenleib, R.: 2004, *Research for Mitigating Syndroms of Global Change. A Transdisciplinary Appraisal of Selected Regions of the World to Prepare Development-oriented Research Partnerships*, Perspectives of the NCCR North–South, Geographica Bernensia, 468pp.

Ives, J.D. and Messerli, B.: 1990, Progress in Theoretical and Applied Mountain Research, 1973–1989, and Major Future Needs, *Mt Res Dev* 10, 2, 101–127.

Leemans, R. and Costanza, R.: 2005, *Integrated History and Future of People on Earth (IHOPE)*, IHDP, UPDATE 02.

LUCC: 1995, *Land-Use and Land-Cover Change. Research Plan*, IGBP Report 35, IHDP Report 7, IGBP Secretariat, Stockholm, 32pp.

Messerli, P.: 1983, The Concept of Stability and Instability of Mountain Ecosystems Derived from the Swiss MAB-6 Studies of the Aletsch Area, *Mt Res Dev* 3, 281–290.

Messerli, B.: 1984, Critical Research Problems in Mountain Ecoystems (Africa, Alps, Himalayas). In: F. Di Castri, F.W.G. Baker, and M. Hadley (eds), *Ecology in Practice* (1981), UNESCO, Paris, pp. 234–242.

Messerli, P.: 1986, *Modelle und Methoden zur Analyse der Mensch-Umwelt-Beziehungen im alpinen Lebens- und Erholungsraum. Erkenntnisse und Folgerungen aus dem Schweizerischen MaB-Programm 1979–1985*, Bern, Bundesamt für Umweltschutz; MaB, 176pp.

Messerli, P.: 1989, *Mensch und Natur im alpinen Lebensraum. Risiken, Chancen, Perspektiven*, Paul Haupt, Bern, 368pp.

Messerli, B. and Hurni, H. (eds): 1990, *African Mountains and Highlands, Problems and Perspectives*, Geogr. Institute, Univ. of Berne, 450pp.

Messerli, B. and Ives, J.D. (eds): 1997, *Mountains of the World. A Global Priority*, Parthenon, Carnforth-Lancs, 495pp.

Messerli, B. and Messerli, P.: 1978, Wirtschaftliche Entwicklung und ökologische Belastbarkeit im Berggebiet, *Geogr Helv* 4, 203–210.

Messerli, B., Hurni, H., Wolde-Semayat, B., Tedla, S., Ives, J.D., and Wolde-Mariam, M.: 1988, African Mountains and Highlands, *Mt Res Dev* 2/3, 93–100.

Messerli, P.D., Wiesmann, U., and Hurni, H.: 2005, The Mountains and Highlands Focus of the Swiss National Centre of Competence in Research (NCCR) North–South, *Mt Res Dev* 25, 2, 174–179.

Moser, W.: 1975, *Einige Erfahrungen mit dem Tourismus in den Alpen – das Oekosystem Obergurgel*, Schriftenreihe des Alpeninstitutes München 3.

O'Sullivan, P.: 2004, Global Change and the Global Economy, *Global Change Newsletter* 58, IGBP Secretariat, Stockholm, 16–18.

Price, M.F.: 1995, *Mountain Research in Europe. An Overview of MaB Research from the Pyrenees to Siberia*, Parthenon, UNESCO, 230pp.

Steffen, W., Sanderson, A., Tyson, P.D., Jaeger, J., Matson, P.A., Moore, B., Oldfield, F., Richardson, K., Schellnhuber, H.J., Turner, B.L., and Wasson, R.J.: 2004, *Global Change and the Earth System. A Planet Under Pressure*, Springer, Berlin, 336pp. (Executive Summary, IGBP Secretariat, Stockholm, 40pp.).

Thompson Klein, J., Grossenbacher-Mansuy, W., Häberli, R., Bill, A., Scholz, R.W., and Welti, M.: 2001, *Transdisciplinarity: Joint Problem Solving Among Science, Technology and Society. An Effective Way for Managing Complexity*, Birkhäuser, Basel-Boston-Berlin, 332pp.

UNESCO: 1999, *World Social Science Report*, UNESCO Publishing/Elsevier, Paris, 351pp.

UNESCO-ICSU: 2000, *World Conference on Science. Science for the Twenty-First Century. A New Commitment*, UNESCO, Paris, 544pp.

Wasson, B. and Underdal, A.: 2002, Human–Environment Interactions: Methods and Theory, *Global Change Newsletter Nr. 49*, IGBP Secretariat, Stockholm, 22–23.

Wiesmann, U.: 2001, Umwelt, Landwirtschaft und Tourismus im Berggebiet – Konfliktbearbeitung im Leitbild Grindelwald 2000. In: Oesterreichisches Studienzentrum für Friedens- und Konfliktforschung (ed.), *Die Umwelt. Konfliktbearbeitung und Kooperation*, Bd. 7, pp. 237–249.

# Chapter 4
# Sustainable River Basin Management in Kenya: Balancing Needs and Requirements

**Boniface P. Kiteme and Urs Wiesmann**

**Abstract** Many research initiatives worldwide are established with the aim of explicitly contributing to sustainable development efforts within their respective contexts. Transdisciplinarity is one of the basic principles used to achieve this aim. In practice, however, this has remained a daunting task for many institutions, especially those focusing on sustainable natural resources management. This is partly due to a number of factors: a limited timeframe of operation; limited scope of work and implementation methodology, which are critical to ensuring broad disciplinary coverage and multistakeholder involvement; and use of approaches that do not guarantee integration of, or ensure responsiveness to, changing ecological processes and socio-economic, cultural and political landscapes.

This chapter presents an example of transdisciplinarity based on 25 years of continuous research in the upper Ewaso Ng'iro north basin on the north-western slopes of Mount Kenya. The chapter describes how this research programme, which aimed to contribute to integrated river basin management, was furnished with structures and apparatus that helped to navigate the transdisciplinarity process through the limiting factors listed above. The chapter shows how the research facility evolved from a district and project planning support institution, based on needs, to a regionally oriented and integrated facility addressing the requirements of sustainability. Projecting on the basis of long-term implications of changes in ecological processes and socio-economic and institutional dynamics on water availability, research activities were embedded in a multilevel, multistakeholder transfer strategy to ensure integration of scientific and local knowledge systems and long-term ownership of preferred interventions. In conclusion, the chapter lists seven recommendations on salient issues of transdisciplinary research.

**Keywords:** Needs · Requirements · Sustainability · Transfer · Water

✉ B.P. Kiteme,
Centre for Training and Integrated Research in Arid and Semi-arid Lands Development (CETRAD), Nanyuki, Kenya
e-mail: b.kiteme@africaonline.co.ke

## 4.1 Background

The upper Ewaso Ng'iro basin is part of the Mount Kenya highland–lowland system located to the north–west of Mount Kenya; dropping from 5200 m to an average elevation of 1000 m at the Laikipia Plateau and the Samburu Lowlands, it covers an area of about 15,200 km$^2$ and has a population of about 800,000 (Kiteme and Gikonyo, 2002). The basin falls within an overlapping administrative structure consisting of three provinces and seven districts under a multitude of decision making structures in a framework of complex and weak institutional arrangements. The basin is characterised by highly diverse ecological systems and zonal differentiation with a steep ecological gradient: precipitation levels drop from over 1000 mm in the tropical rainforest belt to below 500 mm, less than 30 km to the north (Liniger et al., 1998; Kiteme and Gikonyo, 2002).

The basin has, in the past century, experienced unprecedented population dynamics and dramatic land use transformation: pastoralist use at the beginning of the 20th century was superseded during colonial times by large-scale farming and ranching (Kohler, 1987; Kiteme et al., 1998a) causing the population to drop by almost 50% (Wiesmann, 1998). After Independence in 1963, agropastoral communities from high-potential areas immigrated into the region, causing the population to increase by a factor of 10 over the next 40 years (Wiesmann, 1998). At the same time small and medium urban centres grew in importance, and in recent years large-scale horticultural enterprises have been established in the footzone (Kiteme and Gikonyo, 2002; Liniger et al., 2005). These dynamics resulted in (1) settlement processes that gave rise to a multistakeholder society and (2) human induced ecological processes leading to an increasingly degrading natural resource base, as exemplified by chronic water scarcity and the resultant recurring (sometimes violent) user conflicts experienced in the last decades.

The evolutionary process of transdisciplinary research in our project can be traced through six distinct phases defined by the main research focus and key stakeholders involved in each case:

- The first was a pre-phase, preceding Laikipia Research Programme (LRP) that ran from 1976–1984 and consisted of explorative research and baseline surveys in an attempt to 'crack the surface of the nut'. Research at the time focused on the Mount Kenya region and was purely an academic venture serving university interests. However, a thinly veiled ethical impetus by some of the participating scientists triggered the question of the application of research findings.
- Based on the results of the prephase, Phase 1 (1984–1988) marked the formal establishment of the Laikipia Research Programme as an arm of the Laikipia Rural Development Programme (LRDP), the key stakeholder defining research and development needs at the district level. Needs from other stakeholders were reflected through the LRDP. This phase involved more baseline surveys leading to a good understanding of people's development needs and constraints.
- The second phase (1989–1993) was characterised by a 'breakaway' from the LRDP and a significant shift in focus from needs to requirements for

sustainability. This shift resulted in reorganising the approach and scope of research, based on conceptual considerations of sustainability. While the Laikipia district was retained as the main area of focus, a regionally oriented and integrated research approach with a broader topical scope was adopted. The highlight of this phase was the balancing of needs and requirements for sustainability through the formulation of the Water Development Plan for Laikipia district. The spectrum of stakeholders broadened as well to include more government and university departments, development agents, and the civil society.
- The third phase (1994–1998) focused on consolidation of research results (synthesis and transfer), coupled with additional in-depth studies in natural sciences (Natural Resources Management, NRM) and social sciences (Actors Strategies and Perception in Natural Resources Management, ASP) to help answer emerging new research questions, especially at the basin level. Results from these studies, in turn, created a better understanding of the potential for a new cycle of synthesis, as well as areas for training and transfer. The main transfer strategy was the basin-wide water awareness creation campaign targeting key actors in the upper Ewaso Ng'iro north basin.
- The fourth phase (1998–2002) continued the efforts initiated in the third phase, primarily balancing needs with the requirements for sustainability by embedding the campaign strategy into grassroots institutions in the form of Water Users' Associations in the basin, developing structures to close the loop to university training, international exchange, new syntheses in the basin, and extension of lessons learnt and of approaches to other regions in the country. The fourth phase was therefore a phase of transition towards national research and training.
- The fifth phase (post 2002) began with the establishment of the Centre for Training and Integrated Research in Arid and Semi Arid Lands Development (CETRAD) in 2002, with the mandate to build technical and professional capacities, and develop an integrated database for sustainable development in the Arid and Semi-Arid Lands (ASALs) of Kenya.

These phases will be discussed in the following sections under the points of focus of (1) who defines research, (2) how it is linked to society, and (3) how it is used.

## 4.2 Genesis of Research in Mount Kenya Region: University Research Related to Global Change

Research activities in the north-western slopes of Mount Kenya date back to the mid-seventies when the initial explorative investigations were conducted by research scientists from the Institutes of Geography of the Universities of Bern and Nairobi. These explorative studies covered areas such as forest resources, historical and natural differentiation of rural space around Mount Kenya, afro-alpine and periglacial belts, and soil temperatures and precipitation, as a basis for climatic–ecological zonation of tropical mountains (Kohler, 1977; Winiger, 1979), among others. This explorative work was driven by the need to understand the trends of global change, and the research agenda was therefore set at senior research level.

Further studies of hydrology, land resources, land use and land ownership in the early 1980s continued to raise the issue of a potential unsustainability problem in the region, triggering the question of how to apply the research results in order to forestall the problem (Speck, 1982; Pestalozzi, 1983; Leibundgut, 1983; Decurtins, 1992). The study results were synthesised into a comprehensive documentation of the ecological and socio-economic situation in Laikipia district. This provided the initial basis for systems knowledge in the region. At this level of research, students shifted their focus to 'real world problems'.

## 4.3 Research for District and Project Planning

Study results from the prephase contributed to the formulation of a project outline for a rural development project west of Mount Kenya (Laikipia district), and formed the basis of a preliminary report to the mission of the Swiss Technical Cooperation in early 1983 (Winiger, 1983), which in turn, informed the decision to initiate technical cooperation between the governments of Switzerland and Kenya by establishing the Laikipia Rural Development Programme (LRDP) in 1984.

Experience from the initial operations of the LRDP in an area characterised by rapid ecological and socio-economic dynamics finally shaped the decision to establish a research facility to support project planning and implementation in the district (Wiesmann, 1991); this led to the establishment of the Laikipia Research Programme (LRP) in 1984. The research agenda of the LRP was thus naturally determined by the needs of the LRDP as the primary stakeholder client. Research orientation and the research approach were embedded in the LRDP's key areas of focus: *water; agriculture (crops and livestock – small ruminants); small-scale enterprises and village polytechnic; self help groups and social services.* Integrated regional baseline studies were carried out to provide baseline information on topography, agro-climatic zones, rainfall, hydrology, soils, vegetation, dominant systems of land use, the road network and other infrastructure, rural and urban centres and their functions, self-help groups and their characteristics, and other attributes. This database further enhanced existing systems knowledge and helped to create a better understanding of people's development needs and constraints (target knowledge).

Considering the impacts of the drastic land use transformation on natural resources, and owing to the great importance of water problems in Laikipia, water became the main issue on the development agenda in the district. Consequently, the awareness creation campaign initiative launched on the basis of information generated from the baseline studies targeted water undertakers and user groups at the institutional and community levels. The campaigns, implemented jointly by the research team and personnel from the Department of Water, succeeded in raising people's awareness of water resource problems in the district, thereby persuading decision makers to redirect development efforts towards the water sector. Research interventions therefore focused on water development planning for Laikipia district, with an emphasis on enhancing water supply systems to meet the rapidly growing demand for domestic, agricultural and urban development.

## 4.4 Shift from Need Oriented to Requirement Oriented Approach

Before going into the discussion of how the focus of the approach shifted from needs to requirements, the terms 'needs' and 'requirements' shall be briefly clarified. On the one hand, 'needs' are understood as development goals that are explicitly or implicitly expressed by one or several stakeholder groups in the context of sustainable river basin management; for example, water needs of different users for domestic, agricultural, livestock, or industrial purposes in the basin can be quantified and specifically stated. On the other hand, 'requirements' are development goals that are normally not expressed by any single stakeholder group, but derived from conceptual and ethical considerations and negotiations. This can be exemplified by considerations to achieve equity of water distribution among different user groups from a spatial perspective; or the requirement that optimal water use fulfil current needs without compromising the capacity to fulfil the needs of future generations. Therefore, sustainability can only be realised if the focus is on both 'needs' and 'requirements'; focusing on one alone is not enough.

Having clarified the definition of the two terms we now continue to elaborate the developments of the second phase. In this phase, baseline studies (systems knowledge) created a better understanding of the problem and people's development needs and aspirations (target knowledge), thereby influencing investment designs and priorities (transformation knowledge). This necessitated a Water Development Plan for Laikipia district that was based on the principles of sustainability; however, the realisation of such a plan could not be guaranteed using the research approach and focus as set up in this project phase. Sustainability requirements cannot be based on demand alone; they call for broader conceptual consideration which should also include stakeholders who may not fall within the formal political and local leadership structures. It was therefore necessary to move away from the LRDP's needs orientation and shift to requirements orientation by reorganising the research approach and scope based on conceptual considerations of sustainability: while retaining Laikipia district as the main area of focus, a regionally oriented and integrated research approach with a broader topical and geographical scope was adopted. The spectrum of stakeholders was broadened as well to include more university departments, development agents and the civil society.

As a consequence, the integrated regional database developed in phases 1 and 2 was transformed into a comprehensive Geographic Information System (GIS) with additional baseline information such as population and demographic development, migration trends and household characteristics, water resources and water supply systems, and many others, thereby further enriching the systems knowledge. Meanwhile the LRP was integrated into the district's decision making and planning structures (the District Development Committee (DDC), the District Executive Committee (DEC), and the District Planning Unit (DPU)) in order to provide a platform for knowledge sharing and mutual learning among researchers and development agents in the district. Interaction with government and non-government departments through these institutions continued to provide the research team with useful insights into the nature of development problems in the district that required research intervention.

Laikipia district is very diverse in terms of ecological, socio-economic and cultural characteristics, and the established planning and decision making units (administrative or political units) are not identical with the relevant environmental/ecological units. This called for the definition of a common spatial data structure in order to create a basis for integration and general application. Therefore, the baseline data sets were developed into a database which contains comprehensive and relevant disaggregated baseline information on over 250 planning (spatial) units delineated using an agreed criterion of homogeneity (Wiesmann, 1998). The various baseline data sets were then assigned to the planning units which became important reference units for development interventions.

Using the baseline information and based on the delineated planning units, a District Water Development Plan for Laikipia district was elaborated. The plan identified priority areas where water supply projects could be implemented within three to six years, thereby optimising the impact of water development in the district with a view to long-term rational utilisation of the natural and capital resources for domestic, livestock and agricultural purposes in the rural areas. The Water Development Plan was elaborated jointly by the research team, relevant government ministries and departments (district development, water development, local authorities, local administration), and a support team of hydrology experts. The communities participated in setting the priorities among supply needs for different purposes (Leibundgut et al., 1991). The real innovation in this process was the combination of top–down and bottom–up approaches: the top–down approach provides an instrument of rolling planning, which is supplemented by participatory approaches to planning and realising concrete projects within the identified priority areas. The plan was adopted and the local communities implemented the intervention proposals with support from the government, donor agencies and development partners, such as the Swiss Agency for Development and Cooperation (SDC), Dutch Funded ASAL Programme-Laikipia (DGIS), the Catholic Church Aid (CARITAS), and World Vision, among others.

In this phase, the water development planning process serves to articulate how the three knowledge systems applied in the context of Laikipia district: systems knowledge provided an understanding of the ecological and socio-economic dynamics; target knowledge was used to project water demands for domestic, agriculture and livestock uses; and transformation knowledge was applied when identifying priority areas and designing intervention strategies.

## 4.5 Broadening the Research Scope and Consolidating Outputs for Transfer

### *4.5.1 Broader Topical and Geographical Scope to Respond to Sustainability Requirements*

As alluded to in the concluding remarks of the last section, the Water Development Plan for Laikipia district concentrated on solving existing problems: identifying target areas and setting priorities based on needs, with a view to improving water supply to meet the rapidly growing demand in the district. This approach to water development planning, although comprehensively addressing the problem of water supply as well as sustainability requirements at the district level, did not take account of the potential implications of hydro-meteorological and socio-economic dynamics on long-term water availability at the basin level. Three key research questions emerged from this realisation: (1) What are the long-term effects of climate change and land use changes in the upper reaches of the catchment on hydrological conditions in the entire basin? (2) What are the strategies and perceptions of the different actor groups in the regional contexts? (3) What are the existing institutional arrangements for resource management in the basin?

Answering these questions required further consolidation of previous research outputs through synthesis and transfer, and more in-depth studies and training in natural sciences and social sciences; essentially, the situation called for disciplinary and interdisciplinary research to help answer these questions in the regional context. This necessitated restructuring of the LRP in order to broaden and reorient its topical focus and its approach, broaden its geographical scope to include the upper Ewaso Ng'iro basin (15,000 km$^2$), and support further awareness creation and negotiation processes necessary for the implementation of preferred intervention strategies at different levels, as appropriate. Consequently, the sequence of research thrusts and projects was deliberately structured to address (1) natural resources monitoring, modelling and management (through natural sciences research), (2) actors' strategies and perceptions regarding natural resource use and management (through social sciences research), and (3) the synthesis and transfer of research outputs with a view to initiating policy dialogue, strategy development, and technology adaptation (see Fig. 4.1).

This broadened scope also implied an expansion of the institutional network and partnerships. The range of participating university departments was expanded to include Agricultural Engineering, Planning, Range Ecology, Soil Science, Sociology, Economics, Meteorology, and Anthropology of the Universities of Nairobi and the Institutes of Geography and Sociology of the University of Bern. This multidisciplinary nature of implementation was mirrored in the involvement of several funding agencies. These included the Swiss National Science Foundation (SNSF), the Rockefeller Foundation (RF), Research Department of Swedish International Development Agency (SIDA-SAREC), the Dutch funded Traditional Techniques of Microclimate Improvement (TTMI) Project, for scientific research; and the Swiss

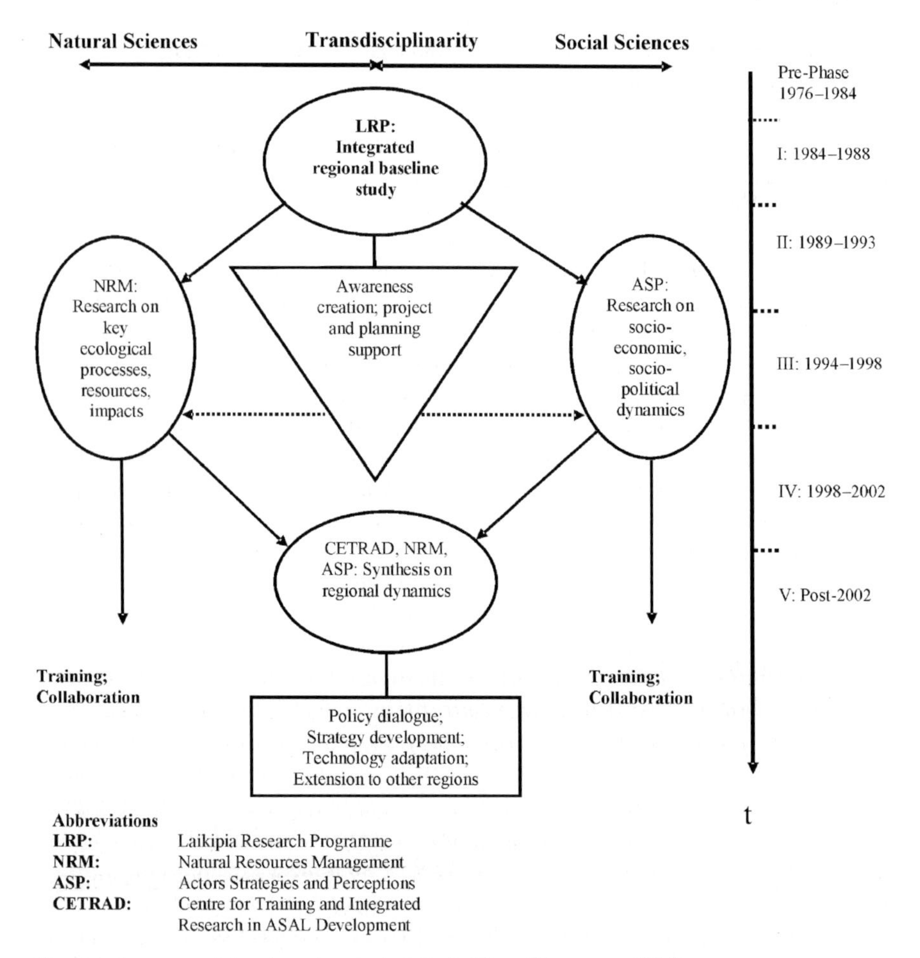


**Fig. 4.1** Sequence of thrusts and projects (adapted from Wiesmann, 1996)

Development Cooperation (SDC), Dutch Funded ASAL Development Programme (DGIS), and Government of Kenya for applied research and transfer work. In addition to government partners at the district level, the Ewaso Ng'iro North Development Authority (ENNDA) became an important transfer partner.

## *4.5.2 Consolidating Research Results: Synthesis on Regional Dynamics*

Broadening the topical and geographical scope led to a deeper understanding of the main challenge facing sustainable river basin management: the problem of balancing ecological sustainability and short-term needs (Wiesmann, 1998) against a backdrop of rapidly degrading natural resources, fast growing population, settlement processes resulting in a multistakeholder society, increasing water demand, scarce water resources with limited potential for expansion, and a framework of complex and ineffective institutional structures.

The comprehensive database containing long-term hydrological and meteorological data from the monitoring network in the basin dating back to 1960, along with data on land use and land cover changes and land management practices, were used to understand aspects such as rainfall characteristics, river regimes, water resources availability, water balance, river water abstractions and their effects on dry season flow (Gathenya, 1992; Liniger et al., 1998), water use conflicts (Wiesmann et al., 2000; Kiteme and Gikonyo, 2002; Liniger et al., 2005), and soil and water conservation (Njeru and Liniger, 1994; Njeru, 2005), among other important ecological aspects. With regard to socio-economic research, general surveys (addressing land use systems, household characteristics, etc.) and in-depth household surveys from selected test areas provided a better understanding of smallholder migrant peasants and their livelihood strategies in the new settlement conditions (Huber and Opondo, 1995; Kuenzi et al., 1998; Wiesmann, 1998). Further elaboration was made on problems, opportunities and options of resource use, management and planning at actor category levels (Betschart, 1996; Flury, 1996; Sottas et al., 1998).

### *4.5.3 Transfer of Results through Awareness Creation Campaigns and Policy Dialogue: Implementing a Multistakeholder and Multilevel Strategy for River Basin Management*

Enhanced systems knowledge regarding the ecological, socio-economic, and socio-political realities in the basin provided useful insights with regard to designing an appropriate strategy towards a more sustainable natural resources management, especially water management, in the basin. This knowledge facilitated designing a multiscale, multistakeholder, multisectoral strategy that not only concentrates on the problems of overutilisation of water resources but also addresses the severe problems of securing a livelihood, as well as the various conflicts directly or indirectly related to water use (Wiesmann, 1998; Wiesmann and Kiteme, 1998; Kiteme et al., 1998b; Wiesmann et al., 2000; Kiteme and Gikonyo, 2002).

The target specific transfer strategy that was developed aimed to reach all relevant stakeholder groups in the water sector within the basin and create awareness among them. The strategy was structured around five thrusts that included (1) district and project planning support, (2) core issue of water use and management directed at reduction of river water use and the development of alternative water resources, (3) land use planning to regulate smallholder settlements in marginal zones of the basin, (4) peasant land use practices, and (5) urban settlements and infrastructural development. The strategy was implemented within the broad framework of a Water Awareness Creation Campaign, and was structured according to three levels: (1) national level, (2) regional/district level, and (3) community level.

The main event at the national level was a workshop and basin-wide tour involving policy makers and decision makers in government ministries and departments, as well as donor institutions and NGOs facilitating water development interventions in the basin; including politicians representing different communities in the region at the constituency level. The primary purpose of the campaign at this level was to solicit the necessary political and policy support for successful implementation of the campaign at the lower levels. In particular, the participation of the Minister and the Chief Executive Officer of the Ministry of Water Development, along with senior officers from the Office of the Attorney General and Office of the President, helped to initiate policy and legislative reforms in the water sector, besides lending political influence to the entire campaign process. The reforms have recently been concluded, and new policy and legislative instruments, along with corresponding institutions (better defined mandates and functions) that promote participatory and integrated approaches have been put in place to oversee water resources development and management (GoK, 1999; GoK, 2002; Liniger et al., 2005). On the other hand, donor agencies and other development agencies helped to promote the need for coordination of external interventions and embedded such interventions into the framework of the campaign strategies. A donor forum coordinated by the LRP was established at the basin level for this purpose.

At the regional level, the campaign focused on government technical and administrative staff dealing with community mobilisation, agriculture and livestock development, and water resources development (allocation, user control and management). The target group also included the seven District Water Boards (DWBs), the Ewaso Ng'iro North Catchment Board (ENNCB), Ewaso Ng'iro North River Development Authority (ENNDA), as well as NGOs and Community Based Organisations (CBOs) in the area. The aim at this level was to solicit support for resource mobilisation and implementation of the campaign strategies at community level, as well as to articulate policy and legislative issues for intervention at the national level, streamline water allocation procedures, and enhance law enforcement. Representatives from these departments formed a Campaign Coordination Committee consisting of 12 members, which was responsible for designing the campaign methodology and materials and coordinating the actual campaign activities through seminars, workshops, and field days at the community level. The committee also facilitated community mobilisation and the formation of Sub-Catchment Clusters which were used as entry points for launching the campaign at the community level. More concretely, the district and catchment water boards helped to streamline and depoliticise water allocation procedures and strengthened surveillance in order to enhance law enforcement.

Stakeholders at this level also initiated policy dialogue and negotiations to influence changes in land tenure systems, land subdivision and smallholder settlements' land use and farming systems (Huber and Opondo, 1995), as well as promote alternative coping strategies among smallholder agropastoralists in subdivided marginal areas of the basin. Although the authorities finally refused to adopt the settlement guideline map due to the political consequences it implied, the map nevertheless created considerable awareness of the need to regulate new smallholder settlements in marginal lands further north.

At the community level, the campaign targeted individuals, corporate institutions/organisations and self-help groups that were using water resources, especially river water, for domestic purposes, crop and livestock production, and 'industry', at different scales in the region. The intervention therefore involved pastoralists, smallholder agropastoralists, small-scale irrigation farmers, large-scale horticultural enterprises, and urban municipalities. The aim of the campaign at this level was to enhance participatory water resources management through community policing, catchment protection, conscious application of soil and water conservation initiatives, dryland farming, and formation of grassroots institutions to provide platforms for negotiations and dialogue and ensure sustainability of the structures already put in place. Through the Campaign Coordination Committee, communities implemented various rainwater harvesting activities at the household and farm levels and introduced drought resistant, early maturing crop species, as ways of enhancing water use efficiency and reducing water demand for crop production. Furthermore, through the sub-catchment clusters, and with the facilitation of the Campaign Coordination Committee, they participated in Field Farmer Schools (FFSs), seminars, and inter/intra sub-catchment exchange visits for experience sharing and exchange of skills. The large-scale horticultural enterprises enhanced rainwater harvesting to supplement river water during dry seasons and installed state of the art irrigation technology for efficient water use.

## 4.6 Consolidating the Gains of the Campaign Initiative and Extending to Other Regions: Transforming the Research Program into a National Research Centre

### *4.6.1 Consolidating the Gains of the Campaign Initiative*

The fourth phase (1998–2002) continued the efforts of the third phase, primarily balancing needs and sustainability requirements by embedding the campaign strategy into grassroots institutions in the form of Water Users' Associations in the basin. The campaign initiative started to manifest positive effects especially at the institutional level: for example, the districts and catchment boards streamlined water permit issuance and became stricter in terms of law enforcement in order to curb water misuse in the catchment. More significantly, a common understanding among the different user groups began to emerge, revealing evident changes in their perceptions and attitudes. It was therefore necessary to develop structures to consolidate and sustain these gains. Consequently, based on the existing community water projects and large-scale horticultural farms in the respective river sub-catchments, the sub-catchment clusters were transformed into Water Users' Associations (WUAs) at the sub-basin level, covering catchments of tributaries of the Ewaso Ng'iro (Kiteme and Gikonyo, 2002; Liniger et al., 2005). The departments of Culture and Social Services and Water mobilised the communities and facilitated registration of the WUAs as self-help groups in conformity with the law. On the other hand, the large-scale horticultural farms provided initial financial and logistic support to facilitate the functioning of the WUAs until they can stand on their own. The more than 13 WUAs formed so far have become useful platforms for negotiations regarding water use and conflict resolution between different stakeholder groups in the basin. Through the efforts of the WUAs, large-scale horticultural farms and some progressive smallholder farmers have intensified rainwater harvesting and increased their storage capacities by means of earth dams and water tanks on their farms. Supported by the LRP/NRM3 and the Water Department, some WUAs have embarked on community based water regulatory devices in several abstraction points to ensure a minimum guaranteed water flow to downstream users.

### *4.6.2 Extending to Other Regions*

Meanwhile, the LRP was approaching the end of its duration. The main challenge was now to formulate follow-up structures to come full circle back to university training, international exchange, new synthesis in the basin and extension of lessons learnt and approaches to other regions in the country. The fourth phase was therefore a phase of transition towards such structures. This required the establishment of an institutional structure with a national mandate and means to implement similar research arrangements. To achieve this, it was necessary to transform the LRP into an institution with a mandate to implement research and training in order to build technical and professional capacities, and develop an integrated database for sustainable development especially in the arid and semi-arid lands (ASALs) of Kenya.

Laikipia district and the upper Ewaso Ng'iro North region remained the most suitable test area because it offers a highland–lowland system with the most dramatic socio-economic and ecological dynamics in terms of land cover, land use and land ownership, population and demographic development, ethnic and cultural diversity, and natural resource sharing arrangements, among other things. Therefore, although the system has unique dry and cold, semi-arid ecological conditions, it nevertheless offers a good representation of other semi-arid environments and highland–lowland systems in the country that are experiencing similar problems of unsustainability. It was therefore desirable to extend into other ASALs in order to capitalise on the long-term experience and lessons learnt.

While continuing with data elaboration for the new areas on a transitional basis, negotiations with the relevant government authorities initiated the process of establishing the necessary structures under the name of Centre for Training and Integrated Research in ASAL Development (CETRAD). Extensive consultations were also held with local and international research and training institutions to help position the new structure in a proper contextual perspective.

## 4.7 Centre for Training and Integrated Research in ASAL Development

Building on LRP structures, CETRAD is making syntheses on regional dynamics and engaging in sustained policy dialogue, strategy development, technology adaptation and extension of lessons learnt and experiences, to other regions with similar clusters of sustainability problems. To implement baseline studies and academically oriented research in the new areas of focus, CETRAD hosts international programmes like the Swiss National Centre of Competence in Research (NCCR) North–South Programme, through which research funds can be obtained. Other programmes such as Eastern and Southern Africa Partnership Programme (ESAPP) and UNDP's Global Environment Facility/Small Grants Proramme (GEF/SGP) Community Management of Protected Areas Conservation (COMPACT) provide core funding and support projects to promote sustainable regional development and sustainable land management.

## 4.8 Recommendations

The foregoing discussions lead to some key recommendations which can be grouped into three categories as elaborated below:

### *4.8.1 Transdisciplinarity as an Iterative Process*

- *Iteration between different types of knowledge*: In the three phases of transdisciplinary research (problem identification, problem structuring, and implementation), all three types of knowledge (systems knowledge, target knowledge, transformation knowledge) are relevant to all issues elaborated in the context of sustainable river basin management. The case of water development planning for Laikipia district and the subsequent awareness campaigns has clearly demonstrated this.
- *Iteration between transdisciplinary, interdisciplinary and disciplinary research*: Experience in each of the research phases, as presented in this paper, has also shown that in transdisciplinarity; disciplinary and/or interdisciplinary approaches will apply iteratively in response to needs for in-depth investigations to answer continuously emerging new research and development questions. This means that while transdisciplinarity is one of the basic principles used to deal with the concerns of different groups affected by unsustainability problems, the alternating use of transdisciplinary, interdisciplinary and disciplinary forms of research is crucial with regard to specific implementation (Hurni et al., 2004).

### *4.8.2 Stakeholder Involvement in Transdisciplinarity*

- *Requirement driven involvement of stakeholders*: The transdisciplinary process discussed in this paper is characterised by a multitude of stakeholder groups all contributing to and impacting in different ways on its implementation. For example, smallholder peasant farmers were the most critical stakeholders when identifying priority needs for water during the water development planning process. On the other hand, district based government departments helped to mobilise local resources and coordinate campaign activities at the community level while the district and catchment water boards helped to streamline water allocation procedures and enhanced law enforcement. At the national level, the Ministry of Water and donor agencies supported water policy and legislative review processes. This shows that the range of actors and institutions involved in a transdisciplinary process can be so broad and diverse, and they can vary to such an extent in relevance and significance depending on what aspect is in focus, that they cannot always be equally covered. Therefore, stakeholder selection in transdisciplinarity should be requirement driven and not just serve political interests or respond to different water needs.

- *Training as a key component of transdisciplinarity*: Based on the need to maintain the iterative process of disciplinary and interdisciplinary research, training in integrative issues becomes part of the sustainability approach, and is therefore a crucial component in a transdisciplinary process. This underlines the importance of university collaboration – in terms of providing a platform for MA/MSc and PhD research and training – as elaborated in the paper.

### 4.8.3 Imperatives of Transdisciplinarity

- *Imperative of need and requirement orientation*: The implementation strategy based on a multistakeholder, multilevel approach to regionally oriented water development planning was embedded in a transfer strategy that was target specific due to topical complexity and stakeholder diversity. This implies that in order to make target knowledge relevant in transdisciplinarity, the target must be defined from a broad perspective of sustainability in order to embrace need oriented target knowledge as well as requirement oriented target knowledge.
- *Imperative of a transparent and accessible information base and communication structures*: Experience cited in this paper shows that a requirement oriented database is an important steering and working tool of transdisciplinarity. On the one hand, it helps to define a common structure of spatial data and information reference (planning units), without which there is a risk of applying spatial frames of reference (catchment or district) that are inappropriate for achieving the required optimisation of resource inputs on the basis of needs, and which more often than not tend to encourage distribution of resources based on special interests and political representation (Wiesmann and Kiteme, 1998). On the other hand, such databases help to forge trust and confidence among stakeholders, especially during policy dialogues and negotiations. Having been scientifically generated, they seem to be less personal, less political and target oriented. Therefore, decisions based on such information systems are widely accepted by the different stakeholder groups and enjoy their support.
- *Imperative of long-term involvement*: This paper finally concludes that long-term research involvement is a prerequisite for transdisciplinarity. The research evolution process described in this paper and the experiences that justified a shift

from one phase to the other, culminating in the transformation from a research programme to a research institution, would not have been possible within a short term.

## References


Betschart, C.E.: 1996, Deciding Actors' Strategies and Perceptions for Sustainable Resource Management and Planning. In: B. Sottas (ed.), *Actors' Strategies and Perceptions – Contributions to the ASP Workshop 1996*, Laikipia – Mt. Kenya Papers – C4, Laikipia Research Programme, Nanyuki, Kenya, pp. 9–12.

Decurtins, S.: 1992, *Hydrogeographical Investigations in the Mt. Kenya Subcatchment of the Ewaso Ng'iro River*, African Studies Series A10, Geographica Bernensia, University of Berne, Berne, 150pp.

Flury, M.: 1996, Development Co-operation and Resource Management – Towards Enhanced Actor Orientation. In: B. Sottas (ed.), *Actor Orientation in Resource Management*, Laikipia-Mt. Kenya Papers – C5, Laikipia Research Programme, Nanyuki, Kenya, pp. 11–17.

Gathenya, J.M.: 1992, *Water Balance of Sections of the Naro Moru River, MSc Thesis*, Department of Agricultural Engineering, University of Nairobi, Kenya, 199pp.

Government of Kenya: 1999, *Sessional Paper No. 1 of 1999 on National Policy on Water Resources Management and Development*, Government Printer, Nairobi, Kenya, 31pp.

Government of Kenya: 2002, *Water Act 2002, Kenya Gazette Supplement*, Government Printer, Nairobi, Kenya.

Huber, M. and Opondo, J.C.: 1995, *Land Use Change Scenarios for Subdivided Ranches in Laikipia District, Kenya, Laikipia Mt. Kenya Reports 19, Laikipia Research Programme, Nanyuki and Group for Development and Environment*, Institute of Geography, University of Berne, Switzerland, 72pp.

Hurni, H., Wiesmann, U., and Schertenleib, R.: 2004, *Research for Mitigating Syndromes of Global Change. A Transndisciplinary Appraisal of Selected Regions of the World to Prepare Development-Oriented Research Partnerships. Perspectives of the Swiss National Centre of Competence in Research (NCCR) North–South*, Geographica Bernensia, University of Berne, Berne, 468pp.

Kiteme, B.P. and Gikonyo, J.: 2002, Preventing and Resolving Water use Conflicts in the Mount Kenya Highland–lowland System Through Water Users' Associations, *Mt Res Dev* 22, 4, 332–337.

Kiteme, B.P., Wiesmann, U., Kuenzi, E., and Mathuva, J.M.: 1998a, A Highland–lowland System under Transitional Pressure: A Spatio-temporal Analysis, *Eastern and Southern Africa Geographical Journal* 8, Special Number, 45–54.

Kiteme, B.P., Liniger, H., Mathuva, J.M., and Wiesmann U.: 1998b, Multi-level Approach for Enhancing Ecological Sustainability: Applications within a Dynamic Regional Context, *Eastern and Southern Africa Geographical Journal* 8, Special Number, 91–96.

Kohler, T.: 1977, *Forests and Forestry in Kenya – The Example of Mt. Kenya Mimeo*, unpublished, University of Bern.

Kohler, T.: 1987, *Land-use in Transition*, African Studies Series A5, Geographica Bernensia, Berne, Switzerland, 122pp.

Kuenzi, E., Droz, Y., Maina, F., and Wiesmann, U.: 1998, Patterns of Peasant Livelihood Strategies: Local Actors and Sustainable Resource Use, *Eastern and Southern Africa Geographical Journal* 8, Special Number, 55–65.

Leibundgut, C.: 1983, A Contribution to the Surface Run-off of Mt. Kenya, Project-draft and Preliminary Results of Hydrologic Investigations to the Water Balance of a Tropical High Mountain and its Surroundings, *Jahrbuch der Geographical Gesellschaft von Bern*, 54/1980–82, 215–242.

Leibundgut, C., Kabuage, S.I., Moser, T., and Wiesmann, U.: 1991, *Water Development Plan for Laikipia District, Kenya, Ministry of Reclamation and Development of Arid*, Semi-arid and Wastelands, Nairobi, Kenya.

Liniger, H.P., Gichuki, F.N., Kironchi, G., and Njeru, L.: 1998, Pressure on Land: The Search for Sustainable Use in a Highly Diverse Environment, *Eastern and Southern Africa Geographical Journal* 8, Special Number, 29–44.

Liniger, H.P., Gikonyo, J., Kiteme, B.P., and Wiesmann, U.: 2005, Assessing and Managing Scarce Tropical Mountain Water Resources. The Case of Mount Kenya and the Semiarid Upper Ewaso Ng'iro Basin, *Mt Res Dev* 25, 2, 163–173.

Njeru, L.: 2005, *Monitoring and Modelling of Crop Growth, Water Use and Production under Dryland Environment N-W of Mt Kenya, Ph.D. Thesis*, Institute of Geography, University of Berne, Switzerland.

Njeru, L., and Liniger, H.P.: 1994, *The Influence of Vegetation on the Water Resources of the Naromoru Catchment – A Water Balance Approach, Laikipia – Mt. Kenya Papers – D17*, Laikipia Research Programme, Nairobi, Kenya.

Pestalozzi, P.: 1983, Historic and Recent Changes of the Agricultural Structures in the Area of Mount Kenya, *Jahrbuch der Geographical, Gesellschaft von Bern*, 54/1980–82, 265–287.

Sottas, B., et al.: 1998, Dilemmas of Deciding Stakeholders: Governance and Open Access to Common Property, *Eastern and Southern Africa Geographical Journal 8*, Special Number, 67–75.

Speck, H.: 1982, The Soils of Mount Kenya Area. Their Formation, Ecological and Agricultural Significance. The Soils of Mt. Kenya Area (1:125000), Reconnaissance Soil Map of the Nanyuki-Naromoru Area (1:100000), *Mt Res Dev,* 2, 2, 201–221.

Wiesmann, U.: 1991, Research with the Aim of Assisting Development. The Example of Laikipia Research Programme. In: F. Ojany et al. (eds), *Proceedings of the International Workshop on Ecology and Socio-Economy of Mount Kenya area*, Group for Development and Environment, Institute of Geography, University of Bern, Switzerland, March 1989, pp. 175–188.

Wiesmann, U.: 1996, Poster Presentation during the CONTICI Group Meeting held in 13th July–15th July, 1997, at the University of Berne, Switzerland.

Wiesmann, U.: 1998, *Sustainable Regional Development in Rural Africa: Conceptual Framework and Case Studies from Kenya*, African Studies – 14, Geographica Bernensia, Berne, Switzerland.

Wiesmann, U. and Kiteme, B.P.: 1998, Balancing Ecological Sustainability and Short-term Needs: A Regional Approach to Water Supply Planning, *Eastern and Southern Africa Geographical Journal 8*, Special Number, 77–90.

Wiesmann, U., Gichuki, F.N, Kiteme B.P., and Liniger, H.: 2000, Mitigating Conflicts over Scarce Water Resources in Highland – Lowland System of Mt. Kenya, *Mt Res Dev* 20, 1, 10–15.

Winiger, M.: 1979, Bodentemperaturen und Niederschlag als Indikatoren einer klimatisch-ökologischen Gliederung tropischer Gebirgsräume. Methodische Aspekte und Anwendbarkeit, diskutiert am Beispiel des Mount Kenya (Ostafrika) (Soil temperatures and precipation as indicators of a climatic-ecologic zonation of tropical high mountains), *Geomethodica* 4, 121–150.

Winiger, M. (ed.): 1983, *The Actual Ecologic and Socioeconomic Situation in Nanyuki (Laikipia District, Kenya): A Preliminary Report prepared for the Mission of the Swiss Technical Cooperation, Laikipia Reports*, Institute of Geography, University of Berne, Switzerland.

# Chapter 5
# Designing the Urban: Linking Physiology and Morphology

**Peter Baccini and Franz Oswald**

**Abstract** In the 20th century a dramatic change occurred in urban development. From a physiological point of view the large scale exploitation of fossil energy and the technical inventions and innovations in the transport infrastructure allowed on one hand a rapid liberation from the limits of renewable biomass and on the other hand a rapid exchange of mass goods over large distances. From a morphological point of view the same factors led to a 'dilution' of urban settlement from dense centres into a network, of highly variable nodes and connections (the Netzstadt). The distinct separation of rural and urban segments within a cultural landscape disappeared. This new urbanity is a relatively young phenomenon. According to the criteria of a 'Sustainable Development', such a culture is not apt to survive on a long-term and global scale. It follows then that the urban systems of the 20th century have to be reconstructed.

'Designing the Urban: Discovering a Transdisciplinary Method' describes the research process of two groups, rooted in very different academic cultures: one in architecture and urban planning, the other in environmental sciences and engineering. At the beginning (in 1993) there was neither a clear concept of how to tackle the methodological problems in order to investigate this new phenomenon nor a reflected concept of transdisciplinary work. The case study presented in the following chapter gives first, a report on 'identifying and answering questions' connected with urbanity and second, some illustrations of the answers found ten years later as a result of the transdisciplinary approach.

**Keywords:** Urban design · Netzstadt · Urban reconstruction · Participatory process · Synoikos

## 5.1 Background

Questions that lead to research projects have their own intellectual history. They are sometimes influenced by the personal biography of the authors asking the questions.

✉ P. Baccini
ETH Zurich, Zurich, Switzerland
e-mail: peter.baccini@emeritus.ethz.ch

G. Hirsch Hadorn et al. (eds.), *Handbook of Transdisciplinary Research*,

This is the case in the present contribution. The two authors, one a scientist and the other an architect, met accidentally at a crucial moment of their academic work. The scientist, a professor in environmental science and technology, had just completed research projects on the metabolism of urban regions. Thus, he was looking for new scientific challenges in which the experiences of the previous research could find a logical application as well as a continuation. The architect, newly appointed as professor for architecture and urban design, was promoting the research idea based on aesthetic, in combination with ecological, considerations regarding the urban landscape of the Swiss Midlands. Thus, he was looking for scientific partners. After the first meeting both were convinced of the relevance of the envisioned subject, the personal integrity of representing two different disciplines and the warm sympathy to each person.

The beginning of the cooperation was marked by the need to tackle two tasks: one, to learn the terms of the other's language and two, to understand the perceptions hidden in the same words used by two persons from totally different backgrounds, e.g. words like 'landscape', 'urban', 'project' or 'process'. We agreed to concentrate on a typical area in the Swiss Lowlands – the so called 'Kreuzung Schweizer Mittelland' (crossing Swiss Lowlands) – and to undertake daylong field trips, where each described to the other the people, things and movements he observed, how and why he perceived them in the specific way he did at the time. This was a guiding experience for the later work with our collaborators, too.

The commonly shared experience of the field trips was also helpful in defining the goals of the project. To both authors it became evident that the scientific findings would have to be convincing and helpful for professional and political actors. These are the people who can influence the processes in planning, building, managing or transforming the human habitat. So, we agreed on the initially vague notion to develop methodological instruments that could help the practice of planning, designing and engineering professions as well as the political decision makers.

In conclusion we started with the hypothesis that the receivers and users of our results base their work on urbanity in the context of the normative issues of sustainable development. To this rather ambitious goal we were lead by the conviction, that the enormous and rapid changes, which had occurred in our biographical period since the 1950s, had to be handled with new, coherent perceptions and instruments towards future life styles and improved environments. These instruments could not be the same ones that had produced the present state of urban culture with little satisfaction for a majority of inhabitants.

## 5.2 Problem Identification and Problem Structuring

At the beginning of the 1990s urban development concepts were based mainly on the paradigm of 'core cities' and their peripheral agglomerates. Territorial management strategies in Switzerland and elsewhere in developed countries emphasised the necessity to strengthen the dense centres and to fight the urban sprawl. However these strategies, implemented in territorial management laws and ordinances, could not prevent a transfer of the majority of the population from a rural type of life, and from the core city life, to a new urban lifestyle, which is mainly defined by the accessibility of urban commodities and services. From former physiological studies (Baccini and Oswald, 1998) it became clear that such an urban development was only possible with a consequent installation of all the infrastructure independent of localities. These include input infrastructure, namely the supply of energy, water, food and construction materials and output infrastructure such as effective waste management, as well as the buildings and connections between them that allow a high mobility of persons, commodities and information. These studies also revealed the high degree to which Swiss settlements depended for energy and materials on the 'globally distributed hinterlands'. It became clear that the key factor for urban development was the built-up area that already existed and not that which was yet to be built on open land.

Therefore a study area was chosen to focus on urban development outside the traditional urban centres established over previous centuries. This decision, taken in 1993, only one year after the UN Conference on Environment and Development in Rio de Janeiro, also involved an awareness of the need to answer the following questions:

- Do these new urban systems meet the criteria of 'sustainability'? In other words: Are they fit for survival in the context of the 'hinterlands' in various scales, from local to global?
- In the case that they do not meet these criteria, what type of reconstruction process should be initiated?

As the first project year showed, the initial concept for the participation of many different academic disciplines was far too ambitious. The project had to be radically reduced in scale without changing the aim. The scholarly dialogue between architects, natural scientists and engineers only got underway at the end of 1994, when all participants were able to agree on one fundamental rule: all members of the research group have 'equal scholarly value'. No discipline has any *a priori* leading function. No method brought to the research takes precedence. The tools for the collective studies should be developed collectively.

From a methodological point of view the following questions were raised:

- What models help us understand the phenomenon of urbanity?
- What metaphors were or are helpful?
- What methods can be used to analyse and design urban systems?
- Is there a generally accepted theory of urban genesis or city transformation?

The answers are given as follows:

- There is not (and probably never will be) a generally accepted theory of 'the city' that can help us distinguish the 'good' premises and methods from the 'bad' ones – good in the sense of being more useful for analysis and synthesis.

- A subjective manner of perceiving human settlements determines the examination method chosen. The different premises result in a subjective selection of questions relevant to the city. The observers' premises and objectives also influence the tools they deploy. The methods used, usually with a scientific foundation, define the settlement type, which is thus a construct of the observer.
- Architecture, history, sociology, political science, anthropology, the natural sciences, or any other discipline is not predestined to play a central role in urban development. In all probability, the best method for defining the 'good city' is based on participatory, transdisciplinary work. We can only learn what a 'good city' is in the relevant social context.

In conclusion, the research process was organised as follows: the scientific competences were concentrated in two domains, a 'morphological group', formed by architects and urban planners, and a 'physiological group', consisting of natural scientists, engineers and one economist. However all decisions with regard to the relevant questions, the tools of investigations, the validation of findings, the selection of models and the next steps, were taken in the plenary, i.e. by all participants in the project.

From the beginning, personalities of the study region (mainly leading figures in politics, economy and culture) were integrated as partners for critical discussions in each phase of the project. This parallel and synchronised process led eventually to the so called Synoikos method.

Complementary to these partners outside of academia and professional practice, professional peer-reviewers were also involved from the beginning. They participated at workshops which periodically served as platforms for presenting the intermediary results of the research work.

Already from an early stage of the project it became clear that an appropriate way to work in mixed groupings (morphological and physiological competences) is the 'activity approach'. The urban system to be elucidated and designed in a transdisciplinary process can be structured around four basic activities (to nourish, to clean, to reside and work, to transport and communicate). From this approach, not foreseen at the beginning, the Netzstadt model and eventually the Netzstadt methodology evolved.

## 5.3 Problem Investigation

### *5.3.1 The Hidden Agendas of the Various Disciplines*

At the beginning the architects considered themselves as the urban planners and designers who are able to take the lead in such a project, due to their competence as generalists. In their view engineers, economists and natural scientists were important suppliers of facts, figures, specific models and tools to support the various architectural blueprints. The natural scientists, particularly, were seen as species mainly interested in theories of cognition and not in realising concrete projects. There was a certain mistrust with regard to scientific methods, because a reductive procedure was thought to do no justice to complex urban phenomena, and to hinder, or even kill any creative synthesis. Designing the urban is closer to an art. Creative individuals or teams lance their piece of work in a competitive process within the forum of a public debate usually more on the basis of normative ideas than of described empirical findings. It is the quality of the product within this process that counts, not the methods that have helped or may support the making of the product. Therefore the group of 'morphologists' within the project were suspicious of scientific research that did not find a conclusion in practical actions. In addition, the morphologists did not bring an established toolbox or the expressed intention to create one in the project.

The 'physiologists' (natural scientists, engineers, economists) were not familiar with the design culture. At the beginning they got lost in the diversity of metaphors the architects were constantly increasing and changing. Scientists are used to the relatively strict vocabulary of their discipline that gives for each notion a clear definition (*terminus technicus*). They were unable to orientate themselves in the given complexity. In their view the morphologists were just surfing on an envelope wrapping the phenomenon 'urbanity'. They accused the architects of acting as chameleons by performing constant label changing. The physiologists had serious difficulties in bringing their tools into a prospective position.

All of the participants had more or less developed values with regard to the 'good city'. After various discussions on target issues it became clear that these values were more or less independent of the participant's discipline. The spectrum was very broad, ranging from dominantly ecologically oriented to scientifically motivated, socially driven, and purely aesthetically based positions. In other words the first layout of norms revealed a chaotic mix. Therefore it became necessary to uncover these hidden agendas in order to come to a first agreement with respect to (1) the target issues and (2) the boundary conditions with regard to the methods to be developed and applied in the project.

### 5.3.2 *From a Chaotic Mix of Norms to a Common System Approach*

After about two years of intensive searching to understand the various backgrounds, competences and goals of the participants, the following framework and structure was agreed upon:

1. Eventually we defined 'urbanity' as follows (Baccini and Oswald, 1998):
   The urban system is a large system made up of geogenic (emergent with the earth) and anthropogenic (emergent with mankind) subsystems. It covers an area of tens of thousands of square kilometers and has a population density of hundreds of inhabitants per square kilometre. It is an all-encompassing three-dimensional network with diverse social and physical links. A relatively high concentration of people, goods and information exists at the nodes of this

network, and there are massive flows of people, goods and information between the nodes. Colonized agricultural and forestry ecosystems and waterways are integrated components of the system.

2. We tried to treat the complexity of our study object, our urban system, with a limited set of elements that allowed us to do morphological and physiological analysis and synthesis. This set consisted of:

   - Six territories
   - Four key resources
   - Four network elements (nodes, connections, borders, scales)

   The urban system is being generated and transformed by *four basic activities* (to nourish, to clean, to reside and work, to transport and communicate). These allow comparisons of human settlements independent of their cultural existence in time and space.

3. We defined essential system qualities to be considered in urban design and not specific final urban states to be attained.

   - Identification
   - Diversity
   - Flexibility
   - Degree of self-sufficiency
   - Resource efficiency

4. The goals for a concrete urban project have to be clarified by a participative political process and should not be given a priori by the professional urban designers. Therefore urban design should consist of three steps:

   - the analysis elucidating the essential properties of the urban system in question, helping the inhabitants to formulate the strengths, weaknesses, opportunities and threats (SWOT) of their urbanity (it is the challenge to develop tools)
   - the sketching of the long-term orientated urban goals by the inhabitants
   - the design for reconstructing urban systems to meet the target qualities of step two, starting from the state defined in step one.

### *5.3.3 From an Interdisciplinary Approach to the Awareness of Transdisciplinarity*

Although the authors did not start the project in 1993 with an explicit transdisciplinary concept, the first steps in 1994 did follow an interdisciplinary direction. At that time the academic discussion about the idiosyncrasy of transdisciplinarity was relatively young. Following this debate and reflecting the progress of our urban research project we started to realise that we met most of the criteria that were selected by the community framing the emerging field of transdisciplinary approaches.

The first monograph of the project was published in 1998 (Baccini and Oswald, 1998) with the German title: 'Netzstadt – Transdisziplinäre Methoden zum Umbau urbaner Systeme' (Transdisciplinary methods for reconstructing urban systems). The title reflects the following important processes:

The formation of four 'mixed' research groups (morphologists and physiologists), one for each activity led to a first synthesis postulating the 'Netzstadt-Hypothesis'. It was a very rough sketch of a model not yet validated and tested in practical urban design.

The process chosen led to a new perception of the phenomenon 'urbanity', expressed with a new vocabulary. This led eventually to a new problem in the communication with the established urban and regional planners. Netzstadt drops the paradigms of classical urban and regional planning. Working in such a research project is an iterative process, passing from subgroups, that are formed temporarily, to the plenary where the decisive conclusions are drawn, leading to new sub-teams (Müller et al., 2005).

With the new approach the project team had created a new identity and a strongly shared conviction that new 'design' tools (methods) had to be developed. This helped to integrate new project collaborators who extended the team, or substituted members who left. It became clear after three years which type of person suits such a project. The person needs a combination of above average social competence, a strong motivation for methodological work on the borderlines of disciplines and the mental and physical stamina to give them the patience for time-consuming workshops.

In the second phase, lasting from 1999–2003, two new project teams centred around the two authors, their assistants and advanced students, elaborated on one hand the Netzstadt model and methods and on the other hand the Synoikos method (for the participative processes). Here the first implementation experiences were gathered.

## 5.4 Implementation

On the basis of two different, but complementary sets of experiments – one in teaching Urban Design with an upper class of students, the other, in producing participatory workshops with participants representing the heterogeneous population of the five municipalities of the 'Kreuzung Schweizer Mittelland' – the hypotheses for designing the urban were tested. The tests consisted of working out practical results, e.g. projects for improving a chosen urban site and its context. The methodology and results were evaluated by participants and critics from outside, finally modified and completed by the researchers for publication. They are published as 'Netzstadt method' (Fig. 5.1), and 'Synoikos method' (Oswald and Baccini, 2003), and can be applied separately or together.

The two methods are based on the same hypotheses for designing the urban and for shaping 'good' urban form:

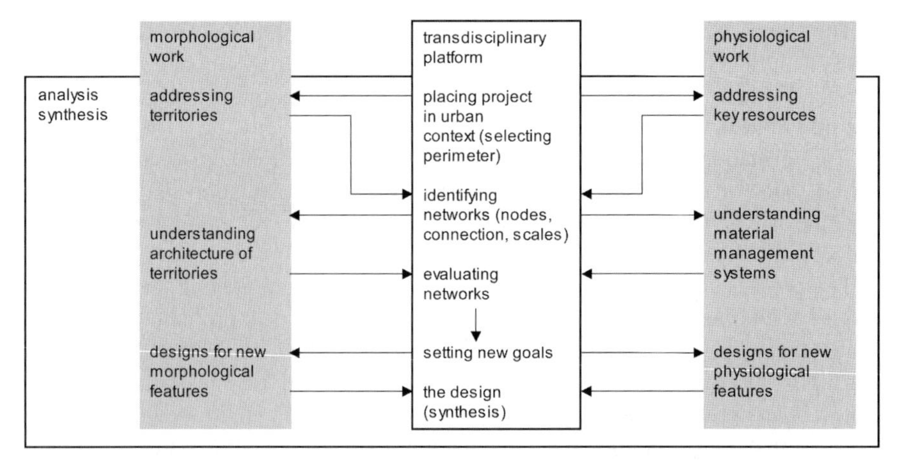


**Fig. 5.1** Schematic overview of the Netzstadt method: the work proceeds along the paths between the disciplinary workshops and the transdisciplinary platform (from Oswald and Baccini 2003, Fig. 3.59; with kind permission of Birkhäuser Verlag, Basel)

- Relative advantage is a result of synergy.
- Synergy is a result of the cooperation of all relevant urban actors.

The Netzstadt method offers a transdisciplinary platform and procedure for designing urban projects. These projects are focused on the improvement of urban qualities and on transforming the existing conditions within a defined context. This includes the reducing of the complexity of the mutual relationships between human activities, territories and resources. These relationships must be prepared with reference to the rational and transparent decision making process within the transdisciplinary design team. The art of designing consists of progressive steps and feedbacks starting with the description and analysis of the existing condition (Fig. 5.2). It is practical to begin this procedure, as proposed by the Netzstadt method, with the morphological, following up with the physiological investigation of the identified issues and problems. According to early feedback since publication, other academic groups have applied this method to their projects.

The Synoikos method, a procedure for participatory workshops, is complementary to the Netzstadt method. The goal is to arrive at proposals for practical actions or an urban strategy and to include specified measures for their implementation with regards to improving urban qualities. In addition, the workshops have to create a substantial majority of participants who actively support the proposed actions and the urban strategy. The workshops, thus, are focusing similarly on designing with the Netzstadt method, on synergies through mutually relating the strengths, weaknesses, opportunities, threats and on generating cooperative majorities for shared target qualities. The processes leading towards these goals are politically meaningful and strengthen publicly the responsibilities of individuals within the open, democratic culture. After the workshops with participants in the study region,

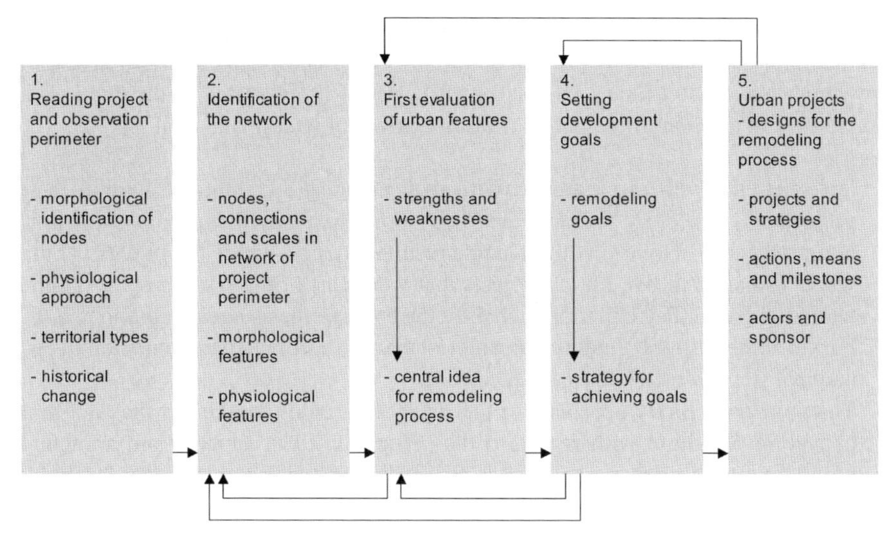


**Fig. 5.2** Designing the urban: applying the Netzstadt method – five procedural steps (from Oswald and Baccini 2003, Fig. 4.4; with kind permission of Birkhäuser Verlag, Basel)

'Kreuzung Schweizer Mittelland', other Swiss regions have applied the Synoikos method.

It is important to state that the Netzstadt method itself does not generate urban designs. It supports analysis and structures the work of designing. The Netzstadt method does not develop new quality objectives for urban systems. For this the Synoikos method was developed. It is step four in the design process (Fig. 5.2).

The fact that the same team who had thoroughly investigated and evaluated the existing conditions in 'Kreuzung Schweizer Mittelland' was also involved in designing scenarios, which served as illustrations for differing future developments of this area, produced synergetic and positive emotional effects during the research. The scenarios served, at the same time, as stimulation for the participants who had to find target qualities.

## 5.5 Recommendations

After more than ten years of experience with our project we can draw the following main conclusions with regard to the transdisciplinary process:

- Within the university borders such projects were and still are quite difficult to realise, due to two important barriers:
  - The established academic evaluation period of three to four years for such research projects is too short.
  - The means of evaluating such projects are very rare or simply non-existent in most research boards in academia.

Our project profited from the fact that the two project leaders had established credentials and that a university provost believed it necessary to risk long-term research investment to support them. The Netzstadt project was evaluated at the end in the framework of a research project on transdisciplinarity (Bergmann et al., 2005).

- For scientists with a research culture that include Ph.D. students within their project, it is necessary to frame the Ph.D. topics in a way that permits time limited projects (3–4 years) with valuable results independent of the process of the whole project. All five Ph.D. projects that were part of the research programme were successful. Participation increased the average thesis period by about 20%. However the doctoral students considered this extraordinary supplementary educational step a worthwile experience.
- The transdisciplinary approach to the topic 'Designing the Urban' has led to a change in paradigm with regard to the perception of urbanity. From an actual point of view this new paradigm has primarily reached some avant-gardes in the academic world. In the biotopes of practice the corresponding methods have not yet found a positive response. In our opinion this is due to the fact that the Netzstadt method uses a new language and as yet lacks a dictionary to help potential users translate the findings into their own idioms of urban and territorial planning.

## References

Baccini, P. and Oswald, F.: 1998, *Netzstadt – Transdisziplinäre Methoden zum Umbau urbaner Systeme*, vdf Hochschulverlag an der ETH Zürich, Zürich, 252pp.

Bergmann, M., Brohmann, B., Hoffmann, E., Loibl, M.C., Rehaag, R., Schramm, E., and Voß, J.-P.: 2005, *Quality Criteria of Transdisciplinary Research. A Guide for the Formative Evaluation of Research Projects.* ISOE Studientexte Nr. 13. ISOE, Frankfurt.

Müller, D.B., Tjallingii, S.P., and Canters, K.J.: 2005, A Transdisciplinary Learning Approach to Foster Convergence of Design, Science and Deliberation in Urban and Regional Planning, *Syst Res Behav Sci* 22 (3), 193–208, Retrieved June 11, 2007, from www.interscience.wiley.com.

Oswald, F. and Baccini, P.: 2003, *Netzstadt – Einführung in das Stadtentwerfen* (Netzstadt – Designing the Urban), Birkhäuser, Basel, 303pp.

# Chapter 6
# CITY:*mobil*: A Model for Integration in Sustainability Research

**Matthias Bergmann and Thomas Jahn**

**Abstract** The project 'Strategies for a Sustainable Urban Mobility' ('Stadtverträgliche Mobilität – Handlungsstrategien für eine ökologisch und sozial verträgliche, ökonomisch effiziente Verkehrsentwicklung in Stadtregionen') is a good example of successful integration work. Researchers from a number of different disciplines, as well as planners from two model cities, formed the research group CITY:mobil and cooperated within this project to develop innovative research methods, on the one hand, and planning tools aimed at a more sustainable mobility in cities, on the other. The project was designed to integrate planning and technical aspects as well as economic, ecological and social goals. Thus, the rather complex task of knowledge integration and social integration by the project team was one of the central challenges within the research process. Therefore, after introducing the details of this specific research project, we conclude with a universal model for a transdisciplinary research process. This model can support researchers in planning and conducting the complex integration demands to meet the dual targets of integrated research results for the area of interest (i.e. the societal problems being the starting point of the research project), and of new interdisciplinary or disciplinary results (e.g. methods, concepts and theories).

**Keywords:** Sustainable mobility management · Cognitive integration · Boundary object · Problem transformation · Social-ecological research · Societal relationships to nature

## 6.1 Background

Objects of transdisciplinary research are not automatically specified: they are the result of a first transdisciplinary research effort at the beginning of the project. This constitution of the research object of 'Strategies for a Sustainable Urban Mobility'

✉ M. Bergmann
Institute for Social-Ecological Research (ISOE), Frankfurt am Main, Germany
e-mail: bergmann@wiko-berlin.de

was based on an intense initial phase where the common research tasks between the participating disciplines were established and a comprehensive integration and cooperation concept for the whole research process was developed.

The project design was constructed through a process of intense discourse among the nearly 20 participants from various disciplines and the model cities who made up the CITY:*mobil* research group. This discourse started with the theoretically and methodologically guided process of 'translating' the various aspects of the societal problem concerning unsustainable development of urban regions into scientific research objects. The shared vision in conceiving the project, the common tasks and the cooperation were the main reasons for the successful integration of the results from the several subprojects into comprehensive and integrated problem solutions for city planners and politicians, as well as into the development of new interdisciplinary and disciplinary scientific tools, methods and theoretical concepts.

When the project was conceptualised in 1993 the term 'transdisciplinarity' was hardly in use in Germany. The project was therefore called interdisciplinary, problem and actor oriented. It was run within the framework of the research programme 'Urban Ecology' of the Federal Ministry of Education and Research (BMBF).

Two research interests provided the impetus for the research project:

- The research programme, as well as the participating institutes, worked on developing sustainable mobility strategies that were problem and actor oriented (rather than oriented towards technical feasibility). Consequently, such strategies had to respect the interests of the urban population (attractive city, urbanity, mobility, ecology), and also the interests of the local government (attractive city, urbanity, mobility, ecology *and* economic feasibility in times of an empty public purse). These manifold interests had to be represented by the cooperating disciplines and actors using adequate integration tools and methods to provide actor and problem oriented solution strategies.
- At the beginning of the conceptualisation process it became clear that there were no appropriate research methods to meet all the societal demands (not just the improvement of the technical standards) using an integrating perspective. Consequently, the development of such scientific methods became one of the most important driving forces and targets for the project.

In compliance with the two project foci (actor oriented problem solutions, and scientific method development) the knowledge generated had to be primarily (in the terminology of this book) system knowledge (how to properly investigate the problem) and transformation knowledge (how to initiate societal transformations).

The project was realised in the interdisciplinary research cooperation CITY:*mobil*, and incorporated engineers, planners, sociologists, and economists. It was supported by the local governments of the German cities of Freiburg and Schwerin, at the same time being the addressees for the implementation of the project results. The experience their institutes had gained within the research network *oekoforum* gave the researchers an advantage in interdisciplinary cooperation. The researchers came from planning, ecology, economics, sociology, engineering, and computer sciences. The representatives of both cities came from the departments for city and transportation planning. A moderator was appointed to the research team to help with cooperation, coaching and supervision, the integration workshops, and to support the project manager.

## 6.2 Integrated Research Methods and Planning Tools

We display a descriptive and analytical model for the transdisciplinary research project CITY:*mobil* that can help with the construction of projects of a similar design (Fig. 6.1). The model we use is developed in Jahn [2005] where he discusses, in depth, the use and meaning of the term transdisciplinarity. This model represents a typical and idealised transdisciplinary research project, which can be regarded as a sequence of three project phases:

- *Phase A*: construction and description of the project and the research team; constitution of the common research object; analysis of the problem dimensions to be treated
- *Phase B*: research process in the four 'subprojects' (SP1–SP4) and the 'main project' (knowledge building, mutual learning and linking between disciplines, researchers and project parts)
- *Phase C*: transdisciplinary integration, product/publication design and impulses for transformation/innovation/implementation.

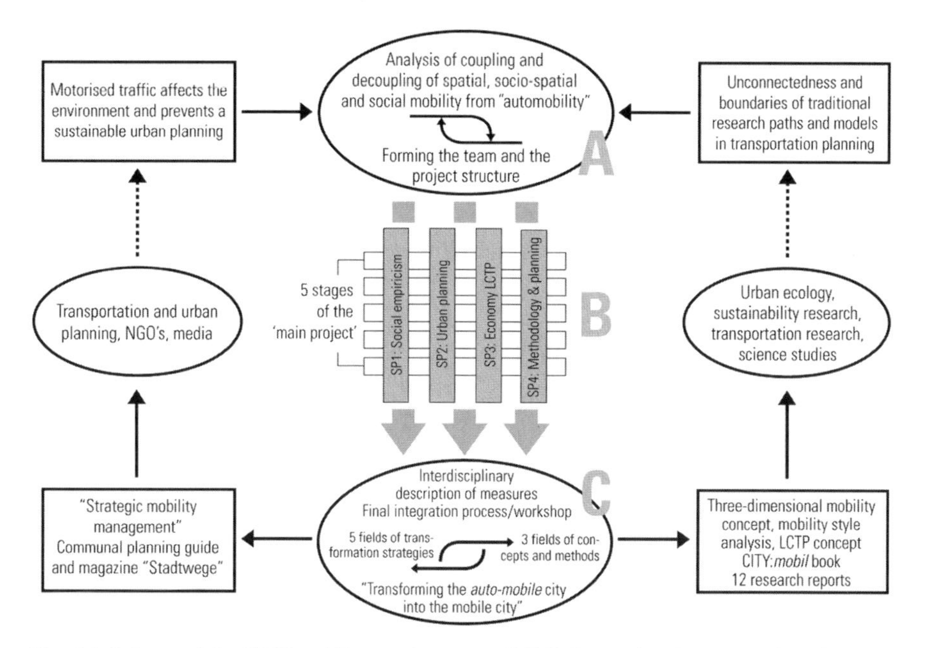


**Fig. 6.1** Scheme of the CITY:*mobil* research process (LCTP: Least Cost Transportation Planning)

In the first stage of the project (Phase A) contradicting normative scientific and political claims of importance and relevance had to be integrated into a research objective that was accepted as *the* focus for the research by all participating disciplines. This was done by mutually agreeing on a number of problem descriptions and resulting analytical questions, definitions of related notions and concepts.

The societal problem was described as an increasing volume of (motorised) traffic, generating a number of ecological and economic problems, impeding sustainable urban development. The scientific problem was summed up as the lack of appropriate disciplinary problem approaches in the conventional transportation research as well as the lack of methods to integrate the motivation for peoples' transportation behaviour – including the aspect of the symbolic dimension of mobility.

Consequently, the research work was not organised in separate project parts dealing with a disciplinary oriented analysis of ecological problems arising from increasing automobile traffic in cities. In fact, the project was carried out under the following principal analytical question that had to be answered by all its subprojects: 'How can the strong coupling between mobility and "auto-mobility" (the use of cars for most mobility needs, and specific car use patterns) be decoupled?' The research path had to transform the transport system and behaviour in those cities from the hegemony of the auto-mobility to a multi-optional and multi-modal mobility – meaning that mobility can be realised in many different ways (normative target of the project) (Jahn and Wehling, 1998; 1999). This notion of a multi-optional mobility also includes special services that respond to the population's needs (delivery services, car sharing and cycle rental, services for cycles/cyclists etc.) and does not only focus on the movement of people from one point to another.

A detailed understanding of the term mobility was an extremely important precondition for the research process, keeping in mind that not only the physical movement from point X to point Y, and the reason for this movement (the social context and its constraints), but also social mobility in terms of social flexibility and societal inclusion, can be regarded as determinants for mobility behaviour. Therefore, the term mobility was defined by the three dimensions of spatial mobility, socio-spatial mobility and social mobility.

This three dimensional concept of mobility was the core concept of the theoretical framework from the very beginning of the project (Phase A) up to the twofold integration at the end (Phase C). It rests upon two theoretical fundamental assumptions: (1) To make a distinction between mobility as motion/movability, as a basic societal relationship to nature (Becker and Jahn, 2003; Becker et al., 2006), and mobility as transportation, as a technical realisation of transport/movements and their infrastructural organisation and aggregation in the physical space. This enables an analysis of the social causes and effects within traffic and transportation as a genuine problem dimension. (2) To distinguish between the socio-spatial and physical dimensions of mobility, thus enabling all participating disciplines – all with equal status – to bring their specific insights, methods and findings into the common research process. Within this theoretical framework, the use of 'mobility' as an overarching integrative term, helped to avoid the weak bonds that often link disciplines in a multidisciplinary setting.

Since the project aimed at new strategies for the cities as well as at new research methods the researchers formulated three strategic goals: (compare Bergmann and Jahn, 1999)

- *Differentiation*: There was no abstract metatheoretical subordination of the participating disciplines under a leading discipline or theory. The cooperation was shaped by a mutual respect and full awareness of the usefulness of the differences between the working and scholarly cultures. (This differentiated approach and its aspects of a new interdisciplinary cooperation was part of the recommendations to the cities advising about cooperation between the authorities responsible for city, traffic and budget planning).
- *Integration*: Respecting the concept of differentiation, the project aimed at combining all findings into an integrated result. Supported by the knowledge of all the participating city administrators and planners, this result had to be transformed into comprehensive and actor oriented strategies. The concepts should be transferable to other cities.
- *Intervention*: The project and its actors aimed to use the project results to influence the current practical and scientific discourse on urban transportation policy and planning. Therefore, a continuous communication process was important within the research cooperation, and with actors of communal governments and associations, on the one hand, and with actors of the scientific community concerned with the problem addressed by the project on the other hand. The project and its actors also aimed to influence the conventional practice of transportation research with the new idea of multi-optional, social-ecological oriented mobility research.

Consequently the disciplinary perspectives had to be linked and the researchers had to cooperate as often and as closely as possible – a demanding effort due to the fact that the participating research institutes were located at five places within Germany and Austria.

Before starting the research work, a comprehensive network plan was mapped, providing a detailed design of the connections between all subprojects and the main project. The project was coordinated by one of the researchers who was supported by a co-chair and a coach/moderator. The coordinator was responsible for the integration of all elaborated knowledge as well as managing the team of about 20 individuals. The coordinator and the leading researchers from the participating research institutes made the decisions that were relevant to all participants respectively to the over all project process.

## 6.3 The Transdisciplinary Research Process

Having described the project construction phase we now detail the research and cooperation process (Phase B).

Based on the theoretical framework of the three dimensional concept of mobility, the research group developed various research perspectives to address the primary question: how can mobility be disconnected from its predominating form ('auto-mobility')? All aspects of the investigation aimed to develop shaping knowledge for transformation processes – in the field (the cities) as well as in the participating disciplines:

- The determining factors for mobility behaviour and orientation were investigated from *the social science perspective*. Orientation in this context means the predominant means of transport used by a group of individuals. Beyond the usual questions about mobility behaviour, like 'how far?', 'how often?', 'by which means of transport?' and 'where to?', the main focus in this subproject was on the lifestyles, circumstances and the attitudes of the urban population and how they are related to mobility behaviour – represented by the question 'why?'.
- The researchers with *the planning perspective* developed a computer learning model with which it is possible to display not just the transportation networks but also the mobility behaviour, the population distribution within the housing areas, as well as the workplace distribution in the cities and their surrounds.
- Along the principle of least-cost planning as used in developing sustainable energy supply strategies, the researchers working from *the economic perspective* developed a comprehensive information and valuation tool with which *all* traffic related costs and incomes of the cities can be displayed – even the hidden ones.
- From the *methodological perspective* the focus was on a critical analysis of research questions and methods in the traditional research on transportation planning. Based on this analysis a new research approach was developed using the idea of an ecology of transportation ('Verkehrsökologie') and a traffic genesis analysis ('Verkehrsgeneseforschung').

These research approaches were investigated in four subprojects.

To obtain the best possible mutual understanding and adaptability between the subprojects a sophisticated system of co-production, interfacing workshops and exchange of findings was established.

Even though there were four subprojects it was possible to achieve successful integration. The cooperation of a number of researchers from several institutes and different disciplines in each of the four subprojects was one of the most important factors for the successful integration of research. The fact that the disciplines and institutes did not work separately, but together, in each of the subprojects became one of the success stories of the project. Since the integration process occurred not only on the main integrating path of the project (so called 'main project'), but also at the level of every sub-project, debate on the mutual learning, alien methods from other disciplines, and strange research 'cultures' etc., it had to be conducted from the beginning and at every level. This prevented conflicts resulting from a huge, rather synthetic phase of integration at the end of the project. The establishment of relationships between disciplines and researchers during the research process, and the links between the various areas of knowledge, was obviously an aid to integration.

In retrospect, one of the most important features of the project construction was this collaboration in the subprojects and the numerous workshops, meetings etc. by the subproject members and by the whole team. This considerably enhanced the integration demanded by the projects' aims.

The project structure also facilitated the cognitive integration of the findings. The researchers working in the subprojects interacted with those from other subprojects in a way that promoted the integrative path of the project. This path – the main project – followed the rationale of a rather conservative research pattern and included six phases:

1. recording an urban inventory of the transportation system and its ecological impacts
2. working out empirical based options for and restrictions to the mobility behaviour, the transportation infrastructure and the city development
3. mapping scenarios concerning the development of the transportation system and its ecological, economic and social impacts for both the model cities
4. working out an integrated impact assessment of the strategies developed
5. conducting an intensive phase of transdisciplinary integration (see Section 6.4)
6. and finally publishing a guide for communities and a book on the general findings aimed at the scientific community.

Thus, the cooperation throughout and the integrative path of the main project provided a solid basis for integrating the multiple kinds of knowledge resulting from the subprojects as well as the main project.

## 6.4 The Twofold Transdisciplinary Integration

The process related integration was facilitated by the integration of the following areas of the conceptual framework:

- the three-dimensional mobility concept,
- the concept to decouple mobility from auto-mobility,
- the integrating management and cooperation concept etc.

Consequently, the task of the final integration phase (Phase C) of the project was to connect and to condense the respective results from the subprojects and the main project into an overall perspective.

It was the aim of the final process to integrate corresponding, as well as differing, findings from all project parts into relevant conclusions for the practitioners in the cities and for the scientific community. This twofold outcome is a significant aspect of transdisciplinary research projects: it has to be a concentrating process, targeting two very different groups and their questions and needs. The integration process was therefore divided into:

- integration at the level of concepts and methods, and
- integration at the level of the communal transformation strategies.

Transdisciplinary research projects should aim to differentiate between and bring together the different perspectives of practitioners and scientists, who also learn from this process. In the CITY:*mobil* project the first (scientific) integration level, for example, determined factors affecting mobility behaviour, bridging concepts for developing questions involving the actors and the transportation system as well as research and transformation. These integrated results had consequences for research methods and designs, for theoretical concepts and new research questions.

The goal of the integration process in developing communal transformation strategies was to catalogue strategic measures for key actors in the transportation sector and to evolve planning and management tools. The approaches include strategies to decrease the number of cars, to improve the performance of the public transport system, to promote cycling, to change space/time patterns and to develop new planning tools.

In order to guide the reader in the field of transdisciplinary integration we list the most important factors that helped ensure successful integration. In each of the three stages of a research project (see Section 6.2) there are special working procedures, measures or attributes that aid transdisciplinary integration.

The following list is one of the results of the recent project 'Evaluation Network for Transdisciplinary Research (Evalunet)' in which a comprehensive set of evaluation criteria for transdisciplinary projects (and hints for their development) was identified in an analysis of research projects. This analysis resulted from an intensive discourse between the researchers of the transdisciplinary projects and the evaluating experts (Bergmann et al., 2005).

## 6.4.1 Phase A

- *Formulation of a common research object*: The societal problem has to be reformulated as a scientific research object that fulfils the function of a boundary object, which frames the process of identifying the main research questions. This should be arrived at in a way that all participants (researchers as well as practitioners) accept it as a guideline for their work within the project. With the formulation of the common research object a very important cognitive basis is provided for the integration and for the 'commonality' (Klein, 2003, p. 107) of all participants in the research process.
- *Differentiation and the integrating potentials of differences*: The research team has to collectively identify and discuss the differences between the specific disciplinary approaches, methods and theoretical concepts and decide on the research paths and integration of the various results. The research assignments of subprojects (or other structures of single research tasks or paths) should not be separated into disciplinary oriented tasks – in the mistaken hope that the overall integration at the end of the project will close all gaps between them.

- *Structuring and mixing the team*: The strategy regarding the research personnel should – according to the subproject strategy mentioned in Section 6.2 – be conducted with the idea of the best possible mix of participating disciplines, methods etc. Therefore, the team should avoid splitting the research tasks among the participating institutions. A (sub) team composed of researchers from different institutions and fields should cooperate in each of the subprojects (or other structures of single research tasks or paths). This strategy facilitates the mutual understanding and learning and ensures that the integration process is an integral part of the subproject.

### 6.4.2 Phase B

- *Integrative structure planning*: A detailed network plan has to be undertaken at the beginning to create a realistic schedule for the research work and – above all – to be able to correlate the different research paths and their results at the right moment of the process. This is even more important when one subteam has to build on the results of another and is dependent on this preparatory work.
- *Tools and working procedures*: There are a number of procedures, tools and measures that can be used to organise the cooperative research work between the disciplines to ease the mutual understanding and integrating process for the whole group and the subgroups. There are partnerships between researchers from different disciplines for mutual control of the comprehensibility and integration potential of results or publications; common systems of categories for the description, analysis and forming of the research objects; multi-criteria assessment methods and moderated integration workshops. This seems to be a promising procedure for cooperation and mutual learning in interdisciplinary settings.

### 6.4.3 Phase C

- *The twofold transdisciplinary integration process*: The ideal type of transdisciplinary integration can be regarded as a success when the generated and integrated knowledge (integrated by means of the instruments mentioned in 6.4.1 and 6.4.2) can be incorporated into practical problem solving strategies as well as into new interdisciplinary knowledge. Ideally, both scientific innovation and societal transformation should be gained. New integrated objects of a second order can be the result – problem solving strategies addressing key actors and management tools, as well as theoretical concepts, methods and new research questions. Dependent on the integration steps undertaken during the project planning and the research process, this final phase of integration can be more or less laborious.
  Most transdisciplinary research projects do not manage to come up with both, readily achievable strategies for social transformation processes and innovative scientific methods, tools and theories. This is also true for the CITY:*mobil* project. It demonstrates how high the standards are set in transdisciplinary

research programmes which demand equally successful results in both the practical and the scientific sphere.
Thus, both the 'arms' shown at the end of the research process in Fig. 6.1 demonstrate the ideal of such project results – which can very seldom be matched.

### 6.4.4 Integrated Model Phase A to C

Transdisciplinary integration is not a task for the very end of a project. In fact the integration has to be in process during the whole of the project: from the preparation, through phases A to C.

To meet this challenge it is helpful to have a model that can be used as a practical planning and evaluation tool. On the one hand, it functions for different research settings on the basis of its universality. On the other hand it is so differentiated that it can help to identify essential points in decision making processes in project routines. Such a model has to be practical, but also – in the sense of understanding science as a scientific problem transformation – has to allow self-reflection. The model aims at supporting research projects in the planning phase (*ex ante*) and allows quality assurance during the process itself or through external or internal evaluations (*ex post*) (Fig. 6.2).

Obviously this model is used to describe the concrete research process of CITY: *mobil* shown in Fig. 6.1. With this methodological tool it is possible to plan or

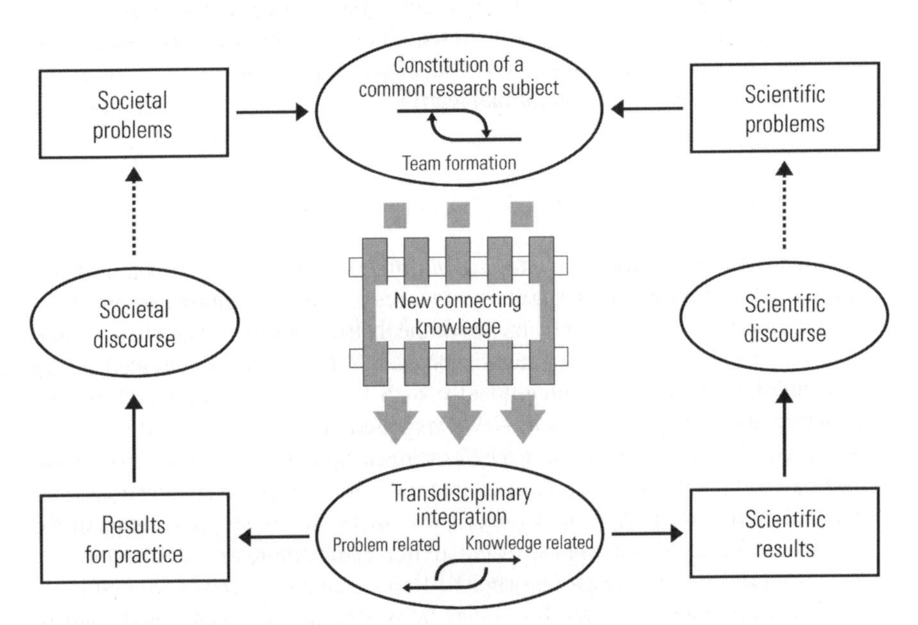


**Fig. 6.2** Model for transdisciplinary research processes (source: Becker and Jahn, 2006. p. 325, translation by the authors)

classify the different types of transdisciplinary research projects along the lines of chosen research settings and project goals. Of course, not every project has to go through the entire cycle of the model and its individual paths with the same intensity – this model represents an ideal.

## 6.5 Project Results

The publications resulting from the project were clearly divided into those for the communal planners and politicians and those for the scientific community. The 'Guide for Planning a Sustainable Urban Mobility' ('Planungsleitfaden für Stadtverträgliche Mobilität', CITY:*mobil*, 1998) obviously belongs to the first category while the book 'Sustainable Urban Mobility' ('Stadtverträgliche Mobilität', CITY:*mobil*, 1999) recapitulates the project findings for a scientifically oriented readership. However, it is not easy to assign each of the project results to only one of these two domains.

The main instruments and measures elaborated are:

- An empirical market research and marketing instrument, specifically targeting groups with different mobility behaviours and attitudes. This aims at shifting the group members from the car to other mobility modes and encourages them to stay with public transport or cycling etc. This new research focus and method is called *mobility-style research* (Götz, 1999; Götz and Jahn, 1998). It is a new research approach, which gives important information to transportation planners and companies. This instrument has subsequently been further developed and has been used in a number of different interdisciplinary and transdisciplinary research projects. It increasingly became a substitute for official transportation research and planning systems (used for surveys for transportation planning), which do not take into account attitudes or life-styles etc. in their survey design.
- A computer learning model of the cities (MOBI*DYN*) combines three modules: transportation networks; allocation of numerous types of different housing areas and work places; and mobility behaviour. It helps planners to understand how such parameters influence the results of their city and transportation planning projects. This model has been enhanced to become a regular planning tool for planning engineers. It is also a new tool for researchers.
- With the aid of the newly developed information system called *Least-Cost Transportation Planning* (*LCTP*) planners and politicians were able to receive comprehensive information about all expenses (including hidden expenses) for the transport sector and its sources. This is a new comprehensive information base on which to make project decisions. It was further developed, in later research projects carried out in cooperation with representatives from the planning departments of a number of German cities, as a standard planning tool for communities, especially as an interdisciplinary information tool for transportation and budget

planners. This project result therefore became an instrument for further research, as well as for concrete urban planning.

- The *'Action Impact Assessment' ('Handlungsfolgenabschätzung')* was developed and put to the test as a new integrated method for the *ex ante* evaluation of transportation projects and their ecological, economic and social impacts, and the problems faced when putting them into public policy. This instrument is primarily a useful scientific tool. It was designed and used to assess the developed results, strategies and measures under the two different heuristic research perspectives 'target group behaviour oriented intervention' ('Zielgruppenorientierte Verhaltensintervention') and 'systemic mobility management' (Bergmann et al., 1999).
- Most of the *28 measures* for a new mobility management in cities were scrutinised in an interdisciplinary organized description course – the 'Action Impact Assessment'. This assessment takes into account topics such as 'The transport company as a user oriented mobility provider', 'The integration of the bicycle', 'The strategic mobility management – the C:M2 process' and 'The assessment of impacts of planning action'.

## 6.6 Recommendations

First of all, one of the main recommendations resulting from this project may sound slightly banal but is nevertheless extremely important for successful complex transdisciplinary research: success or failure in the research process with its transdisciplinary integration requirements depends on the fundamentals being established in the initial phase of planning, formulation and construction of the project. This is essential for achieving a common conception of the work with its cognitive, social, communicative and technical aspects. It will be difficult to straighten out mistakes or omissions made in the initial phase during the later research process.

The success of a project depends largely on an awareness of the importance of defining achievable project goals and being conscious not to measure project success solely on the implementation of its results in the real world. To measure the success of transdisciplinary projects it is more practical and effective to define project goals oriented with the three phases of transdisciplinary research processes and the envisaged integration level. The CITY:*mobil* project, owes its success to the intense cooperation within the research team (about 20 individuals) from the start and to their distinct integration skills. The results were achieved due to the degree of interest in collaboration and the newly developed methodological integration procedures. Implementing the measures would need a period of 10 or 20 years because mobility is a highly 'charged' topic in our society, so an evaluation of these measures would be hard to achieve in the short term.

Moreover, the cooperation of science and practice requires a consciousness of the problems faced by others, which has to be kept alive during the whole research process. It would need another article in this handbook to reflect on the reasons for the problems that arose in the cooperation with the model cities. The impacts of this project are more an influence on planning tools, methods and policies, than artefacts in the model cities. This is due to important reasons such as: political changes during the five years of the project; a lack of readiness by the city governments and departments to cooperate across the department boundaries; and the researchers' preoccupation with overcoming a lack of appropriate research methods. There were considerable consequences in the scientific fields engaged in planning, measuring and surveying in the areas of transportation and ecology. The knowledge base for planning sciences and their curricula has changed as a result of this project.

By the way: the integration in the project was to such an extent that an individual's contribution to the main publications at the project's conclusion were not acknowledged: they were published in the name of the research network CITY:*mobil* [1998, 1999]. It was difficult for participants to be denied adding a named publication to their list. The system of reputation in the scholarly world relies on revised articles but does not acknowledge highly integrated papers published under a group name.

## References


Becker, E. and Jahn, Th.: 2003, Umrisse einer kritischen Theorie gesellschaftlicher Naturverhältnisse (*Societal Relations to Nature. Outline of a Critical Theory in the Ecological Crisis*). In: G. Böhme and A. Manzei (eds), *Kritische Theorie der Technik und der Natur*, München, Wilhelm Fink Verlag, pp. 91–112. Retrieved May 29, 2007, from http://www.isoe.de/ftp/darmstadttext_engl.pdf.

Becker, E. and Jahn, Th.: 2006 (eds), *Soziale Ökologie – Grundzüge einer Wissenschaft von den gesellschaftlichen Naturverhältnissen*, Campus, Frankfurt.

Becker, E., Jahn, Th., and Hummel, D.: 2006, Gesellschaftliche Naturverhältnisse. In: E. Becker and Th. Jahn (eds), *Soziale Ökologie – Grundzüge einer Wissenschaft von den gesellschaftlichen Naturverhältnissen*, Campus, Frankfurt, pp. 174–197.

Bergmann, M., Brohmann, B., Hoffmann, E., Loibl, M.C., Rehaag, R., Schramm, E., and Voß, J.-P.: 2005, *Quality Criteria of Transdisciplinary Research. A Guide for the Formative Evaluation of Research Projects*. ISOE Studientexte Nr. 13. ISOE, Frankfurt.

Bergmann, M. and Jahn, Th.: 1999, "Learning not only by doing" – Erfahrungen eines interdisziplinären Forschungsverbundes am Beispiel von "CITY:*mobil*". In: J. Friedrichs and K. Hollaender (eds), *Stadtökologische Forschung – Theorie und Anwendungen*, Analytica, Berlin, pp. 251–275.

Bergmann, M., Schramm, E., and Wehling, P.: 1999, Kritische Technologiefolgenabschätzung und Handlungsfolgenabschätzung – TA-orientierte Bewertungsverfahren zwischen stadtökologischer Forschung und kommunaler Praxis. In: J. Friedrichs and K. Hollaender (eds), *Stadtökologische Forschung – Theorie und Anwendungen*, Analytica, Berlin, pp. 443–463.

CITY:*mobil*: 1998, *Stadtwege, Planungsleitfaden für Stadtverträgliche Mobilität in Kommunen*, Freiburg, Frankfurt.

CITY:*mobil* (ed.): 1999, *Stadtverträgliche Mobilität*, Analytica, Berlin.

Götz, K.: 1999, Mobilitätsstile – Folgerungen für ein zielgruppenspezifisches Marketing. In: J. Friedrichs and K. Hollaender (eds), *Stadtökologische Forschung – Theorie und Anwendungen*, Analytica, Berlin, pp. 299–326.

Götz, K and Jahn, Th.: 1998, Mobility Models and Traffic Behaviour – An Empirical Social-Ecological Research Project. In: J. Breuste, H. Feldmann, and O. Uhlmann (eds), *Urban Ecology*. Umweltforschungszentrum Leipzig UFZ, BMBF. Springer, Berlin, pp. 551–556.

Jahn, Th.: 2005, Soziale Ökologie, Kognitive Integration und Transdisziplinarität, *Technikfolgen-abschätzung – Theorie und Praxis* 14, 2, 32–38.

Jahn, Th. and Wehling, P.: 1998, A Multidimensional Concept of Mobility – A New Approach to Urban Transportation Research and Planning. In: J. Breuste, H. Feldmann, and O. Uhlmann. (eds), *Urban Ecology*, Umweltforschungszentrum Leipzig UFZ, BMBF. Springer, Berlin, pp. 523–527.

Jahn, Th. and Wehling, P.: 1999, Das mehrdimensionale Mobilitätskonzept – Ein theoretischer Rahmen für die stadtökologische Mobilitätsforschung. In: J. Friedrichs and K. Hollaender (eds), *Stadtökologische Forschung – Theorie und Anwendungen*, Analytica, Berlin, pp. 127–141.

Klein, J.T.: 2003, Thinking about Interdisciplinarity – A Primer for Practice, *Colorado School of Mines Quarterly* 103, 1, 101–114.

# Chapter 7
# Shepherds, Sheep and Forest Fires: A Reconception of Grazingland Management

**Bernard Hubert, Michel Meuret and Joseph Bonnemaire**

**Abstract** New research relating to Mediterranean grazingland management issues, e.g. fire hazard control, biodiversity conservation, vibrant rural areas, etc. addresses the role that livestock farming systems can play. Established technical knowledge about livestock farming cannot contribute to these issues because it relies on standardised animal feeding techniques, focusing on cultivable forage areas and distribution of feed. These have led to the decline of the herders' knowledge of grazing management and of herding practices. Thus original research works were needed. The works have been carried out by an interdisciplinary team of researchers in animal sciences, ecology and social sciences, from an INRA laboratory in Avignon (France).

The research focuses on how the herders interact with ecosystems and landscapes through the management of their flocks, and investigations deal with the question of 'how can one steer grazing for animal production and environment maintenance?' (i.e. control scrub development, maintain biodiversity and promote the habitat of particular populations of plants and animals in need of protection). The paper presents the chronicle of this research over twenty years. It asks how the problematic evolved as a result of the accumulated scientific knowledge and the shift in societal questioning in relation to the rise of environmental concerns and the changes in rural and residential issues. Field observations, surveys and inquiries, public policies analysis, farm monitoring and systems modelling were used, combined or successively, in the course of five different research stages.

The authors lead up to an analysis of the different epistemological standpoints of disciplines and types of actual action representing a transdisciplinary approach. This approach is founded on a distributed view of knowledge, articulating different forms of scientific knowledge – according to the set of involved disciplines – and lay person knowledge and know-how in order to produce relevant categories and rules of action.

✉ B.Hubert
Institut National de la Recherche Agronomique (INRA), Paris, France;
Ecole des Hautes Etudes en Sciences Sociales, Paris, France
e-mail: hubert@avignon.inra.fr

G. Hirsch Hadorn et al. (eds.), *Handbook of Transdisciplinary Research*,

**Keywords:** Problem oriented research · Knowledge and action · Forest fire hazards · Grazing and herding practices · Mediterranean areas

## 7.1 Introduction

A knowledge production activity such as research, when carried out outside the boundaries of conventional academic frameworks in order to deal with complex 'objects', challenges the forms of organisation that commonly define this activity. This is a dynamic that furthers science by inducing a reflexive process in the research activity. This type of situation, i.e. research on issues relating to practice, is carried out through partnership work between researchers and other, non-researcher, actors. Contemporary issues generate such situations in abundance!

The 'reflexive modernity' approach (Beck, 1992) invites us to deal with such issues by learning to be reflexive together. This includes the people who pose the problems, those who are implicated in the problems and those who help deal with them. Many environment and farming related issues are closely tied to current or past practices and founded on considerations of risk and accountability. Can one fight scrub encroachment on pastoral areas without questioning the livestock system models developed over the past 40 years, i.e. standardisation of animal feeding techniques, concentration on cultivable forage areas and feed distribution which attended the decline of herders' knowledge? This is a major change of perspective which fully justifies systems approaches and actions: the purpose is not, or no longer only, to analyse systems but to act systemically. A move advocated by Peter Checkland (1993) and Jean-Louis Le Moigne (1983) many years ago.

In conventional approaches, researchers start with the tools and concepts of their discipline, which they will then try to adapt and extend. For instance they will seek to define the nutritive and feeding values of plants so that livestock farmers may feed their herds or flocks by grazing them on lands whose forage value is better known, while at the same time better controlling invasive scrub vegetation. In short, the issue tends to be reduced to discipline-specific 'objects', e.g. 'animal feeds', which refers back to problems of fibrosity, digestible nitrogen, essential amino-acid rates, etc. through nutritional physiology and microbiology. Outputs are expressed as 'Tables of Feeding Values' of feeds for domestic animals and partly reflect present French livestock systems. However, in the case of complex plant covers, which offer animals simultaneously several dozen grass species, lianas, and a range of tree and shrub leafage during each meal; and when at the same time management objectives grow more complex depending on whether the primary aim is to restore, protect or preserve habitats; researchers are progressively driven to develop and define new objects, starting with a rigorous analysis of the situation or of the issue as it appears to the field actors.

This unavoidably means identifying the objects which are the focus of action on ecosystems or landscapes, i.e. the categories used by the livestock farmers. The problem then reformulates as: how can one use a herd or flock to act on an 'environment' – or subsets of an environment – in order to boost herd or flock production, control scrub development, maintain biodiversity and promote the habitat of particular populations of plants or animals in need of protection? As a result the research question is transformed from 'producing knowledge on the use of an environment' to 'how can one steer grazing for dual objectives', i.e. animal production and environment maintenance? The question has shifted from the relatively conventional objects of a discipline ('intake behaviour in animals') to the objects of action and management ('steering intake at grazing').

In addressing such issues, research must therefore unavoidably engage with other occupational sectors who have their own (political, economic and social) objectives, their own concerns and constraints, different emergencies and paces, and more often than not, different interests, standards, norms and values from those of researchers irrespective of their discipline. Problematics such as 'decision support' associate researchers who produce models and tools and actors who implement them in real situations in order to solve their own problems.

How complex problematics lead to testing the disciplinary frameworks is illustrated by a 20-year research programme on grazing in rangeland areas of southern France. We shall then analyse research attitudes and standpoints raised by cooperation between researchers and non-researchers and among researchers from different disciplinary backgrounds. Finally we introduce an epistemological representation of trajectories in transdisciplinary research.

## 7.2 Background

To illustrate our purpose we selected the example of research on grazing in the Mediterranean rangelands of southern France subject to scrub encroachment. In the course of these studies we attempted to apply scientific methods to new objects in which biological processes are coupled with the strategies farmers apply to manage these processes. The material foundation of such objects is clearly biological, but it assumes new 'qualities' all along the processes which organise production. Rather than recording facts that have been evidenced, these studies aim to highlight the interconnections that give sense to the production systems and the objectives of the people who manage them. This enables the identification of categories that are both relevant for action and scientifically meaningful.

This research programme has been carried out by a multidisciplinary research team, the Ecodevelopment Unit of INRA in Avignon, in collaboration with farmers, cooperatives and technical advisory bodies. The Ecodevelopment Unit was established in 1982 to participate in research on the contribution of small ruminant husbandry to the prevention of forest fires in Provence. The research problematic has evolved over the past twenty years as a result of (i) the accumulated scientific knowledge produced both in individual disciplines and as a result of confrontation and interaction between disciplinary fields – animal sciences, ecology and social sciences, (ii) a shift in societal questioning, the rise of environmental concerns, the growing concern for the biodiversity of Mediterranean rangelands and for an environmental rather than purely technical approach to issues linked to forest protection.

For brevity's sake and to illustrate our approach, the only results presented here are those relating to our study of herding practices. Other studies explored the technical livestock systems in extensive situations, their management and outputs, the design and implementation of public policies regarding the environment and forest fires, the methods of controlling fuel biomass and the forms of collective action among heterogeneous stakeholders concerned with a shared problem on these various issues.

Table 7.1 shows the main stages in the construction of this approach which was initiated some 20 years ago and in which we individualised 5 stages, each characterised by different study objects. For each stage (related to a time period in Column 1) we indicate in six columns:

- the issue as identified in the field
- its translation into a research question and its expression as a study object
- the main disciplines involved
- the research objects designated as 'Level 1' around which transdisciplinary research organises itself
- the main outcomes of these studies
- the 'Level 2' research objects, which express a more formal return of the questions raised within the corresponding conventional disciplinary fields.

Table 7.1 illustrates the process by which a complex issue (Column 2) is translated into a collective and multidisciplinary research question – the central stake of such research programmes – through the construction of a 'study object' (Column 3) detailed into several research objects (Columns 5 and 7) that are relevant from the standpoint of each discipline (Column 4), sometimes even of discipline couples. Outputs, including scientific results (Column 6), issued from interactions between the research team and stakeholders (Hubert and Bonnemaire, 2000). What matters is the return contribution of these different outputs to the original issue through the development of new action categories and rules. These cannot be formulated without some interaction with the 'result users', i.e. the actors who contributed the initial question from which the research problematic was derived. Having been thus collectively identified (Column 3), the research question corresponds to types of objects that have been co-constructed, shared, and dealt with in interdisciplinary programmes. Sub-phases outputs within each stage (as used in Fig. 7.4) are quoted within brackets in Column 6.

## 7.3 Problem Investigation

Broadly, these five stages succeeded one another, albeit with some overlapping since the beginning of the research by INRA's Ecodevelopment Unit. The first stage embraced the beginning of the programme, which was initiated at the request

**Table 7.1** Table 7.1 shows the main stages of the research process. For each stage (related to a time period in Column 1), it indicates in 6 columns: the issue as identified in the field (2); its translation into a research question and its expression as a study object (3); the main disciplines involved (4); the research objects designated as 'Level 1' around which transdisciplinary research organises itself (5); the main outcomes of these studies (6) numbers within brackets relate to Fig. 7.4; the 'Level 2' research objects, which express a more formal return of the questions raised within the corresponding conventional disciplinary fields (7)

| | Problem in the field | Research question/Study object | Scientific disciplines | Research object 1 | Outputs | Research object 2 |
|---|---|---|---|---|---|---|
| 1982–1987 | How can domestic small ruminants be managed to help reduce shrub fuel biomass for improved control of forest fire hazards? | Contribution of woody plants to diets of sheep and goats feeding on grazinglands<br>*How can intake nature and amount in animals feeding on grazinglands be assessed?* | Animal production science<br>Animal ecology<br>Animal nutrition | Intake behaviour<br>Assessment of food offer (woody and herbaceous plants)<br>Assessment of food requirements and their satisfaction through grazing/browsing | Intake capacity of animals double that reported in reference models (1A)<br>Wide diversity of plants in diets (1B)<br>Ability to influence food choices as a result of environmental structuring<br>Highly satisfying zootechnical performances (1D) | Diet diversity<br>Practices and strategies to raise herds/flocks in those conditions |
| 1988–1993 | How can livestock farmers be induced to use scrub-invaded land? | Dual purpose grazingland management: animal output and spatial maintenance<br>*Steering of ruminant intake at pasture* | Animal production science<br>Animal ecology<br>Anthropology of techniques<br>Sociology | Feeding motivation to eat scrub vegetation<br>Methods to stimulate intake on grazinglands | 'MENU' model for herding (2A)<br>Design of fenced heterogeneous paddocks (2B)<br>Feed supplement 'Spécial Parcours®' (2C) | Meals<br>Intake kinetics<br>Changes in definition of the value of these areas according to land use types and techniques |

**Table 7.1** (Continued)

| | Problem in the field | Research question/Study object | Scientific disciplines | Research object 1 | Outputs | Research object 2 |
|---|---|---|---|---|---|---|
| 1994–2002 | What role can scrub-invaded areas play in feeding herds or flocks? | Expression of food choices in complex habitats<br>*Variation in intake behaviour of paddocked herds/flocks* | Animal ecology | Feeding strategies of ruminants in complex environments | A temporal organisation within meals and along days that stabilizes intake (3A)<br>Functional foods based on structure and mass (3B)<br>'GRENOUILLE' model to adjust resources inside the paddocks (3C) | Food diversity approached as 'functional feed' categories<br>Spatial structures adapted to temporal organization of meals |
| 1996 to present | How can livestock farmers be induced to respond to agri-environmental policies aimed at preserving nature? | Design and implementation of arrangements developed in the name of biodiversity<br>*Stakeholders and knowledge production and circulation within these arrangements* | Sociology<br>Anthropology<br>Economics<br>Animal production science<br>Ecology | Design of agri-environmental zoning<br>Development of guidelines and specifications<br>Origins and stabilization of scientific and technical knowledge | Debate on procedures implemented (4A)<br>Conditions for collective action and learning (4B)<br>Handbooks to help design guidelines for grazing/browsing management (technical leaflets) (4C) | Design of public policies and implementation means (ecologization)<br>Role of scientific knowledge in environmental policy emergence<br>Modes of collective action for heterogeneous actors |

**Table 7.1** (Continued)

| | Problem in the field | Research question/Study object | Scientific disciplines | Research object 1 | Outputs | Research object 2 |
|---|---|---|---|---|---|---|
| 2003 to present | How can grazing/ browsing be used to simultaneously ensure the renewal of forage resources and conservation of vegetation structure beneficial to biodiversity? | Interaction between ruminant feeding strategies and dynamics of plant communities that include a woody species having a high dominance potential. *The plant community herd/flock interface in complex environments* | Plant ecology Animal ecology | Motivation of animals to eat plant organs of the dominant woody species relative to the array of foods present in the plant communities. The population dynamics of the dominant woody species and the new development categories induced by browsing. | 'BALPOP' conceptual model (5A) Scenarios of sequential seasonal grazing/browsing in paddocks to control development of dominant woody species (5B) | Definition of 'target organs' to be browsed on the dominant woody species The assortment and balances of plant structures to be offered to the herd/flock at grazed/browsed plant community scale (the paddock) |

of the various stakeholders of Mediterranean forests (foresters, local councillors, livestock farmers, etc.). It focused primarily on the role that domestic herds and flocks could play in controlling excess inflammable forest undergrowth biomass where fires generally start. This stage involved animal scientists and ecologists who first sought ways to measure the daily intake of flocks in such environments, a step never attempted before in animal feeding science (Meuret et al., 1985). The results confirmed quite early the value of grazing for controlling scrub – showing the importance of flock management patterns and therefore of the farmer's role – and not only the efficiency of one or other animal species or race (Etienne et al., 1990; Joffre et al., 1991). It also highlighted the very high intake capacity of domestic animals used to feeding in such environments where they constantly consume a wide diversity of plants. The recorded intake levels were considered to lie far outside the standards established in experimental situations (Meuret and Giger-Reverdin, 1990). These field observations were subsequently validated by tests in digestibility cages specially designed to partly mimic the diversity of food the animals were used to grazing in the field (Meuret, 1988). The results showed that the feed value of these scrub-covered stretches of land was double that expected until then! At the time the plant aspects could not be adequately dealt with, dominated as it was by the 'Pastoral Value' dogma (Daget and Poissonet, 1971). The Pastoral Value (PV) is actually unable to take into account the relative nature of food preferences which animals express, according to the state of the offered food, its diversity (diversity of species, of plant organs, of phenological stages, etc.) and proximities. Besides PV ignores the feeding value of most shrubs to which it attributes a null value.

The lack of disciplinary knowledge of biological processes, i.e. an exceptional and daily intake of plant mixes by the flocks, triggered the second research stage, which began with enquiries to the practitioners used to manipulating the feeding motivation of their flocks, i.e. the shepherds. These declared from the outset that '*the everyday challenge is to offer the flock the feeding resources within a sequence that constantly revives the appetite*', a standpoint on the intake process which animal nutritionists were quite unable to deal with at the time. Research in this second stage associated animal scientists and anthropologists who carried out observations and in-depth enquiries with a wide range of livestock farmers and shepherds. There resulted a first representation of the 'steering of grazing circuits', the 'MENU' model (Fig. 7.1), based on the criteria of experienced shepherds. MENU makes it possible to design, for the length of a bi-daily grazing circuit, a succession of complementary areas whose coordinated use stimulates intake in a less palatable 'target area' (e.g. an area to be cleared of scrub) (Meuret, 1993). This type of grazing practice confirms that no individual plant or fenced grazing area possesses an intrinsic value that can be defined in the absolute and mapped. They all possess a relative, circumstantial value, linked to the way the shepherd stimulates the feeding motivation of his flock to a lesser or greater degree (Meuret, 1997). The MENU organisation takes advantage of the heterogeneity of rangeland plant resources and encourages the use of 'stimulating feed mixes'. On the other hand, conventional experimental data are based on the evaluation of foods considered individually to produce a metrology of ingested quantities and feeding value of each food.

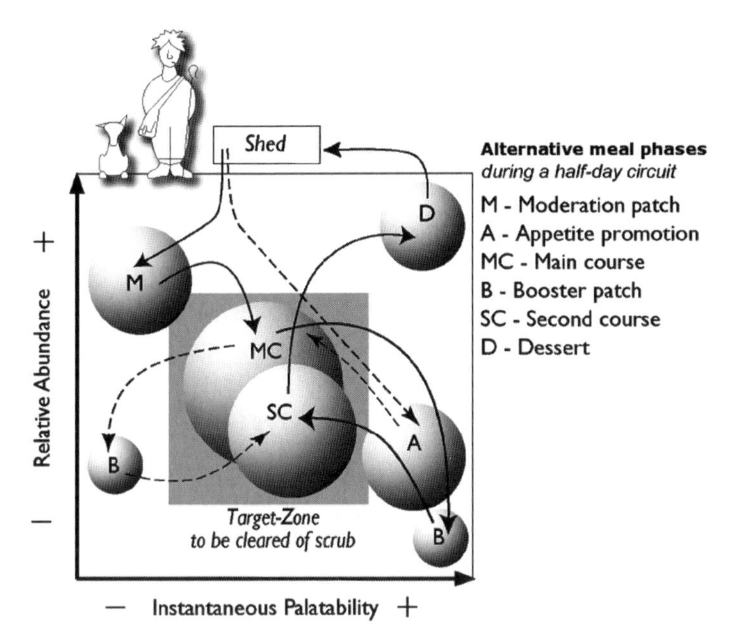


**Fig. 7.1** The MENU model enables a shepherd to stimulate intake on a "target-zone" that would be insufficiently attractive to the herd without a particular organization of the grazing circuit (each corresponding to a half day or a meal). The shepherd takes advantage of spatial heterogeneity of the land in designing the circuits so as to give the herd access to an appropriate succession of contrasting sites (meal phases) in terms of relative palatability and resource abundance (after Meuret, 1997)

The following two stages were almost concurrent. One sought to investigate in depth the intake process of animals on rangelands. The approach associated ecology and animal sciences, in order to better understand and interpret the earlier observations of grazing practices carried out with the contribution of shepherds, and to make them intelligible to the increasing number of actors concerned with grazing on areas designated as remarkable in environmental policies (protected areas, conservatories, Natura 2000 network, etc.). These actors had progressively noted the relative uselessness, in the situations they had to deal with, of the technical recommendations available in livestock science (Chabert et al., 1998). Beside contributing original knowledge of the 'feeding strategies' of animals on diversified grazinglands, which required spectacular advances in the modelling of biological processes with the help of bio-mathematicians (Agreil et al., 2005), this stage also produced a model of the spatio-temporal organisation of grazing on rangelands for the use of livestock farmers and managers of areas of specific interest: the 'GRENOUILLE' model (Fig. 7.2) (Agreil et al., 2004). This organisation model is based on the categorisation of the wide diversity of vegetation occurring in the paddocks to be grazed – at times confusing and therefore disheartening for a farmer – into only a few 'functional foods' fairly easy to identify on the basis of the distinct plant structure offered to the animal: 'large' and 'small' foods which allow a higher or lower intake rate ($\text{gmin}^{-1}$).

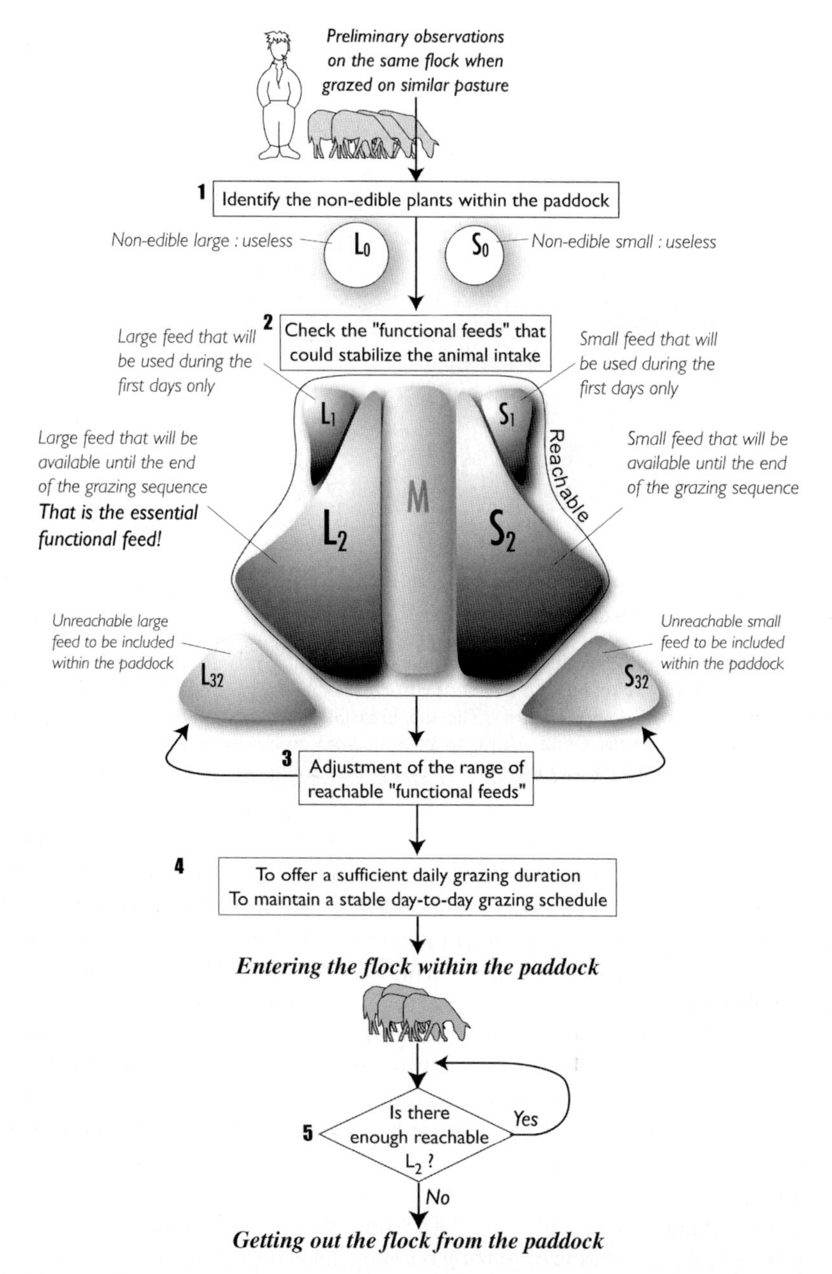


**Fig. 7.2** The GRENOUILLE method enables a sheep farmer to detect and adjust the "functional feeds" to offer within a paddock made on a very diversified vegetation. A minimum of experience or preliminary observations with the same flock are needed for the five diagnostic stages (Nos 1–5). The one indispensable functional feed is the "Permanent large" food (G2). It enables the sheep to stabilize their intake within each meal sequence and between days, both in terms of quantity and quality. The indicator for taking the flock out of the paddock becomes simple: is enough G2 food available (grasses and shrubs whose format enables the sheep to take easily large eating bites) (after Agreil et al., 2004)

The other concurrent stage aimed to better understand the factors underlying the increasing weight of environmental stakes centring on biodiversity preservation. These concerns aimed to change rough grazinglands, that had been marginalised by the technical models developed in the name of agricultural 'modernisation' in the 60s and 70s and ignored by grassland research, into areas of specific interest listed in international agendas and for which new forms of public incentives were devised and implemented in order to maintain them in their existing state, thus avoiding their reversion to an ecologically more banal forested state and contributing to landscape uniformity that makes them more susceptible to the development of major forest fires ravaging hundreds or even thousands of hectares (Deverre et al., 1995; Hubert et al., 1995; Alphandéry and Billaud, 1996). Researchers in sociology, agro-economics, management sciences, ecology and animal production science are currently involved in this work. The social sciences were thus induced to take into account the facts of ecological functioning, of know-how and skills and of the devices used to act on certain dynamics and steer them towards states considered to be more favourable (for the feeding of animals, protection against forest fires, biodiversity management, landscape conservation, etc.).

Finally, the most recently initiated stage in the research centres on interactions between grazing and plant dynamics at the level of plant communities, in situations where feeding herds and controlling scrub species with a strong dominance capacity detrimental to biodiversity, is the principal objective. This stage, which capitalises on newly acquired knowledge on the 'feeding strategies' of grazing and browsing animals, involves the participation of ecologists specialising in plant population dynamics. Their purpose is to reconsider accepted views on the diversity of plant species that compose these heterogeneous environments and the effects of the animals' selective feeding choices on the dynamics of this diversity. In this innovative transdisciplinary articulation, plant ecologists attempt to progressively approach their biological 'object' from the animal's standpoint and its motivation to consume complementary plant structures, i.e. 'large' and 'small' foods, in the course of a meal. An initial conceptual model, BALPOP, has been proposed (Magda et al., 2005) which articulates the very short time span of feed choices with the longer time span of plant community dynamics. As an output it proposes a couple of multi-annual scenarios of field utilisation patterns. These make it possible to interrupt or limit the reproductive cycle of over-invasive shrubs through browsing of a series of 'target plant organs', which moreover correspond to the 'functional feeds' of animals.

## 7.4 Problem Rebuilding

In field recordings emphasis is therefore, logically, on the diversity of the resources consumed and their organisation within meals (Agreil et al., 2004). One no longer seeks to fine-tune herding practices on a daily basis food 'offer' and 'demand', as is the case in conventional trough ration feeding, or even on artificial pastures, on the basis of a few parameters selected because they are the only ones that can be measured in the paradigm of conventional animal science. The land to be grazed or browsed having lost its negative qualifier of 'heterogeneous grazingland' is now viewed as a set of potential resources, within which one must encourage interactions that can improve the animals' appetite while steering their impact on the plant cover, given the available knowledge on feeding strategies in animals and the response of plants to grazing and browsing.

Several types of research outputs have been achieved: organisation models, such as MENU and GRENOUILLE, developed from the joint modelling of scientific and livestock farmer knowledge, a commercial feed that boosts intake at pasture 'Spécial Parcours®' (De Simiane and Meuret, 1994), and handbooks and recommendation leaflets on the use of different environments according to the objectives assigned to grazing (Colas et al., 2002). MENU and GRENOUILLE have benefited from recent theoretical advances in animal nutrition (in particular the concept of 'instantaneous palatability' of foods) and enable livestock farmers and environmental managers, as well as specialists in ruminant feeding, to discuss innovative management patterns for herds and flocks induced by environmental issues. By drawing in and organising appropriate concepts (the successive 'courses' of a menu-circuit, the 'meal', 'large' and 'small' foods, etc.) combined with identification in the field of different categories of functional foods (depending on the species present and their phenological stages), these models introduce new categories for action (new for both the scientist and the shepherd), which the farmers appropriate by discussing them among themselves as well as with the researchers.

The rich diversity of such areas defined as 'environments' is in fact expressed as the components of an organised action and not as different categories of plant covers. As such, it can become a resource to be valorised through innovative herding practices. Contrary to the notion of 'heterogeneity' – a synonym of imposed constraints in the eyes of animal feeding science and agronomy – these grazing practices are designed to 'build up' the feeding value of such environments by organising a feeding context that takes advantage of the herd/flock's feeding habits, the time patterns of plant offer, resource encounter over time, and spatial polarisation, etc.

This research dynamics opens the way to developing Level 2 'research objects'. These are closer to the organised disciplines as they give meaning to new questionings. They provide a base for the current elaboration of new models in animal feeding through grazing and browsing: plants, plant mixes capable of playing a role in 'reboosting appetite' in the course of a meal as well as for new reflection on environmental issues (Meuret, 2004). These Level 2 research objects will cause animal production science to evolve (Bonnemaire, 2001; Bonnemaire and Osty, 2004) as they serve as a common working base involving the promoters of the approach described above, animal production scientists and animal nutrition specialists (Meuret and Dumont, 2000) as well as sociologists and management scientists who investigate the forms of representation and styles of management of ecological dynamics in settings that result from public policies (Deverre, 2004; Teulier and Hubert, 2004; Hubert, 2005).

As a result, this process may cause disciplines to evolve by having to deal with objects that have been developed from problems encountered in the field of practice, and not only from theoretical scientific frameworks. Under this dual influence, disciplines will evolve inevitably. We experienced it with animal nutrition science taking into account the diversity of food items and intake rhythms and with sociology taking into consideration the specificities of biophysical processes. The history of science provides ample examples of how disciplines have evolved through scissiparity or through cloning (by changing their theoretical framework for example), whereas in the process described here scientific disciplines change through their contact with problems to be solved in the real world. What we witness here is the joint production of categories that are new for the researchers and of rules that are relevant for action (as in 'MENU' and 'GRENOUILLE' models). These differ from the categories initially formulated by the actors as they are also the result of the researchers' work and specific goals, and have been co-constructed through concrete approaches and cross-learning by the researchers and the stakeholders.

## 7.5 Problem Structuring

As we can see, in situations involving researchers in partnerships with non-researchers, the research questions follow actual trajectories during the conception and the management of the research operations that they generate. These trajectories express more particularly the evolution of the research programme and of the partnership situation: they are a constituent part of the objects themselves and of the research results. But they also are dependant on the diversity of viewpoints that need clarifying and the scientific postures of the researchers involved.

### *7.5.1 Problem Structuring According to the Ways Research Work is Performed*

The notion of 'standpoint' is used here in the sense given by Legay (1993, 1997), Pavé (1994), Bonnemaire (2001) and Bonnemaire and Osty (2004) to define research approaches to complex systems, which privilege a specific entry chosen for its relevance to the objective defined. The standpoint is a scientific one, i.e. a constructed, chosen standpoint, contrary to standpoints in our daily life which are not deliberately constructed. The latter are defined, according to semiologist L.J. Priéto (1975) by the activities of the subjects and their position in relation to a reality. Several standpoints are of course possible, as are the different '*Weltanschaungen*' mentioned by Checkland (1993).

A range of disciplines have thus been brought to play in the different stages of the programme, each contributing its paradigms, its concepts as well as its theoretical and methodological frameworks, despite their not being always compatible with each other. For this reason it was necessary to identify the forms or articulation between disciplinary postures. A. Hatchuel (2000) identifies three 'models' in research work, each of which corresponds to specific conditions of exercising a professional activity and of involvement with the different partner categories.

- the *'laboratory model'* – most common in experimental research in agronomy, physics and biology, etc., based on the observation that since the world is wide and complex, one needs to isolate a part of it in order to investigate it (the researcher will therefore extract a piece of it and separate it from the rest of the world);
- the *'field model'* – originating in the ecological, social and economic sciences: the world cannot be viewed as a closed entity; it is complex and involves a vast number of interactions. In order to investigate it as it is, be it in nature or society, the researcher needs to build objects or artefacts to portray it;
- the *'intervention research model'* – derived from the management sciences and from ethno-methodological or comprehensive approaches in sociology. This model assumes that researchers do not stand outside of the world, that they are stakeholders in the world which they investigate and should therefore place themselves in situations of interaction by engaging in collective action with the partners who have called on them.

Hence there exists a direct relation between the researchers' scientific posture, the ways in which knowledge is produced and the knowledge itself. The types of knowledge produced will be different according to the work posture adopted, which derives from the disciplinary affiliation of each researcher, as well as their history and culture. This is an important point, since believing that one stands in a paradigmatic context, which isn't the one which one actually stands in, will generate controversies that lead into impasses (as people can no longer understand each other) or end up with unacceptable solutions.

In brief, people act according to the way they understand the world and to the concepts they use to represent it. In this perspective, it is crucial that researchers themselves have a clear view of the status they assign to their research approach, to the type of knowledge they generate, to the form of intervention they implement and to the type of partnership they engage in.

### 7.5.2 Problem Structuring According to the Perspectives these Different Standpoints Assign to the World

The diversity of standpoints cannot be piled up on top of one another in the hope this will provide total understanding of the functionings that are being investigated. Such an all-embracing vision is unreachable. However, it is crucial to learn to approach these functionings in a different way, by associating different and complementary viewpoints through transdisciplinary approaches, as recommended by Richard Bawden (1997) in a paper on agronomic approaches to sustainable development. He invites us to break out of the strictly technological and anthropocentric vision that generally prevails in agronomy, yet warns us against getting trapped in a solely ecocentric analysis, which, conversely, would privilege a strictly naturalist approach. For R. Bawden, these options need clarifying in order to both avoid paralysing confusions and to re-establish the values that convey sustainability to facts within their societal context. How then does one change one's standpoint or create bridges between standpoints? An analytical grid is needed to find one's way through this diversity of perceptions of what appears to be important and the diversity of ways actors position themselves in order to act.

The grid proposed by Richard Bawden and presented by in Hubert (2002, 2005) may serve as a basis for designing such a grid. This consists of four sections or quadrants partitioned by two orthogonal axes: a vertical axis contrasting reductionist visions of the world in the lower part, with holistic visions in the top part; and a horizontal axis distinguishing an objectivist (or positivist) vision to the right, from a constructivist vision to the left (Fig. 7.3). The bottom-left quadrant is not relevant here and shall not be considered. We shall distinguish:

- a techno-centric quadrant (bottom right) centering on technicity as a means of developing and valorising nature: at its core lies the notion of productivity. This corresponds to standardisation, to the production of standards and benchmark data, as well as to prescriptions as a means of transmitting knowledge.

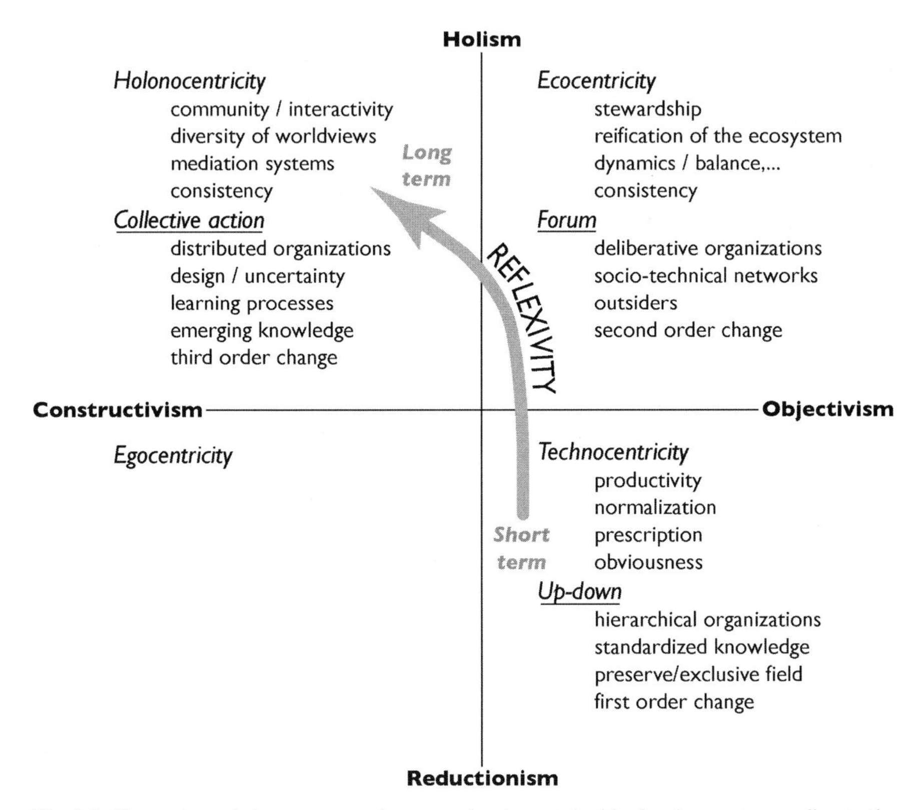


**Fig. 7.3** Three viewpoints on agronomic approaches to sustainable development according to the grid proposed by R. Bawden (1997). Key words characterizing each of the quadrants and related action procedures

- an eco-centric quadrant (top right) in which nature is not solely at the service of humankind who, therefore, must respect its functionings and dynamics. The notion of the ecosystem is central to representing the world and its functioning, cycles, equilibria and dynamics. We are dealing here with coherences, balance sheets, and not with the absolute objectivation of the things that make up the world. There is no longer truth per se, about the complexity of the world, but knowledge that is more or less complete, and therefore uncertain.
- a 'holistic' quadrant (top left) where the focus is on social interactions and 'solidarities' (alliances as well as oppositions). Besides the functionings of nature, what is being considered here are the activities, practices, intentions and objectives of human groups who derive their resources from these functionings. There are no natural systems per se but rather knowledge systems and systems of action on the world. We need to recognise these different systems and the knowledge and concepts underpinning these different thought systems (which include the other quadrants) in order to act on the world: the role of mediation arrangements and the different representations of these thought systems are to enable such compromises. This is a world where emphasis is on coherence rather than objectivity, so long as we accept the existence of a diversity of viewpoints and that each is relevant for those people who carry and express them.

In moving from bottom right to top left across this grid one progresses towards increased reflexivity and recognition of the time span of the temporal processes involved. Hence reflexivity involves greater awareness of the long term dimension. Indeed, the differences between the quadrants are the principles and modalities of the relation to action.

### 7.5.3 Problem Structuring According to the Resulting Types of Action

As regards action procedures (Fig. 7.3), when action takes place exclusively within the 'techno-centric' quadrant, the procedures involved are founded on forms of hierarchical organisation, of standardised knowledge. The model functions on the basis of planned organisation (in the sense given by Dodier, 1997), in which '*the concretization and functioning of objects obeys laws whose designers, the scientists, detain knowledge ensuring their mastery over networks, in so far as the operators observe the plans*'. Networking procedures are considered to be stable, having been defined by the designers who possess the necessary functional knowledge in worlds sheltered from outside intervention. Such has been the case in farming over the past 50 years resulting in the 'modernisation' of French agriculture and the 'Green Revolution'. In such contexts, responses to uncertainty or to environmental change relate to what may be termed 'first-order changes' where solving a problem roughly consists of amending the routines applied, changing the rules or re-adjusting the bodies of references. For instance one may agree to change regulatory standards upwards or downwards on the validated observation that current standards are not applicable or insufficiently relevant. However the actor networks, the procedures that activate them and the ways knowledge is generated, remain unchanged.

In the 'eco-centric' quadrant, procedures involve non hierarchical deliberative organisations that foster debates, such as forums. These deliberative organisations are founded on so-called socio-technical networks in which a great diversity of actors meet to discuss widely heterogeneous entities such as their individual value systems, norms and standards and objects related to the ecological systems and processes in debate (Mormont, 1996; Deverre et al., 2000). Action as well as confrontations and associations among actors are organised within flexible frameworks. These forums are open to outsiders to the domain or system being debated: the argumentative framework defines the right of access to the deliberations on the issue in debate and the boundaries, which may generate controversies. Statements by each participant will thus undergo a translation process within the group, allowing them to be shared and socialised. The process confers equivalent status to heterogeneous resources and *a priori* incommensurable interests that require adjustments among the actors regarding the very definition of what they are (Callon, 1989). The process is dynamic, involves power struggle and negotiation to redirect interests or create alliances on an issue. It also enables the actors to establish links with other situations they know and thus to steer action and formulate a judgment on its execution (Akrich, 1993). Here, adaptations, changes and solutions to problems relate to 'second-order changes'. The whole process does not stop at just changing the rules and standards, but may also question the principles and standards at the source of these rules. In this case the problem to be solved will need to be formulated within a setting that has been remodelled by the extended assemblage of actors and in the new terms ensuing from these successive translations. A given factor or activity examined may prove ineffectual: one will then need to examine other factors, bearing on other linked processes, which had not initially emerged. The objects considered have thus altered the network of actors involved, enabling innovative solutions to be sought for that particular situation.

The third quadrant is the realm of 'collective action', which is more organised than the forum. Here, organisation forms centre on the social actors. The dominant concern is no longer an ecology that prevails in the name of 'nature', but the analysis of relationships with nature based on available, tacit as well as explicit knowledge, and of activities founded on this knowledge and their reciprocal ability to evolve in situations of interaction with other forms of knowledge and other types of activities. Central to this is the in-depth transformation of the knowledge of each party – and not only of their statements as in the above situation – through cross learning promoted by formal interaction arrangements. They are distributed-type organisations in which *a degree of uncertainty as to the functioning of 'objects' is acknowledged as inescapable and an intrinsic part of the technical networks and which accept that uncertainty is managed locally and publicly as events occur by operators considered to be the holders of heterogeneous knowledge that cannot be reduced to an only central body of knowledge* (Dodier, 1997). The forms of knowledge, concepts and activities of the arrangements will allow rapid responses to ceaselessly changing situations by confronting viewpoints: emphasis here is on adaptation capacities to change.

In this quadrant, strong emphasis is laid on design, i.e. construction of the reasoning frameworks that underpin the understanding of the world, learning processes, and the production of new emergent knowledge resulting from interaction among participants. In dealing with problems in this type of setting and with such principles and procedures, changing routines and rules as in the techno-centric quadrant becomes irrelevant, as does challenging the reasonings that underlie these routines as in the eco-centric quadrant: often challenged are the values at the origin of these reasonings, and their underlying principles. This involves changing the value system, i.e. the problem as posed initially may no longer be relevant and the situation which had been posed as a problem must then reconsidered from different standpoints. The problem may no longer exist, or may have shifted to other questions. For instance solving problems posed by biophysical functionings may require inventing and implementing new forms of organisation of the activities that generated these malfunctionings. This involves imagining and implementing novel arrangements charged with managing new objects developed critically in relation to the processes examined initially (see for instance Couix and Hubert, 2000 on prevention of forest fires). The actor networks change along with the forms of their organisation and the state of shared knowledge, producing novel solutions.

### *7.5.4 Research as a Situated Activity*

Figure 7.4 displays the trajectories followed by the different stages of our research programme in these four quadrants. Each stage is signalled by its main outputs referenced as in Table 7.1 (Column 6, within brackets). Most of this research is seen to have combined approaches situated in several quadrants, producing fairly different types of outputs. Some of these are situated at the border between two epistemological standpoints as a result of disciplinary and methodological interaction. These borderline areas are where the social sciences will meet ecology or where animal production science will learn to consider the diets of animals feeding on rangelands. Our initial work was clearly positioned in the techno-centric (zootechnical performances, intake capacity) and eco-centric quadrants (diet diversity). The second stage, made possible by the contribution of social sciences, started with *in situ* interviews and observations to produce quantitative data that could be compared with accepted norms and standards in order to keep in touch with the animal science community. Thus in the following stages aspects raised in the first stage were explored in greater depth by mustering new resources in ecology and social sciences in order to deal with new, shared study objects: the grazed ecosystem, the 'individual' actors, the collective dimension of public policy implementation and new institutional arrangements for field studies.

In measuring the feeding value of forage resources using animals in digestibility cages, the researcher is clearly positioned in a techno-centric approach

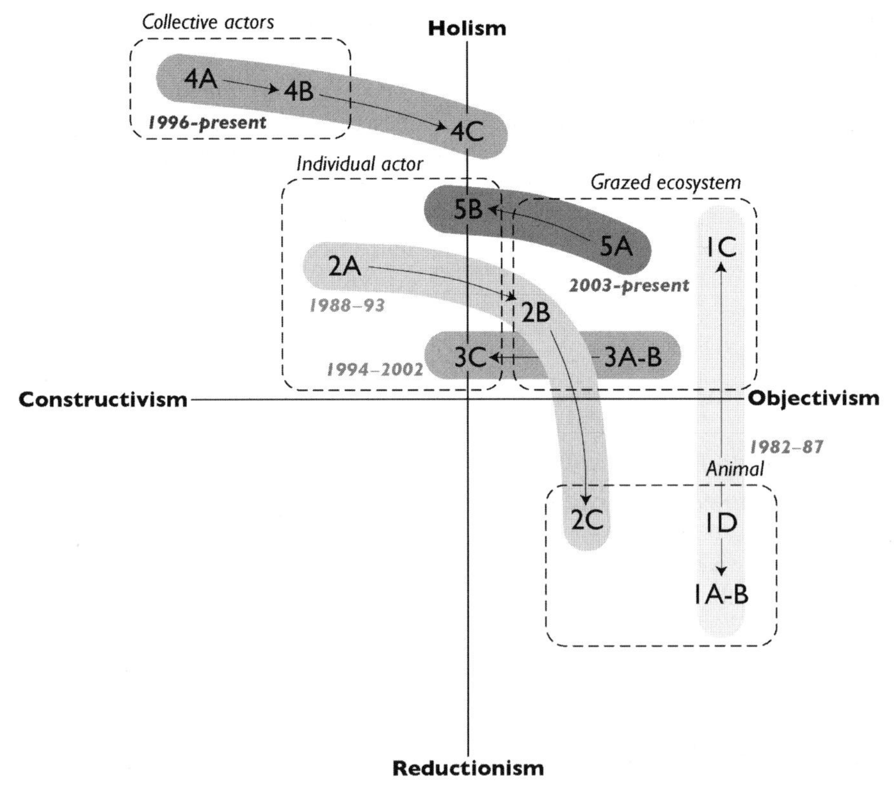


**Fig. 7.4** Trajectories followed by the different stages of the research programme in the three quadrants according Fig. 7.3 setting different viewpoints on agronomic approaches to sustainable development. Each stage is signaled by its number and main outputs as referenced in Column 6 of Table 7.1

(the laboratory model), even when the forage includes holm-oak branches and the animal is a lactating goat, instead of the conventional model in animal production – the castrated male sheep – what is being sought is comparability with zootechnical data. Inversely the 'MENU' model starts from livestock farmer knowledge and seeks to make it intelligible to researchers and technicians. The objective here is to produce generic knowledge from a range of knowledge by mobilising concepts and methods drawn from several quadrants in an intervention research approach. Likewise, whereas in Stage 4 the focus is on knowledge and forms of organisation mobilised in agri-environmental schemes, Stages 3 and 5, concentrate efforts on the processes underpinning the dynamics of grazed ecosystems and use the field model to this end, conscious that these ecosystems are managed by people who mobilise their knowledge in historically and geographically situated contexts.

Collaboration among disciplines thus enables the objects investigated to be studied without reducing their complexity by combining forms of knowledge arising from different epistemological postures in a transdisciplinary approach. In our opinion, this provides a way to manage the tension between scientific exigency and societal relevance of the issues we are ambitious to deal with. The former, demands knowledge outputs that can be validated by the scientific communities concerned (animal, ecological and social sciences), as they have been produced within their respective epistemological frameworks but have been approached with the help of novel research objects and have resulted in original knowledge. The latter expects knowledge outputs applicable to actual situations, i.e. appropriable by several types of actors who question their own knowledge in the face of an evolving context that changes their usual objectives.

## 7.6 Recommendations

### *7.6.1 Research as a Capability to Combine Different Forms of Knowledge Production Activities*

The purpose is not to give pre-eminence to the 'holistic' standpoint (especially as it acknowledges the other standpoints): what matters is at least to clarify and recognise the differences in standpoints and consequently in the related forms of knowledge and types of stakes. They need clarifying so that they can be mobilised in a complementary way, as well as knowing where each fits, in order to move from a short- to a long-term perspective, thus increasing our reflexivity on what is being implemented, or even challenged. The same goes for research postures, which are also included in this categorisation.

In the laboratory model output occurs chiefly within the 'techno-centric' quadrant. The knowledge produced is not readily usable in the 'eco-centric' quadrant as the criteria and modalities in which knowledge is produced in the first quadrant are not directly compatible with ecological approaches. Taking account of ecological processes in an approach that would remain strictly experimental, as in agronomy, requires considerable reductions – if only temporal. The duration of an experiment, for instance, and the moment it represents in wider dynamics – with the risk of overlooking ill-identified but crucial interactions, that is, unless one is content with measuring the effect of a few targeted factors whose contribution to an encompassing functioning would have been explicated earlier. Even in this case work modalities will need changing and shift to observation protocols in the field and to modelling. Likewise, dealing with issues in the 3rd 'holistic' quadrant can only be achieved by getting involved as a researcher, in a collective action with the other partners concerned with a problem, which this situation poses to them. Callon (1998) qualified this approach as a pragmatic communication model linking research and decision processes, which he differentiates from the 'diffusionist' and 'imperfect knowledge' models, both of which are not very distant from our two preceding models. In this model interactions between knowledge systems – the actors among themselves and the actors with the researchers – open new action and relational fields and foster the emergence of new 'communities' of individual and collective actors linked together by scientific and practical knowledge, techniques, standards, cultural preferences and natural realities; causing knowledge, involvements and exchange networks to evolve simultaneously.

Our purpose is not to declare a particular posture to be superior to, or more scientific than another. What matters is to acknowledge that different objectives cannot be reached by just any research posture and that each posture relies on its own rules that must be respected: this is the basis of scientificity. That which makes a difference in the research work are not the objects, but the approaches, which researchers apply, and the methods they use to investigate these objects. This means that at a given time along the research process, some researchers at least will need to engage in interactive approaches with their partners in the projects, thus enabling the identification and construction of socio-technical objects which will then be approached by each researcher from the angle of their disciplinary competencies. The interactions that engendered their construction will guarantee that these objects make sense for both the researchers and their partners. Each one of the quadrants aims at specific outputs, all of which are useful in studies whose objectives are to better take into account the stakes of sustainable development in environmental or agricultural problematics *sensu lato*. We need to be able to recognise the relevance of each posture and its limitations in order to agree on working modalities that will produce specific and complementary outputs through a transdisciplinary process.

### 7.6.2 Research as a Dynamic Exploratory Process

In fact the relevance of the study objects and of the questions that they convey represents a significant driving force in discipline evolution. Research excellence is to be grounded in the quality of the trajectory, which starting from a complex issue taken as such, enables the identification of relevant study objects, then of research objects that hold promising results in each discipline. Hence the importance of making explicit the contents of Columns 2 and 3 of Table 7.1, i.e. the process through which an issue is transformed into research questions and objects, e.g. from how to control scrub encroachment to how to motivate flocks browsing. The dynamic that transforms these objects contributes to shifting the boundary between objects and between disciplines, e.g. animal science, ecology and anthropology (Columns 3 and 4). Disciplinary boundaries are thus put to the test and shift through confrontation with new objects and their investigation. Such an approach consolidates individual disciplines whereas excessive compartmentalisation weakens them.

A collective of researchers should report on this progression and, for their part, scientific bodies should be capable of assessing it and not focus evaluations solely on contributions to the disciplines as is the general practice when keeping to conventional assessment patterns. Neither is it useful to dig out ‘mythical’ syntheses when at best one will find activity reports or research contracts. Their evaluation is quasi impossible and so is their generalisation. The necessary references are lacking when these are aggregates of results forming assemblages of knowledge which are specific to individual situations and not necessarily easily articulated and made explicit. These assemblages of knowledge are unavoidably embedded in different disciplinary paradigms and can be viewed from different standpoints. They therefore can be interpreted simultaneously within different disciplines. The outputs of this kind of contextualised interdisciplinary research are assessed both with reference to the disciplines concerned and to action in terms of relevant categories and rules of action.

Modelling, as a rigorous procedure, provides a particularly appropriate and fertile tool in object construction and formalising, owing in particular to the heuristic value of formalised representations in interdisciplinary dialogue (e.g. 'MENU', 'GRENOUILLE' and 'BALPOP'). It also enables a simulation of alternatives as well as the exploration of wider timescales and the legitimisation of implementation conditions by specifying the validity domain of results. Because of the abstraction procedure involved, modelling allows the generalisation of the results obtained in each case study and the formalisation of time dimensions beyond the knowledge produced in situ in the course of the observation period (Legay, 1997; Hubert, 2001).

A fundamental change is introduced by the reworking of research questions regarding a societal issue into an 'object of study' and its detailing into research objects (move from Columns 2–3) explicitly situated within an interdisciplinary framework that is both operational and coherent. It leads to working with processes, modes of intervention, interactions, coordinations, and flow transfers rather than with descriptions, stocks or balance sheets. In other words the researcher is invited to focus on objects relating to management and action, e.g. to set stock a paddock, to decide to move stock to another one, to organise a grazing 'circuit', etc. No longer can the knowledge produced be separated from the activity concerned.

**Acknowledgments** We are very grateful to Laurence de Bonneval for her scrupulous and faithful although rush mode translation of this paper and to Su Moore for relevant English editing.

## References

Agreil, C. and Meuret, M.: 2004, An improved method for quantifying intake rate and ingestive behaviour of ruminants in diverse and variable habitats using direct observation, *Small Ruminant Research* 54, 1–2, 99–113.

Agreil, C., Meuret, M., and Fritz, H.: 2005, Adjustment of feeding choices and intake by a ruminant foraging in varied and variable environments: new insights from continuous bite monitoring. In: V. Bels (ed.), *Feeding in Domestic Vertebrates: from Structure to Behaviour*, CABI Publishing, Wallingford, pp. 302–325.

Agreil, C., Meuret, M., and Vincent, M.: 2004, Grenouille: une méthode pour gérer les ressources alimentaires pour des ovins sur milieux embroussaillés, *Fourrages* 180, 467–481.

Akrich, M.: 1993, Les objets techniques et leurs utilisateurs. De la conception à l'action. In: B. Conein, N. Dodier, and L. Thévenot (eds), *Les objets dans l'action. De la maison au laboratoire*. coll. "Raisons pratiques", 4, EHESS, Paris, pp.35–58.

Alphandéry, P. and Billaud, J.-P. (eds): 1996, Cultiver la nature, *Études Rurales* 141–142, 233pp.

Bawden, R.J.: 1997, Learning to persist: A systemic view of development. In: F.A. Stowell, R.L. Ison, R. Armson, J. Holloway, S. Jackson, and S. McRobb (eds), *Systems for Sustainability*, Plenum Press, New York and London, pp. 1–5.

Beck, U.: 1992, *Risk Society: Towards a New Modernity*, Sage, London, 260pp.

Bonnemaire, J.: 2001, Enjeux sur les savoirs et les objets de la zootechnie: l'élevage entre science, technologie, nature et société, *C.R. Acad Agric, Fr.*, 87, 4, 237–260.

Bonnemaire, J. and Osty, P.L.: 2004, Approche systémique des systèmes d'élevage: Quelques avancées et enjeux de recherche. *C.R. Acad. Agric. Fr.*, 90, n2 – séance du 11 février 2004, 29 pp.

Callon, M.: 1989, *La science et ses réseaux. Genèse et circulation des faits scientifiques*, Paris, La Découverte.

Callon, M.: 1998, Des différentes formes de démocratie technique, *Annales des Mines*, janvier, 63–73.

Chabert, J.-P., Lécrivain, E., and Meuret, M.: 1998, Éleveurs et chercheurs face aux broussailles, *Courrier Environnement Inra* 35, 5–12.

Checkland, P.: 1993, *Systems Thinking, Systems Practice*, J. Wiley & Sons, Chichester, 330pp.

Colas, S., Müller, F., Meuret, M., and Agreil, C. (eds): 2002, *Pâturage sur pelouses sèches: un guide d'aide à la mise en oeuvre*. Espaces Naturels de France, Fédération des Conservatoires régionaux d'Espaces naturels, Programme LIFE-Nature Protection des pelouses relictuelles de France, 139pp.

Couix, N. and Hubert, B.: 2000, Promoting collective learning in a land-use management project: Thirteen years experience in researcher-technician partnership in the Cévennes, France. In: LEARN (eds), *Cow up a Tree. Knowing and learning for change in agriculture. Case studies from Industrialised Countries. Science Update*, INRA Editions, Paris, pp. 121–140.

Daget, P. and Poissonet, J.: 1971, Une méthode d'analyse phytoécologique des prairies. Critères d'application, *Annales d'Agronomie* 22, 5–41.

Deverre, C.: 2004, Les nouveaux liens sociaux au territoire, *Natures, Sciences et Sociétés* 2, 12, 172–178.

Deverre, C., Hubert, B., and Meuret, M.: 1995, The know-how of Livestock farmers challenged by new objectives for European farming: I. Rangelands usages, greatness and decay. In: *Proceedings of Vth International Rangeland Congress*, Salt Lake City, pp. 115–116.

Deverre, C., Mormont, M., and Selman, P.: 2000, Consensus building for sustainability in the wider countryside, Rapport final à l'Union Européenne, contrat no. ENV4-CT96-0293.

Dodier, N.: 1997, Remarques sur la conscience du collectif dans les réseaux socio-techniques, *Sociol. Travail* 2, 131–148.

Etienne, M., Hubert, B., Jullian, P., Lécrivain, E., Legrand, C., Meuret, M., Napoléone, M., Arnaud, M.-T., Garde, L., Mathey, F., Prévost, F., and Thavaud, P.: 1990, Espaces forestiers, élevage et incendies, *Revue Forestière Française*, n spécial 'Espaces forestiers et incendies', 42, 156–172.

Hatchuel, A.: 2000, Intervention research and the production of knowledge. In: LEARN (eds), *Cow up a tree. Knowing and learning for change in agriculture. Case studies from industrialised countries. Science Update*, INRA Editions, Paris, pp. 55–68.

Hubert, B.: 2001, Une démarche systémique pour une agriculture durable: concilier survie et durabilité des systèmes pastoraux, *Journal for Farming Systems Research-Extension*, special issue, 57–85.

Hubert, B.: 2002, Le traitement du long terme et de la prospective dans les zones ateliers (suite). Les rapports entre chercheurs et acteurs, *Natures, Sciences, Sociétés* 10, 4, 51–62.

Hubert, B.: 2005, L'interdisciplinarité sciences sociales/sciences de la nature dans les recherches sur problème. In: R. Teulier and P. Lorino (eds), *Entre connaissance et organisation: l'activité collective. L'entreprise face au défi de la connaissance*, Colloque de Cerisy, La découverte, Paris, pp. 133–155.

Hubert, B. and Bonnemaire, J.: 2000, La construction des objets dans la recherche interdisciplinaire finalisée: de nouvelles exigences pour l'évaluation, *Natures, Sciences, Sociétés* 8, 3, 5–19.

Hubert, B., Deverre, C., and Meuret, M.: 1995, The know-how of Livestock farmers challenged by new objectives for European farming: II. Reassigning rangelands to new, environment-related usages. In: *Proceedings of Vth International Rangeland Congress*, Salt Lake City, pp. 251–252.

Joffre, R., Hubert, B., and Meuret, M.: 1991, Les systèmes agrosylvopastoraux méditerranéens: réflexions à propos de la gestion de ces espaces fragiles. Dossier MAB 10, Unesco, Paris.

Legay, J.M.: 1993, Questions sur la dynamique de l'exploitation halieutique. Préface. Actes de la table ronde ORSTOM/IFREMER. In: F. Laloé, H. Rey, and J.L. Durand (eds), *Questions sur la dynamique de l'exploitation halieutique,* ORSTOM Editions, pp. 9–11.

Legay, J.M.: 1997, *L'expérience et le modèle. Un discours sur la méthode.* Coll. Sciences en questions, INRA éd., Paris.

Le Moigne, J.L.: 1983, *La théorie du système général. Théorie de la modélisation.*Puf (2ème édition), Paris.

Magda, M., Agreil, C., Meuret, M., Chambon-Dubreuil, E., and Osty, P.-L.: 2005, Managing resources by grazing in grasslands dominated by dominant shrub species. In: J.-A. Milne (ed), *Pastoral Systems in Marginal Environments*, Wageningen Academic Publishers, The Netherlands

Meuret, M.: 1988, Feasibility of in vivo digestibility trials with lactating goats browsing fresh leafy branches, *Small Ruminant Research* 1, 273–290.

Meuret, M.: 1993, Les règles de l'Art: garder des troupeaux au pâturage, *Études et Recherches Systèmes Agraires et Développement* 27, 199–216.

Meuret, M.: 1997, How do I cope with that bush?: Optimizing on less palatable feeds at pasture using the Menu model. Recent advances. In: J.E. Gonda et.al. (eds), *Small Ruminant Nutrition*, Options Méditerranéennes, A-34, pp. 53–57.

Meuret, M.: 2004, Concevoir des habitats pour troupeaux domestiques, *Espaces Naturels* 8, 11.

Meuret, M., Bartiaux-Thill, N., and Bourbouze, A.: 1985, Évaluation de la consommation d'un troupeau de chèvres laitières sur parcours forestier: méthode d'observation directe des coups de dents; méthode du marqueur oxyde de chrome, *Annales de Zootechnie*, 34, 159–180.

Meuret, M. and Dumont, B.: 2000, Advances in modelling animal-vegetation interactions and their use in guiding grazing management. In: *Proceedings of the 5th International Symposium on Livestock Farming Systems. Integrating Animal Science Advances into the Search for Sustainability*, Posieux (Switzerland), EAAP Pub, Wageningen, pp. 57–72.

Meuret, M. and Giger-Reverdin, S.: 1990, A comparison of two ways of expressing the voluntary intake of oak foliage-based diets in goats raised on rangelands, *Reproduction, Nutrition, Développement*, 6, 2, 205s–206s.

Mormont, M.: 1996, Agriculture et environnement: pour une sociologie des dispositifs, *Economie rurale* 236, 28–36.

Pavé, A.: 1994, *Modélisation en biologie et en écologie,* Aléas éd., Lyon.

Priéto, L.: 1975, *Pertinence et pratique. Essai de sémiologie,* Edition de Minuit Paris.

De Simiane, M. and Meuret, M.: 1994, Un complément spécial parcours, *La Chèvre*, 2002, 26–29.

Teulier, R. and Hubert, B.: 2004, The notion of "intermediary concept" contributes to a better understanding of the generative dance between knowledge and knowing. In: *20th EGOS Conference, Ljubljana (Slovenia), 30-06/3-07/2004.*

# Chapter 8
# Fischnetz: Involving Anglers, Authorities, Scientists and the Chemical Industry to Understand Declining Fish Yields

**Patricia Burkhardt-Holm**

**Abstract** The Fischnetz project was initiated for two reasons: fish catches, especially of brown trout, have decreased by 60% over the last 20 years in many rivers and streams and the health status of numerous brown trout populations has been found to be impaired. Several cantons and members of the public requested that the Swiss Federal Institute of Aquatic Science and Technology (Eawag) and the Federal Office for the Environment (FOEN) identify the causes and propose measures for improvement. This bottom–up initiative led to the transdisciplinary project Fischnetz, which started out as a nationwide network and lasted 5 years – from the end of 1998 to January 2004. The name of the network, 'Netzwerk Fischrückgang Schweiz', abbreviated 'Fischnetz', indicated the integrative and communicative nature of this approach. Its main objectives were (I) to document the health status and the fish decline, (II) to identify their reasons and (III) to suggest corrective measures. It was hoped that the fostering of networking would help ensure continued collaboration and information exchange after the end of the project. Broad public acceptance was, in part, reflected by the considerable financial support from all 26 cantons, the Principality of Liechtenstein, federal authorities, the chemical industry, and the Fisheries Association.

In this article emphasis is placed on how the different stakeholders (fishermen associations, national and cantonal authorities, researchers and representatives of the chemical industry) were involved. In particular, building up a network was central to the integration of the already existing data and know-how of all participants. This was a prerequisite for identifying knowledge gaps and initiating research projects. The collaboration throughout the project ensured an efficient exchange of results, ideas and conclusions enabling the setting of new priorities and agreement on further procedure and proposed measures.

**Keywords:** Fish catch · Angler · Aquatic ecosystems · Stakeholder involvement · Water quality

✉ P. Burkhardt-Holm
Programme Man – Society – Environment (MGU), Department of Environmental Sciences, University of Basel, Basel, Switzerland
e-mail: patricia.holm@unibas.ch

G. Hirsch Hadorn et al. (eds.), *Handbook of Transdisciplinary Research*,

## 8.1 Background

Over recent years evidence has shown that fish populations in many Swiss rivers and lakes have experienced serious decline. Corresponding data are primarily based on the yearly records of angler catch, and point to a decrease of up to 60% since the 1980s (Fischnetz, 2004a). Catch decline was recorded for brown trout, roach, grayling, nase and many other species, to varying degrees depending on geographical distribution (Friedl, 1999). However, most data were available for the native brown trout since this is the preferred angler species and most common in Switzerland.

The unexplained decline in fish numbers has caused considerable concern, not least due to the large number of anglers in Switzerland. Approximately 240,000 persons, which is equivalent to 6% of the Swiss population between 15 and 74 years of age, practise angling at least once a year (Klingenstein et al., 1999). Furthermore, the income from selling licenses is important for the cantons. With a decrease of 23% in the numbers of angling permits for rivers and streams and 46% for combined permits for rivers and lakes sold between 1980 and 2000 a significant reduction was already noticed (Fischnetz, 2004b). In addition, fish are a powerful bioindicator. Accordingly, concern was raised regarding fish disappearance as humans fear for their own health. This is underlined by the fact that many Swiss inhabitants obtain their drinking water from water filtered through riverbanks (including, for example, 400,000 people in the canton of Bern). Therefore, an action plan needs to address, at least indirectly, questions of relevance to human health. For example when pathogens or chemicals are detected in fish, the risk of transfer of these compounds via drinking water to humans has to be considered.

The project was situated at Eawag, and defined as a cross-cutting project. This type of project is established at Eawag to address multidisciplinary issues, which are often brought up as a public need. Eawag's mission statement is to bridge the gap between science and society by excellence in research and collaboration with experts outside of academia. In addition, Eawag takes a neutral position and is thus, favoured as a leader in this types of project (cf. cross-cutting project 'Ökostrom', Truffer et al., 2002).

In this study, firstly, some background knowledge on brown trout is given; secondly, the phases of the project are presented; and thirdly, methodological approaches are described.

## 8.2 Brown Trout Life Cycle

Stream-dwelling brown trout migrate upstream during late autumn or early winter to headwaters and smaller rivers in order to deposit their eggs in gravel bed structures where the eggs incubate over winter. Unpolluted water, oxygen-rich water and clean, coarse gravel are a prerequisite for successful spawning and reproduction. The number of eggs produced and deposited per female is an increasing function of body size (a 30 cm female produces between 800 and 1,300 eggs/year). In Switzerland, brown trout do not migrate long distances to spawn, but may prefer to move upstream to smaller rivers that provide more suitable spawning habitat. To survive the incubation period deposited eggs must receive sufficient dissolved oxygen. In Switzerland, oxygen concentration in the running water is generally high. However, transport of oxygen to eggs will be impaired if fine sediments accumulate over the spawning site or if the gravel is clogged by fine sand. Additionally, very high river floods involving sediment transport may uncover buried eggs, washing them away or destroying them. The eggs hatch in early spring, and the so-called 'alevins' remain in the gravel until their yolk-sac is exhausted and then emerge and begin to take external food (~April, May). Accumulation of fine sediments also impairs the ability of alevins to emerge from the gravel after hatching. Soon after gravel emergence, these fry disperse locally and establish territories, which they defend vigorously against other fry, and from which they gather their food (Elliott, 1994). The availability of territories is closely linked to the quality and quantity of available habitat. Suitable territories are provided by pools, riffles, boulders, undercut banks, logs, and a sinuous stream bank (high variation in width). Food resources (e.g. prey fish, insects, drift) are more readily available in streams that provide a diverse river bank and bed structure, overhanging vegetation, low levels of suspended sediment, and low levels of micro-pollutants. Poor water quality, especially high concentrations of nitrogen compounds, can also lead to increased mortality of fry. Similar to eggs, young fry may be washed away by high flows. Young fry are also especially vulnerable to disease and parasites, which are more prevalent in polluted water and warmer temperatures. While in many parts of the world, particularly in lakes, eutrophication (excessive nutrient inputs leading to algal growth) has caused fish death associated with low oxygen, this has not been a problem in Swiss rivers because of low ambient nutrient levels and generally high flow velocity. Since all streams are ultimately limited in their ability to provide territories and food, there is an upper limit to the number of fry that can survive this stage. This upper limit is called the 'carrying capacity' of the stream and is believed to be an important factor in limiting populations. After their first year of life, brown trout are less vulnerable to disease or competition for territories, but their demand for larger territories increases. After about two to three years, depending on the growth rate, juvenile trout become reproductively mature and begin to spawn. It is worth mentioning that brown trout in Switzerland rarely live longer than 5–7 years (Fischnetz, 2004b).

The natural distribution range of brown trout in Switzerland can be estimated according to the fish zone designations of Illies (1978). This designation system uses characteristics such as slope, flow, substrate size, and temperature to estimate the naturally dominant fish species for a stream reach. In addition to designated trout zones in areas of high slope and low temperature, other species' zone designations in Switzerland, including grayling, barbel and bream, may be found in lower densities. In Switzerland almost all rivers, especially those popular with anglers, are stocked with trout. Stocking of fry is most common and may occur in either the spring or autumn. If the densities of native and stocked fry far exceed the carrying capacity, competition may become so severe that the chance of survival of all fry is greatly reduced. This will result in a final number of surviving fish that is actually lower than the carrying capacity. Anglers in Switzerland are required to document, and submit to the appropriate canton, records of the number, size, location, and date of all fish caught and retained. However, the number of unsuccessful trips and the duration of trips are not generally recorded; therefore we do not know whether or how angler catch behaviour has changed in detail over the last decades.

## 8.3 First Phase of the Project: The Start and Formulation of Objectives

In March 1998 representatives of all cantonal fisheries administrations, of the federal administration and of research institutions came together to discuss the extent of the decline in fish numbers and the often quoted impaired health status. Stating the inability to detect the causes at once resulted in the declaration of intent to build up a nationwide network. The objectives of this network were:

1. To collect, synthesise, and critically evaluate the existing data on fish catches, population abundance and health status of fish in Swiss rivers over the last 20–30 years; to develop a basis for standardised field monitoring of fish abundance and health; and to prepare for future investigations on the abundance and health of fish (monitoring of success).
2. To search for causes/causative factors; to identify the most important causes of the present situation; and consider their significance. To approach this aim, twelve hypotheses on potential causes were formulated.
3. To develop and propose options for correction.
4. To provide continuous and unbiased information on the project to the public and the scientific community.
5. To foster synthesis and networking.

## 8.4 Second Phase: Raising of Hypotheses and Putting Them into Research Questions

To structure the search for the causes, twelve hypotheses were developed. These were raised first, in a preliminary form, in March 1998 at a meeting of several cantons (Berne, Aargaue, St. Gallen, Zurich, Vaud), the Centre for Fish and Wildlife Health, the FOEN and the Eawag. Participants ranked the hypotheses individually according to their own personal assessment of the importance of these factors for brown trout decline and fish health impairment. After the start of Fischnetz, these hypotheses were later discussed and partly modified by the project management team. They included cause-effect relationships at multiple levels, with some overlap and interaction between them (Fig. 8.1).

As a consequence, the first four hypotheses comprise effects, the latter eight address potential causes:

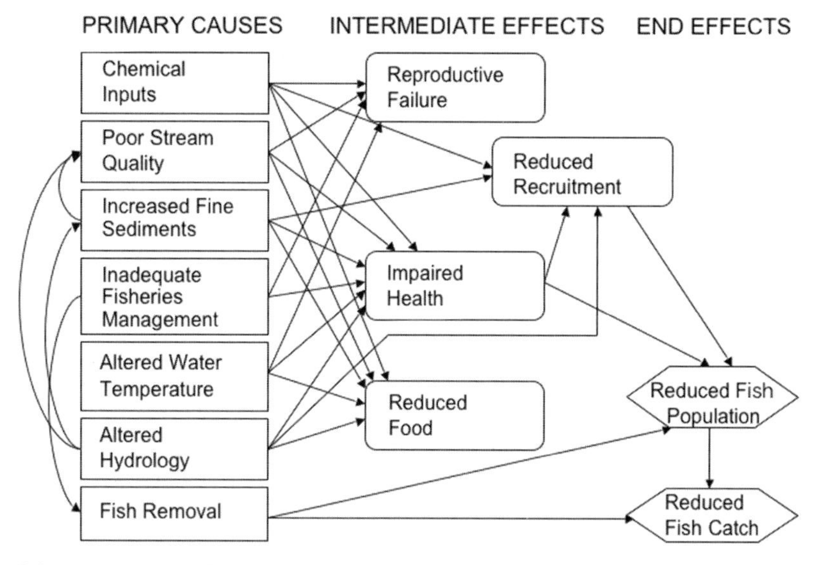


**Fig. 8.1** Hypotheses: relationships between the hypotheses and feed back loops. Hypothesis No. 1, the integrative hypothesis is not indicated by a single box since it is represented by the whole network of relationships. Primary causes, for which control measures are possible, are indicated by rectangles; intermediate causes by rounded rectangles and end effects are represented by hexagons. Arrows indicate the directional influences of one factor on another

- The declining fish catches are due to more than one factor, each possibly having a different regional significance;
- The fish population is affected by reproductive failure of adult fish;
- or from reduced recruitment of young life stages;
- The health and fitness of the fish are impaired;
- Chemical pollution (both nutrients and micro-pollutants) is causing harmful effects;
- Poor morphological quality and longitudinal connectivity of rivers affect fish survival and recruitment;
- The relative amount of fine sediments has increased and led to sediment clogging, which reduces spawning success and disturbs the embryonic development of brown trout;
- The amount or quality of food is insufficient;
- Fisheries management, i.e. stocking and angler behaviour are causing the declining fish catches;
- Excessive removals of fish by birds is responsible for apparent declines;
- Water temperature has changed to the disadvantage of the fish, especially trout;
- High floods in winter and corresponding sediment transport have changed detrimentally.

However, for reasons of transparency and demonstration of our respect for the preceding compilation by the participants of the first meeting, which put the hypotheses together for the first time, we stuck to these hypotheses to a large extent. The hypotheses were put into specific research questions and compared to existing knowledge. Knowledge gaps were filled with results from new projects. The procedure was as follows: According to disciplinary and professional background, each member of the project team was appointed as an expert and was responsible for one to three hypotheses. For some hypotheses two or three persons were responsible. For each hypothesis, a detailed research plan was developed according to the business plan, answering the following questions: Why is this hypothesis reasonable? How important do we judge this hypothesis to be (and why)? What knowledge is already available? What are the open questions and which of them are most important to confirm or refute the hypothesis? What would be the best and most feasible way to raise and collect data? How could this be done and who has the knowledge (regarding data or regarding methodological know-how) and capacity to do this? When should this be done? Which questions are of the highest priority as well as in context to the other hypotheses? What would this cost? At a two-day retreat, all project members came together to discuss and further develop this detailed research plan. In particular, the intention was to put the hypotheses together in a context, to prioritise, to build links between the suggested sub-projects, and to decide on the budget and responsibilities. Decisions were made and out of 22 suggested sub-projects, 13 were initiated for the current year, the others planned to start the following year.

## 8.5 Third Phase: Addressing Research Questions on Documentation and the Search for Potential Causes

The longest phase took almost 4.5 years of the whole project duration. During this time sub-projects were carried out, connections between them were drawn and synergies were created. Fierce debates and discussions on the background, traditional knowledge and interpretation of the results of sub-projects shaped this phase.

The decisions about the type of study and how the studies were conceived were based on feasibility (time, money, personnel resources), the priority given to the corresponding hypotheses and the already available knowledge. Besides internationally distributed knowledge available in the literature, we were aware of the additional knowledge that was not published or only available in grey literature, in particular regarding the Swiss situation. This comprised knowledge of existing problems on one hand (studies on problematic rivers and streams, selected due to unusual low numbers of fish, fish health problems or extreme habitat situations) and on remediation measures on the other hand and rarely, from rivers and their biocoenoses which had been continuously studied with the same methods for a longer time duration.

Accordingly, questions were addressed by Fischnetz research studies that were of high priority, but for which scientifically sound knowledge was rare: for example, there is considerable concern about the potential of both natural and man-made environmental substances which affect fish health and reproduction. In particular, the issue of hormones and hormone-like substances in the rivers found to cause intersex in fish (a mixed gonad, comprised of both, male and female gonadal elements in the same organ) or the synthesis of female egg yolk proteins in male fish raised concern among the scientific community, the policy makers and the public at large – many thousands of research papers in the last 10 years and a great number of newspaper articles have dealt with this issue. Such phenomena were not yet studied in Switzerland and public concern had to be met as well as scientific interest. Consequently, a research programme was started in Switzerland, initiated by the Swiss National Science Foundation (SNSF) and FOEN (http://www.snf.nrp50.ch/), aimed at assessing the hazards and risks that arise when endocrine disruptors are processed through ecosystems which in turn cause human and animal exposure. The fish catch decline and the potential role of loss of habitat and endocrine disruptors were stated reasons for its initiation (nrp50, implementation plan). Accordingly, in the frame of Fischnetz, we started the project 'Endocrine Disruption in *S*witzerland: *A*ssessment of *F*ish Exposure and *E*ffects' (SAFE) investigating the exposure to and effects of oestrogen-like substances in the surface waters of Switzerland on brown trout and its potential contribution to population decline. First results showed a low, but consistent exposure of fish in our rivers to oestrogenic activity, and a synthesis of the egg yolk protein vitellogenin in male fish (Burki et al., 2006; Vermeirssen et al., 2005). As more data on the effects of exposure to oestrogenic activity of critical life stages, such as alevins and maturing fish are collected they will be integrated in a population model, paralleled by a population study to assess potential population effects in an oestrogenically polluted river.

Former studies pointed to a potential contribution to the proliferative kidney disease (PKD) having a considerable effect on the brown trout population. This disease is caused by the single-cell parasite *Tetracapsula bryosalmonae*, which proliferates in the kidney and other organs and eventually leads to death of the host. No treatment or cure is currently available. PKD-induced mortalities in brown trout, of up to 90%, occur when water temperatures surpass 15°C for 2 weeks or more (Wahli et al., 2007; Schubiger, personal communication). Young fish are especially affected and as a consequence fish stocks lack offspring. Under the auspices of Fischnetz the distribution of PKD among river trout was examined at several hundred river sites and fish farms and at more than half of the river sites PKD-positive brown trout were found (Wahli et al., 2007). It was shown that PKD occurred particularly in the waters of the Swiss lowlands where water temperatures higher than 15°C are common in summer. Since the presence of PKD in rivers correlates with reduced fish catch (Fischnetz, 2004b), PKD appears to be a significant factor contributing to the fish decline in Swiss rivers.

Literature reviews (on fine sediments and effect of siltation, on fisheries management or synthetic compilation of former studies on macrozoobenthos) were regarded as adequate to confirm or refute our hypotheses when a vast amount of knowledge was available but not yet compiled in a way to answer our questions. Available studies either monitored selected parameters (e.g. occurrence of specific diseases or changes in water temperature), studied a representative river, or synthesised several projects focussing on the same question or the same geographical region. For example, at the beginning of Fischnetz consequences of global climate change for our native fish populations were even less clear than they are today. Various studies pointed to an increase in water temperature (Bader and Kunz, 1998). Experts from the federal authorities share this view (A. Jakob, Bundesamt für Wasser und Geologie (former name), expert hearing for Fischnetz, May 2002). Such an increase could specifically affect brown trout, due to the requirements of the different life stages: the egg stage is the most temperature sensitive, with optimal development occurring at 5°C, whereas below 1 and above 9°C an increased mortality is observed and temperatures above 12°C are lethal. Maximum growth rates of juvenile and adult brown trout occur at 13°C, while growth ceases below 3°C and above 19°C. River water temperature change was monitored automatically at 1 minute intervals since the 1960s at 95 sampling stations in a Swiss wide project Nationale Daueruntersuchung Fliessgewässer (NADUF) (www.naduf.ch). Daily mean discharge data were also available for the same time period. For the purposes of our study, other physical data, brown trout catch data, data on the occurrence of the temperature-dependent kidney disease PKD and results on preferred temperature range were integrated and served to assess the consequences of climatic change for water temperature and brown trout populations in alpine rivers and streams. We showed that regionally coherent warming of ~1.5°C occurred abruptly in 1987/1988 in the rivers and streams resulting in an upward shift in thermal habitat. Due to physical barriers restricting the longitudinal migration of brown trout, an upward shift of habitats implies an effective habitat reduction. The overall consensus suggests that the decrease in brown trout population is at least partly, a consequence of this change, accelerated by PKD (Hari et al., 2006).

In several cases, the results received did not sufficiently answer our questions. This led to the extension of some projects beyond the project duration (sub-project SAFE, www.snf.nrp50.ch). In other cases, the selection of rivers to study took so much time, due to necessary preceding checks that projects were started with much delay and, thus, extended beyond the project's duration (sub-project test areas).

## 8.6 Fourth Phase: Synthesis and Development of Measures

Results from various studies helped to rule out several of the hypotheses (Burkhardt-Holm et al., 2005). To evaluate the data and consider the aforementioned interactions and feedback loops between the projects, a well-structured synthesis was necessary. In a first approach, we developed a Bayesian probability network as a means for summarising both the qualitative and quantitative information. This type of network has the advantage of making causal assumptions explicit. It is based on a dynamic representation of the brown trout life cycle and is extended to include the effects of natural and anthropogenic factors. Furthermore, conditional probability distributions were also based on carefully elicited judgements of scientists and experts in this field. The relative importance of the various stress factors was assessed by comparing various model scenarios. In four river basins with varying characteristics to represent the range of conditions in the Swiss lowlands, major stress factors have been identified. Selection criteria for the river basins included evidence of a significant brown trout catch decline over the preceding two decades and the discontinuation of stocking for the two years under study, in order to be able to assess the natural recruitment potential. In addition, the river basins are typical in that three of them exhibit a multitude of potential causal factors. In each basin, the brown trout population was studied at three sites. These sites were either separated by barriers or the distance between the sites was great enough that migration was considered to be of minor importance.

The catchment area of the River Emme is characterised by a steep pre-alpine region at the headwaters (1,400 m) with seasonally fluctuating flow due to the melting of snow in spring and where several tributaries and storm events influence the River Emme's flow (especially in the midland reach). A history of flooding led to intensive river management activities in the 19th and 20th centuries with many barriers placed along the river, resulting in separation of tributaries and poor gravel transport. Additionally, water is removed for irrigation purposes. In consequence, the Emme basin exhibits particularly low flow in summer, resulting in low depth and a decreased number of adequate habitats for brown trout. Natural trout habitats are mostly found in the upper portion. Two large wastewater treatment plants (STPs) discharge into the downstream Emme, while a number of smaller STPs discharge into the river's tributaries. The catch of brown trout was found to have declined by 60% between 1989 and 1999.

The Liechtensteiner Binnenkanal (LBK) in the Swiss midlands is a channel constructed in the 1930s for flood control and land usage with a rather constant flow. The only prominent barrier at the mouth of the channel was removed and the stretch was restored in 2000; allowing free migration between the LBK and the Rhine River. Restrictions of natural habitats are mainly due to longitudinal constructions leading to poor variability in width and to regulated flow resulting in high levels of fines and sediment clogging. Only one small community's (4,500 inhabitants) STP discharges into the LBK. The catch of brown trout was found to have declined by 94% between 1981 and 2002.

The Necker, a pre-alpine river with natural, seasonally fluctuating flow, has its source at 1,300 m above sea level. River morphology is hardly disturbed, providing a variety of habitats for all life stages of brown trout. A small amount of wastewater discharges into the Necker (input of four small STPs, with treated waste of fewer than 10,000 people). The catch of brown trout was found to have declined by 58% between 1988 and 2002.

The river Venoge is located in the western midlands of Switzerland and flows into Lake Leman. The flow is influenced by melting snow in spring and increased rain fall at the end of the year. Many vertical barriers were constructed in the last century. Coupled with natural hindrances, they pose a threat to migration. The fish habitats are poor in the lower portion but sufficient in the upper portion. The area of the basin is 231 $km^2$ and land use includes 47% agriculture and 34% forest. Eighteen mostly small (two are 10,000–50,000 people, others $\leq$10,000 people) STPs discharge into the Venoge or its tributaries. The catch of brown trout was found to have declined by 40% between 1987 and 2000. However, due to the classification of the lower sites as grayling or barbel zone, these reaches cannot be expected to support high densities of brown trout.

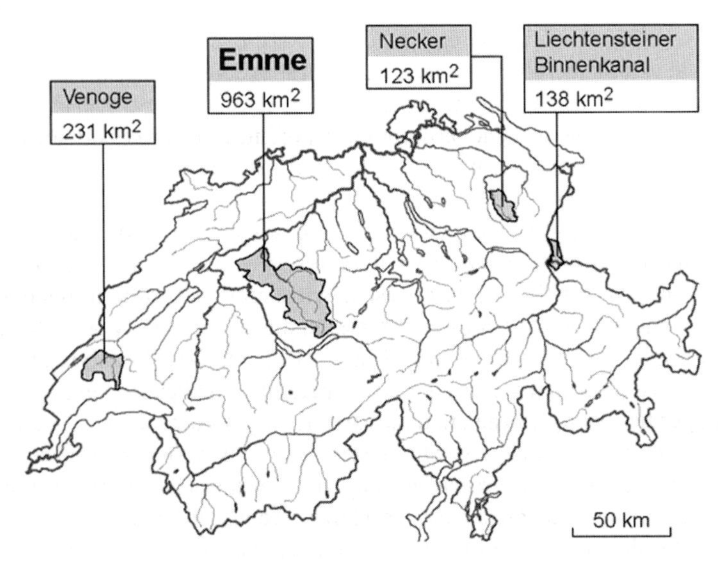


**Fig. 8.2** Location of test rivers in Switzerland with size of their catchments (map © 2006 swisstopo)

Major stress factors were identified: impaired habitat is very important at all except the least impaired sites. Sediment clogging by fines and PKD is also very important at various sites. Wastewater inputs (mostly diluted) are a contributing factor at three of the 12 sites. In some stretches, effects of stress factors are partially offset by stocking. Conclusively, the results of the network model calculations indicated that the relative impact of the different causal factors differs between the locations, depending on the combination of the causative factors found at a specific site (Borsuk et al., 2006).

Another approach we applied to identify likely causal agents that would explain the adverse effects is the weight-of-evidence analysis or retrospective ecological risk assessment (Forbes and Calow, 2002; Burkhardt-Holm and Scheurer, 2007). It aims at evaluating the available evidence as objectively as possible. Although we had sufficient data only for the most thoroughly investigated parameters, the analysis makes the assessment more transparent, and thus significantly facilitates the discussion of the subject, in particular with experts outside of academia. This epidemiological approach asks a series of questions about the occurrence, relevance and interrelationship of potential factors. In our project, these have been answered for the set of the 12 river sites as mentioned above. According to this approach, PKD was the most probable cause in some reaches, whereas it was seen as water quality and temperature for other river stretches, and degraded habitat morphology for other rivers. Several reaches were affected by several factors, and since this approach does not tackle the combination of factors, synergistic effects for instance could not be assessed. Another drawback of this approach appears when knowledge is lacking, for example, on the effects of chemicals.

In conclusion, both integrative approaches show that not a single factor is responsible for the widespread catch decline, but a combination of stressors contribute to the observed negative effects.

Two factors were assessed to be of minor importance for the Swiss wide phenomena: a decline in benthos organisms as food for fish was not confirmed, although, of course, this may play a role in specific river sections. Secondly, the increase of high flood in winter in combination with an adverse effect of gravel transport on brown trout eggs and in-gravel embryos was recorded in only very few rivers out of 41 studied. Fish-eating birds, especially cormorants and goosanders may be an important cause of fish loss in some regions, especially where these birds have increased in abundance. Nevertheless, the hypothesis of fish-eating birds was refuted as cause for national fish catch decrease. For some factors, the database was not sufficient for a proper evaluation: an increase in fine sediments and subsequent increase in siltation harming developing broods of brown trout in the gravel could not be assessed. Similarly, fisheries management was not shown to have affected fish populations. In both cases, more research has to be carried out.

## 8.7 Compilation of Existing Knowledge

In the first phase, the Fischnetz project team appealed to all people who we assumed might have data. We then collected further data via an announcement published in our Fischnetz-info brochure (fischnetz-info, No 1). The survey pool was made up of mainly committed persons from the environment and protection authorities of the cantons or the federal offices, as well as researchers in this field. In response to our appeal we gathered much data from 30 projects.

To structure the corresponding data of so many and very different projects, we developed a *catalogue of criteria*, which were checked for each project. We decided on a systematic scheme for presenting all details and numbered every project, later called sub-projects. One important criteria of Fischnetz stipulated that each sub-project had to match at least one of the Fischnetz objectives; besides, they had to be sufficiently large; the results had to be achieved in the course of the Fischnetz duration time; delivered in writing; and complemented with raw data and scientifically adequate discussion. The above mentioned sub-projects were relevant to Fischnetz in fulfilling its aims. The project members had to agree to actively collaborate with Fischnetz, in particular to help integrate the results in the overall context. Quality criteria had to be fulfilled to guarantee a minimal standard of scientific reliability. However, since these sub-projects were financed by other sources, Fischnetz had no direct means to influence the direction, extent or applied methods in these approaches. As a consequence of the corresponding evaluation, the successful projects were invited to join Fischnetz as independent sub-projects, profiting from exchange of not yet published results, from networks and from the *material – and sample archive*. In this archive all material collected by Fischnetz sub-projects was compiled. Sampling was announced in advance, so that other researchers had the opportunity to collaborate in the sampling or ask for additional material to be collected at the same place and date. By this, material was additionally investigated or sampling was extended in 8 projects resulting in more comprehensive assessments of the respective rivers.

In return, Fischnetz profited from the incorporation of the results of the independent projects and the knowledge of the project collaborators. Further, the list helped to avoid overlapping when planning new projects.

In addition, wherever significant information gaps were identified, which were not expected to be closed by running projects, Fischnetz initiated, financed and carried out new research activities. In total, 19 sub-projects were partly or totally financed by Fischnetz (calculated for cash flow, without personal resources of project team members, which was necessary in all sub-projects). Interestingly, when all sub-projects were taken together and the percentage of the resources investigated was calculated for the different categories of research, it can be summarised that mainly field studies, monitoring projects and synthetic work was funded. In particular the latter was financed largely by Fischnetz, which reflects (I) the specific goal of this project, and (II) the difficulty of getting financial support for inter- and transdisciplinary work.

By this procedure, Fischnetz profited from 77 research projects over 5 years. Each of the members of the sub-projects was coached by a project team member whose responsibility included the exchanging of results and methodological knowledge with the project team and related sub-projects, the presentation of the results, and eventually the invitation of the project leaders to the project team meetings.

## 8.8 Stakeholder Involvement

Stakeholders were involved on all levels of the project organisation, which guaranteed a consistent and continuous representation of the most important interest groups, namely the angler associations, the authorities, the chemical industry and the scientists.

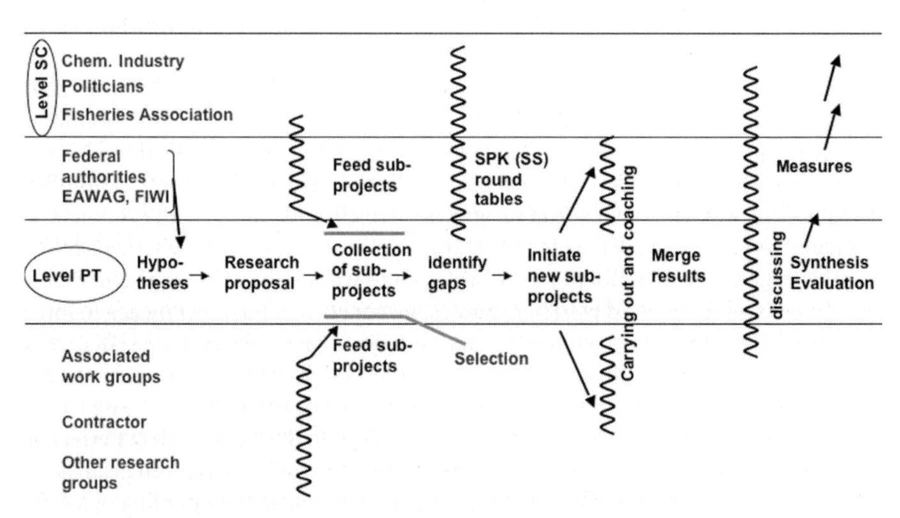


**Fig. 8.3** Participation and evolution of the project; SC: steering committee; PT: project team; SPC: sub-project conferences; SS: special seminars

This involvement was necessary in order to compile the available but distributed knowledge and to build the basis for later implementation. The financial support of the project showed the acceptance of, as well as the high expectations for, this project: all cantons and the principality of Liechtenstein, the FOEN, Eawag, the fisheries associations and the chemical industry supported this project substantially, as well as research foundations and the research institutions involved.

The *steering committee* was composed of six senior members of the civil service and further representatives of administration, fisheries, the chemical industry, and science. It confirmed the aims, monitored the success, and was responsible for the implementation of the relevant political measures. They met once a year and were especially active to ensure the financial support.

The *project management team* was in charge of the detailed planning, implementation and delivery of objectives, the identification of research gaps and the initiation of complementing scientific studies. It was also responsible for the communication and the synthesis of the results. This inter- and transdisciplinary team was comprised of 12 people with expertise in fisheries science, environmental chemistry, ecotoxicology, physiology, population biology, limnology, hydrology and climate change. They also represented various stakeholders, such as authorities, chemical industry, fisheries associations and science. It was the priority task of this team to coordinate the efforts of the different sub-projects, reduce overlaps and use the arising synergies to reach the aims of Fischnetz.

To facilitate the continuous and efficient aim directed process, a *business plan* was established (Burkhardt-Holm et al., 2002). Regular work meetings were carried out once a month, as well as yearly retreats to review the aims and set up the right priorities.

To integrate the knowledge of the external experts (namely the sub-project leader), extra academic and international experts, additional tools were established.

In a first kick-off meeting with the external experts the objectives and the hypotheses were discussed and shaped. Sub-projects already collected were presented. Open questions were collected and a brainstorming of ideas took place on how to solve the pressing questions. This resulted in an initial, rough outline for sub-projects. All the ideas were considered when preparing the detailed research plan. At the same time, all experts and the general public were kept informed of the process: the integration, refining or revising of 'their' ideas at the various meetings, conferences, the Fischnetz homepage and via the Fischnetz-info brochure.

Furthermore, a lot of intrinsic knowledge has been collected over time by fisheries associations. Here, the main task was to clarify the aims of Fischnetz, define the project's limitations and establish a basis of trust, which provides foundation for further collaboration and information exchange. Consequently, (I) representatives of Fischnetz gave numerous seminars at meetings of fisheries and professional associations and (II) invited the representatives of these associations as external experts to discuss specific questions (for example, on the role of macrozoobenthos as fish food; PKD distribution and counter measures; fish-eating birds; distribution and harm; modern basics of fisheries management) at six national *expert hearings* (89 participants). Four international expert hearings (48 participants) were carried out to address specific questions to selected experts, in order to ascertain the current state of knowledge, discuss methodological approaches and gather information and advice concerning further procedure.

Six regular meetings with the sub-project leaders (*'sub-project leader conferences'*), mostly between 25 and 40 persons, provided further platforms for mutual learning. At the beginning of each meeting, selected members of the project team or researchers from other relevant or related projects gave thematic input representations, mostly referring to specific questions and hot topics due to conflicting interests. Subsequently, ample time was given for questions and clarifications. In the second part of each meeting, small group discussions resulted in options for the further procedure of the projects dealing with the named issues. Finally, these suggestions were discussed in the session. This procedure was extremely effective especially for experts with a lot of knowledge but little experience in presenting their ideas to a broad audience. We also had a contrasting experience with the external experts (mostly fishermen) when a project team's suggestions were distributed as texts (although explicitly stated as 'provisional') in advance of the discussions. These ideas were criticised and finally suspiciously refused, most probably because suggestions in cold print were not understood to be a basis for open discussions.

How was the long-term commitment ensured? Facilitating elements were the transparency between partners and the quick information on the progress and outcome of each of the projects. The latter was provided by the newly created journal '*Fischnetz-info*' which appeared regularly every 3–4 months and summarised the most important results in an easily understood form, responded to frequently asked questions and made the planned progress transparent. In addition, we built up a *website*, where we listed all finished and running sub-projects and provided the corresponding intermediate and final reports. We made press releases, press reviews and selected press articles available; published protocols of our meetings; and announced our sub-project leader conferences and public seminars. *Public seminars* were conducted (once a year, about 150–200 participants each) to ensure knowledge transfer.

## 8.9 Implementation

The measures were proposed at the end of the project, and handed over to the senior citizens and politicians of the strategic committee. A final public seminar with more than 230 participants, and a press conference confirmed the high interest and acceptance by the public at large.

Major achievements were attained. The synthesis allowed the assessment of the importance of factors suspected of contributing to fish decline and impaired fish health: although several long range historical data sets are incomplete and many relationships between factors are unknown, concrete measures for improving living conditions of fish were derived from our results. The measures described are of varying effectiveness and act on different levels. Intervention at the source of the problem is particularly effective but sometimes difficult to realise. Clearly, at this stage any proposed measure must be adapted to local conditions and focus on the list of top priority issues.

PKD: PKD is more widely distributed than anticipated. As one of the most important implementation measures with considerable impact, acknowledging the potential impact of PKD on wild populations, this disease was added to the list of notifiable diseases (Tierseuchenverordnung TSV, 1995, revised 2002), in cooperation with cantonal and federal authorities. A further consequence was the development of more sophisticated diagnostic tools to improve monitoring of the disease and its implications for fish health (Schubiger, personal communication). In addition, Fischnetz proposed that stocking with fish from PKD-infected waters into PKD-free or untested waters is to be avoided.

Habitat morphology: the connectivity of rivers and streams has to be improved along their entire length, as well as with their tributaries. The river banks should be remediated and the flow discharge should be improved to simulate more natural situations, wherever possible. These measures together allow an increase in habitat diversity, necessary for the different life stages of brown trout (see above).

Water quality: we have to abide by water quality standards as defined in the EU Water Framework Directive (WFD). Water quality should neither acutely threaten the life of fish or other organisms, nor in the medium or long term induce a negative health effect on a fish or its progeny. Equally important is better supervision and consistent application of the Water Protection Regulations in agricultural areas with an arable area of 10% or more, in order to reduce pollution from pesticides or other harmful substances. Risk assessment, the substitution of substances that are not easily degradable, as well as the remediation of inefficient sewage treatment plants (e.g. increasing the age of activated sludge, improved degradation of nitrogen compounds) is recommended.

Further aspects of future activities were derived: the network built up during the time of Fischnetz, was continued by the establishment of an information centre (paid by Eawag, FOEN and the fisheries associations, established at Eawag) and named FIBER (Fischereiberatungsstelle). Its main task is to help answer anglers' questions. The cantons agreed to pay for the establishment of a further advisory board for the next three years (Fischnetz+, situated at the University of Basel) to advise the local and cantonal authorities. Within this, the authorities of the Swiss cantons are using data from Fischnetz to improve the living conditions of fish in Swiss rivers (www.fischnetz.ch). We hope that these efforts will eventually improve the quality of the aquatic environment and restore the fish populations in Swiss rivers.

## 8.10 Recommendations

The whole project was a response to a societal problem. From this point of view it could be regarded solely as a public service. However, the problem of the potential causal factors for the decline of fish catch and impaired health status raised a lot of questions of basic and issue-oriented research of extraordinary importance for the whole scientific community dealing with anthropogenic effects on fish. A win–win situation was enhanced by the multiple outcomes: a catalogue of measures to implement the raised knowledge on one hand and an extended network between all participants, a number of publications, and Ph.D. and diploma theses. Transparency and fast distribution of results, the planned projects and the consequences of project experiences were among the most important aspects that demonstrated the credibility, efficiency and the speed of progress of Fischnetz.

However, certain aspects – and answering specific research questions – would have to be improved. In particular, the lack of data and the effort necessary to present the data in a compatible form for evaluation was a problem. The quality and quantity of data in several cases was too poor to allow explorative data analysis. In respect to collaboration, shortage of time was a constant problem during the project. Fund raising for research projects was time consuming for all researchers.

Regarding the transdisciplinary aspects of the project, mutual learning was successful. The utilisation of the mentioned methods, in particular the workshops and the issue groups, proved of value. The knowledge transfer and exchange of ideas and opinions contributed to a better understanding between the stakeholders with their different backgrounds. Scientists learned that one of their tasks is to help solve societal problems and practitioners accepted that research has to follow rules in order to produce reliable, reproducible and individually independent results.

One year after the end of the project, an external peer review was carried out to evaluate the project. The project was appreciated as a prototypic transdisciplinary project due to its approach: of taking on the real world problem of fish decline, and the emphasis on collaboration between researchers and stakeholders. The peer review evaluation committee of the project Fischnetz stated in its report that the participative nature was especially reflected by the incorporation of the external partners from the very beginning of the project; and the continuous support and exchange of results, scientific knowledge and the discussions on an equal footing with the project management team. The formulation of the problem was assessed as exemplary, addressing the concerns of all external partners and all important aspects of the issue.

**Acknowledgments** Thanks are due to H. Hoffmann-Riem and G. Hirsch for helpful comments, E. Vogel for language corrections. I am especially grateful to all the people involved in Fischnetz for their enormous efforts. The project was financially supported by all cantons of Switzerland, the principality of Liechtenstein, the Eawag, the FOEN, the Swiss Society of Chemical Industries (SGCI), the Centre for Fish and Wildlife Health at the University of Berne and the University of Basel.

## References

Bader, S. and Kunz P.: 1998, *Klimarisiken – Herausforderungen für die Schweiz*, Wissenschaftlicher Schlussbericht NFP 31, VdF, Zürich, 307pp.

Borsuk, M.E., Reichert, P., Peter, A., Schager, E., and Burkhardt-Holm, P.: 2006, Assessing the decline of brown trout (Salmo trutta) in Swiss rivers using a Bayesian probability network, *Ecol Model* 192, 224–244.

Burkhardt-Holm, P., Giger, W., Güttinger, H., Peter, A., Scheurer, K., Suter, M.J.-F., Ochsenbein, U., Segner, H., and Staub, E.: 2005, Where have all the fish gone? The reasons why fish catches in Swiss rivers are declining, *Environ Sci Technol* 39, 441A–447A.

Burkhardt-Holm, P., Peter, A., and Segner, H.: 2002, Decline of fish catch in Switzerland – Project Fishnet: A balance between analysis and synthesis, *Aq Sci* 64, 36–54.

Burkhardt-Holm, P. and Scheurer, K.: 2007, Application of the weight-of-evidence approach to assess the decline of brown trout (*Salmo trutta*) in Swiss rivers, *Aq Sci*, 69, 51–70.

Burki, R., Vermeirssen, E.L.M., Körner, O., Joris, C., Burkhardt-Holm, P., and Segner, H.: 2006, Assessement of estrogenic exposure in brown trout (*Salmo trutta*) in a Swiss midland river, *Environ Toxicol Chem* 25, 2077–2086.

Elliott, J.M.: 1994, *Quantitative Ecology and the Brown Trout*, Oxford University Press, Oxford.

Fischnetz: 2004a, Short final report 2004 (in German, English, French, Italian), EAWAG Dübendorf, BUWAL Bern, Switzerland, Retrieved July 7, 2006, from http://www.fischnetz.ch.

Fischnetz: 2004b, Final Report of Project Fischnetz (in German, French), EAWAG Dübendorf, BUWAL Bern, Switzerland, Retrieved July 7, 2006, from http://www.fischnetz.ch.

Forbes, V.E. and Calow, P.: 2002, Applying weight-of-evidence in retrospective ecological risk assessment when quantitative data are limited, *Hum Ecol Risk Assess* 8, 1625–1639.

Friedl, C.: 1999, Fischfangrückgang in schweizerischen Fliessgewässern, *Mitteilungen zur Fischerei*, BUWAL, Bern, 32pp.

Hari, R.E., Livingstone, D.M., Siber, R., Burkhardt-Holm, P., and Güttinger, H.: 2006, Consequences of climatic change for water temperature and brown trout populations in Alpine rivers and streams, *Global Change Biol* 12, 10–26.

Illies, J.: 1978, *Limnofauna Europaea*, Gustav Fischer Verlag, Stuttgart, 532pp.

Klingenstein, J.S., Lüthi, B., and Weiss, T.: 1999, Angeln in der Schweiz, Schweizerischer Fischereiverband (SFV), European Anglers Alliance (EAA), Klosters, 1–45.

Truffer, B., Bloesch, J., Bratrich, C., Gonser, T., Hoehn, E., Markard, J., Peter, A., Wehrli, B., and Wüest A.: 2002, Ökostrom aus Wasserkraft. Ein transdisziplinäres Forschungsprojekt. Schlussbericht (1997–2001), *Ökostrom Publikationen*, Bd. 10, November 2002, EAWAG, Dübendorf.

TSV, 1995. Tierseuchenverordnung, *amendment* 3.2001, Federal Veterinary Office, Bern, Switzerland.

Vermeirssen, E.L.M., Burki, R., Joris, C., Peter, A., Segner, H., Suter, M.J.-F., and Burkhardt-Holm, P.: 2005, Characterisation of the estrogenicity of Swiss midland rivers using a recombinant yeast bioassay and plasma vitellogenin concentrations in feral male brown trout, *Environ Toxicol Chem* 24, 2226–2233.

Wahli, T., Bernet, D., Steiner, P., and Schmidt-Posthaus, H.: 2007, Geographical distribution of PKD in brown trout (*Salmo trutta*) of Swiss rivers: An update, *Aq Sci*, 69, 3–10.

# Chapter 9
# Nanoscience and Nanotechnologies: Bridging Gaps Through Constructive Technology Assessment

**Arie Rip**

**Abstract** In the Dutch nanoscience and technologies research and development (R&D) consortium NanoNed, an additional 'Flagship' for Technology Assessment (TA) & societal aspects of nanotechnology was created [www.nanoned.nl/TA]. This has given us an opportunity to experiment in real time with interactions between nanoscientists and technologists, and build on our earlier work in Constructive Technology Assessment. The fact that the consortium was prepared to spend some of its funding on TA and societal aspects is part of larger dynamics: the wish of nanotechnologists (in the US and elsewhere) to avoid the impasse around (green) biotech, by taking into account ethical legal and social issues (ELSI) from the beginning, and taking them seriously.

I will offer an overall diagnosis of the reflexive co-evolution of science, technology and society (Rip, 2002), which will allow me to better position the specific problem structuring that we developed. The 'doubly-fictional' character of TA of nanotechnology is still mostly promise: the societal impacts can only be speculated about. This has led us to further develop socio-technical scenarios and their use in interactive workshops with various stakeholders and other actors. Actors involved in such workshops, or working with the scenarios, are expected to become more reflexive. This may lead to the modification (hopefully for the better) of technological developments and the way these become embedded in society.

**Keywords:** Constructive TA · Socio-technical scenarios · Reflexivity · Nanotechnology · Paths

## 9.1 Background

At a macrolevel, the co-evolution of science, technology and society can lead, and has in fact led, to stable patterns or regimes (sometimes including explicit governance arrangements). The distinction between technology variation and societal

✉ A. Rip
University of Twente, Enschede, The Netherlands
e-mail: a.rip@utwente.nl

G. Hirsch Hadorn et al. (eds.), *Handbook of Transdisciplinary Research*, 

selection (which carries part of the argument about an evolutionary approach) is actually predicated on the existence of an historically evolved regime where technology development is separate from uptake and use (Rip and Kemp, 1998). Variation and selection are institutionally separated – and are bridged by anticipations, experiments, and interactions.

Historically, the separation was part of the industrial revolution. In a further step, during the 19th century, engineers and other technology actors were given a mandate to develop new technologies and confront society with them, as long as this could be presented as progress. This was complemented by the emergence of the idea of 'technology' in general, symbolically linked to ideals of progress (Smith and Marx,1994). The mandate and its symbolic justification offered a macroprotected space. It was maintained until the 1960s (when Harold Wilson, the then UK Prime Minister, could still proclaim a 'white-hot technological revolution'), but then it started to break down.

The existence of the mandate, and the macroprotected space for technological developments, however, had allowed the establishment of institutions and divisions of labour which could not simply be turned back. While shifts do occur, and there are contestations, the overall pattern of a gap between promotion and control and attendant division of labour between promoters of new technology and critical selectors remains strong.

Thus, in our modern societies there is an asymmetry between 'impactors' (those at the source of technological impacts) and 'impactees'. This can be a difference in power, but is always a difference in timing. Initiators of technological development know more, and have invested more, at an early stage, and impactees and spokespersons for society have to wait and in a sense, follow their lead. Collingridge (1980) has identified the knowledge and (public) control dilemma which follows this situation. The problem is already visible within 'innovation journeys' (Van de Ven et al., 1999) as a flexibility dilemma (Verganti, 1999): one has to foreclose options at a moment when not enough is known. In fact, assessment is ongoing, and the challenge is to improve the assessment process by, for example, including more aspects and perspectives at an early stage as Constructive TA proposes (Rip et al., 1995; Schot and Rip, 1997).

The asymmetry has another component: technology developers are insiders and do not necessarily know very much about the 'outside'. However, adoption and diffusion, is up to 'outsiders', who have other interests and expectations. The story of nuclear energy has been one of struggle between insiders and outsiders since the 1960s. On a much smaller scale, the same storyline is visible in the development of cochlear implants for deaf people (Garud and Ahlstrom 1997), especially when it turned out – unexpectedly for the insiders – that the deaf community was negative about taking deaf people out of their own culture by providing them with implants (Blume, 2000; Reuzel, 2004). Garud and Ahlstrom (1997) suggest that technology developers are working in an 'enactment frame', and look at the world as a challenge or barrier to be overcome (while from the outside actors can be selective according to their various backgrounds). The 'enactment frame' then leads to a concentric approach to product development: get the product right, then look at market and regulation, and afterwards worry about public acceptability. Deuten et al. (1997) have mapped this phenomenon for biotechnology products. As they have shown in a narrative analysis of one of the cases the ‘enactment frame’ itself evolves, because, for example, other ‘niches’ emerge (Deuten and Rip, 2000).

For our topic, it is important to note that the asymmetry gives rise to a division of TA labour where insiders articulate ‘promotion’, and outsiders ‘control’. A similar division of labour, at one remove, is visible in the way governments and their agencies handle technology in society. Government technology policy most often focuses on promotion of (selected) technologies, as in the case of stimulating the electronic superhighway. The main challenge for such technology policy is to ‘pick the winners’, at the level of the society. At the same time, other agencies of the same government are occupied with reducing the human and social costs of the introduction of new technology, for example through safety and environmental regulation. This dichotomy between promotion and control of new technology is thus part of the *de facto* constitution of modern industrialised societies, and is reflected not only in the division of labour between government agencies, but also in cultural and political views, as in the assumption that there will be proponents and opponents to a new technology.

The gap is precariously bridged by a variety of instruments and actions, by existing and emerging nexuses and by division of labour in sociotechnical regimes. It is in this world full of asymmetries and gaps that Constructive TA (Rip et al., 1995; Schot and Rip, 1997) is located. Theoretically – it can be seen as a series of bridging events, and when institutionalised, as a nexus; practically – it organises interactive workshops supported by sociotechnical scenarios; and normatively – there is an overall aim to improve technology and society, and a conviction that increasing reflexivity of ongoing co-evolution will help. Democratisation of technology can be associated with Constructive TA, but that is a possible effect, not a goal. In other words, participation is a means, *not* an end. The key point is to enhance reflexivity through anticipation, feedback and learning (Schot and Rip, 1997), and translate this into actions and interactions.

Constructive TA is an attempt to broaden design and development of new technology by increasing interaction and reflexivity. It modulates the co-evolution of science, technology and society from within, as it were (Rip, 2006a); and social scientists, socio-technical scenarios (STS) scholars, as well as reflexive technologists and mediators, provide support and incentives. It follows that Constructive TA must be transdisciplinary, or nondisciplinary. My own definition of transdisciplinarity focuses on nondisciplinarity: a combination of epistemic goals and distributed locations where disciplinarity is not relevant (Rip, 2004). A similar thrust is seen in Turpin’s (1997) evaluation of Australian Collaborative Research Centres: contributions to knowledge independent of disciplinary knowledge, and diffusion through movement to new problems and new contexts rather than through publication in disciplinary journals.

## 9.2 Problem Structuring

There is, by now, a range of approaches to technology assessment. Foresight and future visioning emphasise the open future, and there are now proposals for ‘vision assessment’ (Grin and Grunwald, 2000). At the other end of the spectrum there is the comparison of existing technological options by firms and R&D institutions in order to select the promising ones. Within the range of approaches, a cluster of approaches and methodologies have been developed and piloted over the last 10–15 years, which emphasise real time interaction and learning. There are various labels, including Interactive TA (Grin and Van de Graaf, 1996; Grin et al., 1997), Real-Time TA (Guston and Sarewitz, 2002), and Constructive TA (Rip et al., 1995). We prefer the label Constructive TA to emphasise the goal of contributing to the actual construction of new technologies and the way these become more or less embedded in society – rather than simply waiting for the changes and then trying to map possible impacts.

For new and emerging science and technology like nanoscience and nanotechnologies, Constructive TA has to address what can be called its doubly 'fictional' character: many of the envisaged uses of nanotechnology are still science fiction, and so the study of possible impacts is social science fiction. In our specific situation – the TA and Societal Aspects of Nanotechnology Programme, within the Dutch R&D consortium NanoNed – there is an opportunity: being part of a nanotechnology consortium gives us easier access to address the general Constructive TA goal of interaction with nanoscientists and technologists; and within the ongoing developments of nanotechnology, we can make co-evolution more reflexive. We have gained credibility in the nanoworld, for example in European Networks of Excellence and in international conferences.

In nanotechnology, promises about new technological options abound, but little can be said definitively about their eventual realisation, let alone the impacts on society. As I have already noted, an impact assessment of nanotechnology will necessarily be speculative: one could call it science fiction (about future nanotechnology) combined with social science fiction (about the world in which future nanotechnology would have impacts). Still, it is important to try and anticipate and create controlled speculations about possible futures so as to stimulate reflection and broaden the scope of strategic choices about nanotechnology, and more generally.

While it is too early to expect actual conflicts about nanotechnology (even if demonstrations are occasionally staged), concerns are being articulated, for example about uncontrolled spread of nanoparticles. What happened (and continues to happen) with biotechnology and genomics is often seen as a lesson. Nanoscientists have actually called on governments to help avoid the impasse that has occurred with (green) biotechnology. (An example is Vicki Colvin, in the Hearings of the US Congress concerning the new Nanotechnology Bill, April 2003; see my discussion in Rip (2006b)). Genomics stimulation programmes have an ELSI component: Ethical, Legal and Social Issues, modelled after the Human Genome Programme in the 1990s. ELSI is now taken up, in various ways by European and American R&D stimulation programmes for nanotechnology. To include TA studies in the NanoNed programme is a similar proactive response to possible societal concerns. In practice, this need not imply a willingness to interact: it could work as a division of moral labour, where the scientists can point to the social scientists and ethicists as reasons why they themselves do not have to think about societal and ethical aspects (as happened to some extent with ELSI in the Human Genomics Programme). When there is interaction, it requires ELSI scholars to link up (though not identify) with the perspectives of the nano-enactors.

While the actual future cannot be predicted, the occurrence of socio-technical dynamics and emerging irreversibilities implies that there is an *endogenous future*. Embedded in the present are preferred directions, which imply that a trajectory will be followed (Dosi, 1982). Even if paths are created while 'walking' (Garud and Karnøe, 2001), emerging paths can be mapped, and the dynamics of their emergence can be analysed (prospective technology analysis) – and the results can be fed back to actors. This, by itself, is not new. It overlaps with technological roadmapping, where one reasons forward to desired performances and functionalities, and identifies barriers to be overcome, and then reasons back to the efforts necessary to achieve them. In some cases, such activities have become institutionalised, for example in the semiconductor industry where an international consortium makes such roadmaps and uses them for strategic coordination. Predictions based on Moore's Law serve as a guideline for strategic decisions, and thus become self-fulfilling.

Socio-technical scenarios, in contrast to roadmapping, are open-ended. They branch out into the future, and are structured by prospective technology analysis. An advantage in terms of interaction and reflexivity is the way such scenarios link up with technology enactors. Actors always work with partial and diffuse versions of scenarios to orient themselves and others. A social science supported TA will improve the quality of the scenarios. A structural problem is that enactors tend to project a linear future, defined by their intentions, and use the projection as a roadmap – only to be corrected by circumstances. Mapping tools which force actors to consider the non-linearity of evolution, and accept the complexity, can be developed to make them more effective (if the actors are prepared to accept such social science based support).

The approach combines:

- mapping and analysing the ongoing dynamics of technological development, the actors and networks involved, and the further (and possibly conflictual) embedding of the technology in society, with particular attention paid to emerging patterns, including preferred technological paths and so-called dominant designs, which will shape further co-evolution
- identifying and articulating socio-technical scenarios about further developments, possible impacts, forks in development and the possible choices of various actors. This stimulates technologists and other relevant actors to reflect on their strategies and choices, making these more socially robust
- organising interactive workshops and other 'bridging events' where a (relevant) variety of actors participate (up to critical NGOs), but always including nanoscientists and technologists. Socio-technical scenarios allow them to probe each other's worlds in a structured way.

Not only is nanotechnology at an early stage, it is an enabling technology: the promises of nanotechnology have to be realised through its uptake in other products and services (from faster DNA analysis to improved coatings, sunscreens and drug delivery). Its impact depends on what happens there, and is in that sense co-produced, even if one might consider that it is nanotechnology which has made the difference – thus, the well-known problem of attribution of impacts to earlier actions (Rip, 2001). In our recent work, we therefore also consider socio-technical scenarios for a sector, say food packaging, where nanotechnology might be taken up.

In the past the emphasis was on socio-*technical* scenarios. We are now developing *socio*-technical scenarios as well, e.g. for risks of nanoparticles, for food and nano (see below). Societal issues, which in the *socio*-technical scenarios become visible only at a later stage and which reflect the concentric bias of enactors, and their implications for the governance of new and emerging science and technology will now be at centre stage: risks to health and environment, economic and distributional impacts, equity issues and sustainable development (particularly in developing countries), longer term and ethical issues.

Our approach to prospective analysis and assessment of a technology at an early stage can build on a number of relevant methodologies and pilot studies (including SocRobust, an EU-FP5 project, see Larédo et al. (2002)), as well on recent advances in disciplines such as innovation studies (itself an interdisciplinary domain), sociology of technology, evolutionary economics, industrial economics, industrial ecology and political science. Interestingly, the boundaries between these disciplines are fuzzy, and there is an increasing overlap. It raises challenges for these disciplines, viz. to analyse processes in real time rather than in retrospect, and to do so in interdisciplinary collaborations.

## 9.3 Problem Investigation and Interactive Activities

An initial set of four workshops, addressing the domain of lab-on-a-chip, a microsystem that uses nanotechnology to improve performance and develop new functionalities were held in February 2006. Individual scenarios of the evolution of the domain and the challenges had been drawn up before the workshops, based on interviews by the CTA researcher (PhD student, Rutger) with various actors (including but not limited to the eventual workshop participants), and his independent analysis of the structure and dynamics of the domain. The scenarios provided a platform for the workshop discussions about future developments and strategic choices. Two of the workshops had only insiders as participants. The discussion then centred on the technical challenges that they identified. The other two workshops had insiders and (professional) outsiders as participants. Although broader issues were raised, there was still a tendency to focus on the technical aspects. The concentric view of enactors, and the attendant division of moral labour, dominated. The evaluation of the workshops showed that participants appreciated the opportunity to hear views and visions of other actors, and interact with them. This definitely broadened the strategic reflection of the participants. When asked after two months whether this had made a difference in their own work, they reported that little was visible yet. Ongoing projects would have priority, and only when moments of choice occurred, e.g. a new project, did they expect to be able apply their broader reflection – if still relevant.

In another workshop in June 2006, prepared and led by PhD student Douglas Robinson, the domain of cell (on a chip) analysis was the topic. Insiders as well as professional outsiders participated. The role of the socio-technical scenarios was slightly different: Robinson had created three different futures for the domain (on the basis of his document analysis and interviews), which were circulated beforehand to provide a platform for the workshop discussion. During the workshop, a multipath mapping exercise was conducted (as developed by Robinson and postdoc Tilo Propp) where paths and their related challenges were depicted for the domain, and participants could identify further issues and discuss strategic choices (Fig. 9.1). This led to lively discussions, and to further interactions between some participants after the workshop. One of the participants, the director of a young start-up company developing sensors to measure electrolytes in the blood in real time (which then allows point-of-care diagnosis), picked up on the diagram and filled in the details for his company, adding possible futures. Robinson continues to be involved with the company, and we have been adding to the diagram to show how societal embedment and issues involved in it could be included (still following the concentric bias).

Our interactions with actors in and around the nanotechnology world (scientists, technologists, entrepreneurs, insurance companies, regulators, NGOs) during preparation of the workshops and during the workshops will increase reflexivity of co-evolution of nanotechnology and society. This is further enhanced because the workshops are partly or wholly positioned as part of the activities of EU Networks

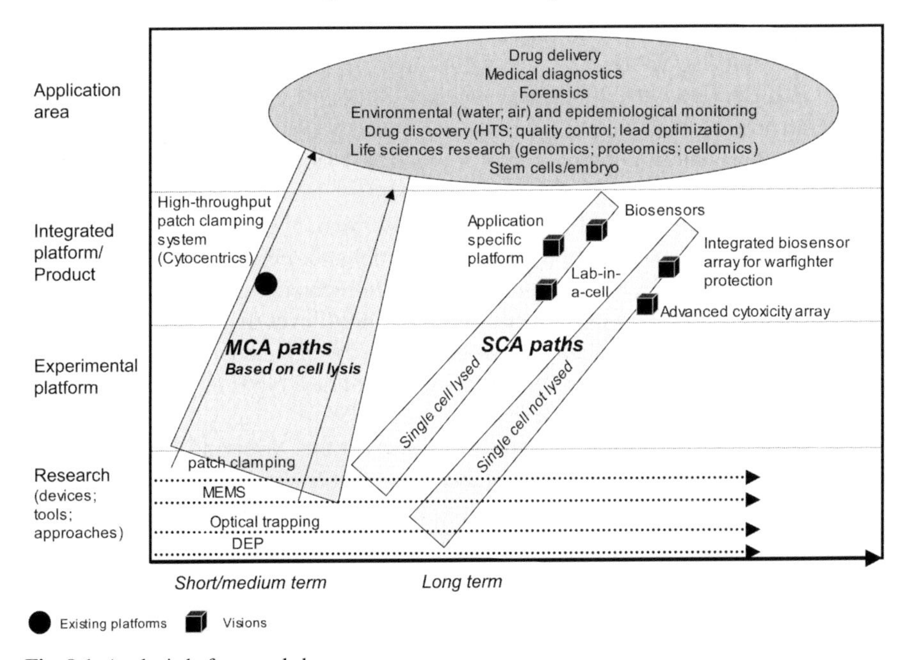


**Fig. 9.1** Analysis before workshop

of Excellence in nanotechnology (Nano2Life, FRONTIERS). Such direct linkages with nano-actors enable productive interactions, but can also be constraining when modernist 'enactment' perspectives are pushed. The junior researchers have to resolve these tensions in their day-to-day work, with support from the senior researchers in the TA NanoNed Programme.

I note in passing that the senior researchers, as well as some of the junior researchers, who move about in the nanoworld speak at conferences, are members of ethics boards, and have informal conversations with nano-actors (nanoscientists and policy makers). Such a presence by itself already stimulates reflexivity. This leads to further, sometimes joint activities. Van Merkerk co-authored an article with a nanoscientist on the need for broader considerations (Van Merkerk and Van den Berg, 2006). Robinson was asked to lead the Ethics Board of the EU Network of Excellence Frontiers, but had to decline – finishing his PhD was more important. Arie Rip was invited to contribute to the International Dialogue on Responsible Development of Nanoscience and Technology, organised by key policy makers from USA, Europe and Japan.

In the TA NanoNed Programme, further interaction is stimulated by inviting PhD students and postdocs in nanoscience and nanotechnologies to study broader, societal aspects of the research project or programme they are working in. Transdisciplinarity then starts from the science side, rather than the social science and humanities side. One three month study has been concluded. Working on basic questions of Scanning Tunneling Microscopy in the liquid phase, the PhD student (Duncan den Boer) envisaged possible, even if long-term applications, such as the synthesis of exotic molecules. He mapped out the potential innovation and value chain for such an application, and interviewed relevant actors about their views on the necessary structures and activities. Such an approach, explored before by social scientists (De Laat, 2000), was new and difficult for the PhD student. There will be an article describing his findings and the experience itself will allow us to create a support and training package for other PhD students who want do similar projects.

Our further socio-technical scenario exercises have not yet been used in dedicated workshops, but they have been part of presentations to nano-audiences of various kinds, inviting them to become more reflexive. One scenario post-doc Marloes van Amerom and myself have developed, shows the multilevel dynamics of the emergence of risks of nanoparticles as a legitimate issue, which leads to further research and proposed regulation, and to further discussion of new types of governance (like 'soft law') and of routes not taken (Fig. 9.2).

An interesting finding that helped us to construct this 'evolving risk landscape' is the way the intervention of re-insurance company Swiss Re, driven by its own concerns about possible financial losses, shifted the question of the risks of nanoparticles from a fringe concern voiced by some critical NGOs, to an issue high on the agenda of national and international authorities. In other words, de facto reflexivity can be induced by third parties from the outside.

The other multilevel diagnosis and subsequent scenario building is about the food packaging sector (which is actually the intersection between the food value

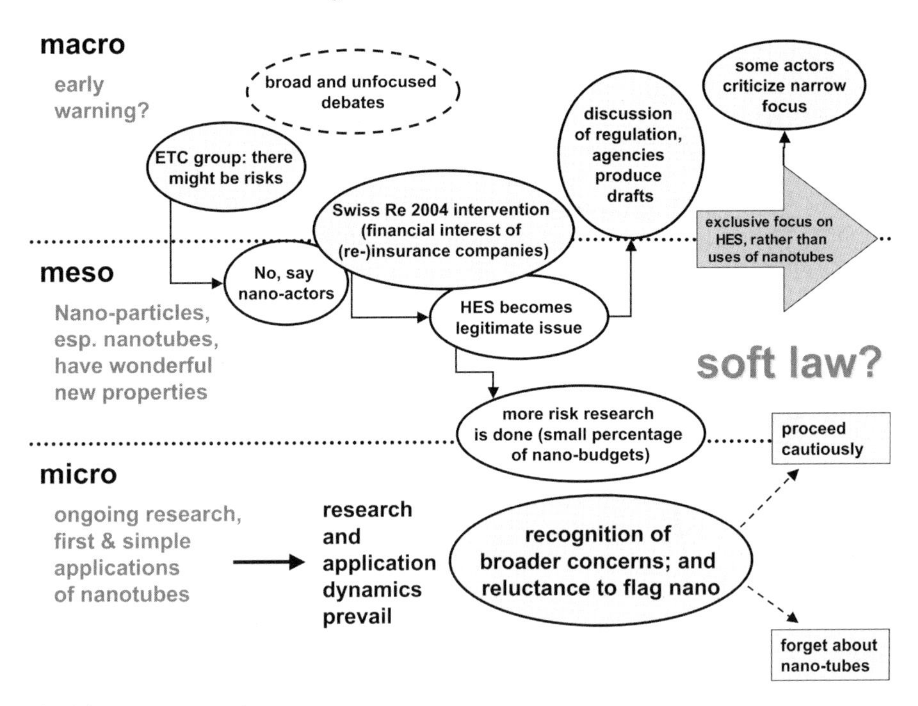


**Fig. 9.2** Macro meso micro

chain and the packaging value chain) and its possible uptake of nanotechnology, and of broader concerns about nanotechnology. The figure (developed by PhD student Haico te Kulve and myself for a presentation at Arizona State University's Center for Nanotechnology in Society, 23 February 2007), outlines socio-technical dynamics of the sector in the Netherlands, and in Europe more generally (Fig. 9.3). For reasons of space, I will not explain the analysis behind the depiction of ongoing dynamics, nor the reasons for choosing the three scenarios indicated in the picture. But there is a theory of co-evolutionary interactions and outcomes behind it, including recognition of the key role of so-called 'linking pin entrepreneurs' (a variation of 'institutional entrepreneur').

There is 'bounded' reflexivity in the nanoworld already: nano-enactors are sensitive about public images of, and public reactions to nanotechnology and can sometimes be overly sensitive. They are interested in results of focus group exercises and willing to experiment with citizen juries, as in the UK [www.nanojury.org]. A symmetrical approach is necessary, and one entrance point is to trace the perceptions and problem definitions of the nano-enactors as well (Rip, 2006b). Feedback from these findings may be unsettling to the actors in the nanoworld, but is necessary to improve reflexivity. Again, senior researchers with some standing vis-à-vis nanoscientists and technologists have to protect the junior researchers, and make sure the message is heard and appreciated. In general, one should not expect interactions to be harmonious. Interactions between nano-enactors and (C)TA agents are important, but their interests are different and may conflict.

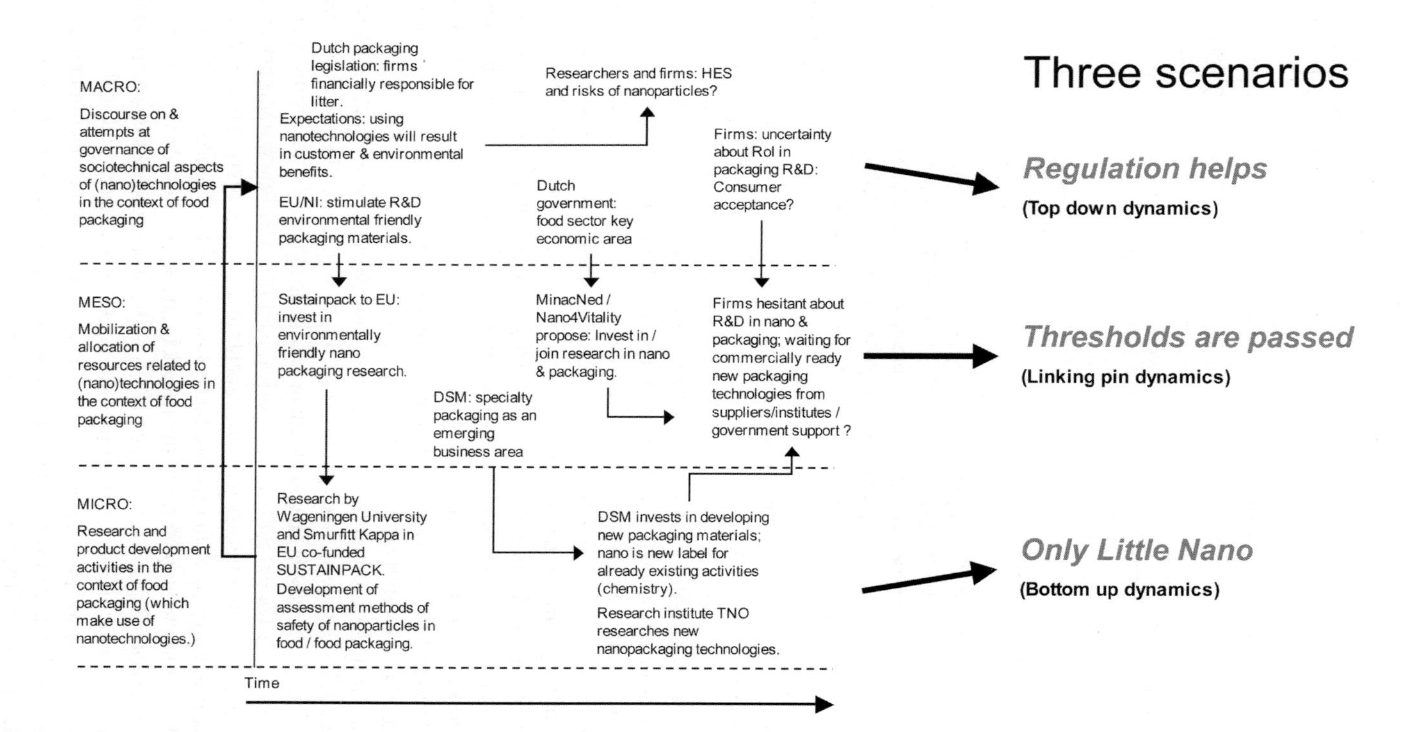


**Fig. 9.3** Three scenarios

Another type of conflict can arise between the social scientists and humanities scholars. Economists (esp. neo-classical economists) and sociologists can collaborate, but can also clash. Increasingly important is the interaction between ethicists (and other normative scholars like political theorists) and empirical scientists/scholars. Although there are clashes, there is also a willingness to listen to the other side. This indicates a latent challenge to transdisciplinarity (latent, because it has not been articulated as such). Transdisciplinarity is linked to real world issues, and cannot escape addressing normative issues; but it is not clear how this can be done in an integrated manner and in a way that avoids having normative issues relegated to the footnotes only, as the personal views of the author(s).

## 9.4 Recommendations

Key recommendations are to use socio-technical scenarios (the combination is important) based on analysis and diagnosis, to support and structure what would otherwise be mainly conversation. Now the focus is on interactive workshops with a requisite variety of participants and a broader use of scenarios. They can also be used as a microcosm, to play out and assess various outcomes.

Our focus has been on the interaction between nanoscientists, technologists and other enactors in and around the nanoworld. Increasing their reflexivity is important because they are in a position to make a difference. It does require the attention of the CTA agents to maintain some independence, and not to become (or be seen to become) public relations officers, or otherwise lubricate the frictions between nanotechnology and society. There is a trade off: if one wants to bridge the gap between innovation and ELSI, one has to associate with both sides. A further point is that there should be a division of labour, with other social science and humanities scholars starting at the ELSI side, and even more importantly, with societal voices being articulated and heard. This is actually the case in the Netherlands, where the Rathenau Institute is focusing on broad societal impacts and the associated debate.

Clearly, the TA and Societal Aspects of Nanotechnology Programme is problem oriented, and while various disciplines are mobilised, there are no (or only incidental) disciplinary products. This has to do with the fact that we cannot define and control our own research object. Instead, we work on the combination of analysis and diagnosis to support reflexivity of co-evolution. It is also clear that transdisciplinarity is more than scholarship unencumbered by disciplines. It is not just one approach to scholarship among others, it is also a key part of an important change in science and society. Thus, one has to learn about the approach ('how to do it') as well as position what one does in relation to broader developments.

One can see the emergence of Foresight, TA, and ELSI, and the interest in 'governance' of science (in Europe and elsewhere) as instances of reflexive co-evolution in a period of punctuated equilibrium. Viewed in this way, they can be assessed together, and the question can be raised of whether they add up to a new pattern – perhaps a new de facto constitution – in which transdisciplinary work would come into its own.

**Acknowledgments** This article draws on research funded by NanoNed [www.nanoned.nl/TA].

## References

Blume, S.: 2000, Land of Hope and Glory: Exploring Cochlear Implantation in the Netherlands, *Sci Tech Hum Val* 25, 139–166.

Collingridge, D.:1980, *The Social Control of Technology*, Frances Pinter, London.

De Laat, B.: 2000, Scripts for the Future: Using Innovation Studies to Design Foresight Tools. In: N. Brown, B. Rappert, and A. Webster (eds), *Contested Futures. A Sociology of Prospective Techno-Science*, Ashgate, Aldershot, pp. 175–208.

Deuten, J.J. and Rip, A.: 2000, Narrative Infrastructure in Product Creation Processes, *Organization* 7(1), 67–91.

Deuten, J.J., Rip, A., and Jelsma J.: 1997, Societal Embedment and Product Creation Management, *Tech Anal Strat Manag* 9(2), 219–236.

Dosi, G.: 1982, Technological Paradigms and Technological Trajectories: A Suggested Interpretation of the Determinants and Directions of Technical Change, *Res Pol* 6, 147–162.

Garud, R. and Ahlstrom, D.: 1997, Technology Assessment: A Socio-Cognitive Perspective, *J Eng Tech Manag* 14, 25–48.

Garud, R. and Karnøe P.: 2001, *Path Dependence and Creation*, Lawrence Erlbaum Associates, Mahwah.

Grin, J. and Grunwald, A. (eds): 2000, *Vision Assessment: Shaping Technology in the 21st Century*, Springer Verlag, Heidelberg.

Grin, J. and van de Graaf, H.: 1996, Technology Assessment as Learning, *Sci Tech Hum Val* 21, 72–99.

Grin, J., van de Graaf, H., and Hoppe, R.: 1997, *Technology Assessment Through Interaction. A Guide*, Rathenau Institute, Den Haag.

Guston, D.H. and Sarewitz, D.: 2002, Real-Time Technology Assessment, *Tech Soc* 24, 93–109.

Larédo, Ph., Jolivet, E., Shove, E., Raman, S., Rip, A., Moors, E., Poti, B., Schaeffer, G.-J., Penan, H., and Garcia, C.E.: 2002, *SocRobust Final Report*, Project SOE 1981126, Management Tools and a Management Framework for Assessing the Potential of Long-Term S&T Options to Become Embedded in Society, TSER Programme of the European Commission, January 2002, Ecole des Mines, Paris.

Reuzel, R.: 2004, Interactive Technology Assessment of Paediatric Cochlear Implantation, *Poiesis & Praxis* 2, 119–137.

Rip, A.: 2001, Assessing the Impacts of Innovation: New Developments in Technology Assessment. In: OECD Proceedings, *Social Sciences and Innovation*, OECD, Paris, pp. 197–213.

Rip, A.: 2002, *Challenges for Technology Foresight/Assessment and Governance*, A report commissioned by the Commission of the European Union, contributing to the STRATA programme area 'Sustainability: R&D policy, the precautionary principle and new governance models', University of Twente, Enschede.

Rip, A.: 2004, Strategic Research, Post-Modern Universities and Research Training, *High Educ Pol* 17, 153–166.

Rip, A.: 2006a, A Co-Evolutionary Approach to Reflexive Governance – And Its Ironies. In J.-P. Voß, D. Bauknecht, and R. Kemp (eds), *Reflexive Governance for Sustainable Development. Incorporating Unintended Feedback In Societal Problem-Solving*, Edward Elgar Cheltenham, pp. 82–100.

Rip, A.: 2006b, Folk Theories of Nanotechnologists, *Sci Cult* 15(4), 349–365.

Rip, A. and Kemp, R.: 1998, Technological Change. In: S. Rayner and E.L. Malone (eds), *Human Choice and Climate Change, Volume 2*, Battelle Press, Columbus, pp. 327–399.

Rip, A., Misa, Th., and Schot, J.W. (eds): 1995, *Managing Technology in Society. The Approach of Constructive Technology Assessment*, Pinter Publishers, London.

Schot, J. and Rip, A.: 1997, The Past and Future of Constructive Technology Assessment, *Technol Forecast Soc Change* 54, 251–268.
Smith, M.R. and Marx, L. (eds): 1994, *Does Technology Drive History? The Dilemma of Technological Determinism*, MIT Press,Cambridge.
Turpin, T.: 1997, CRCs and Transdisciplinary Research: What are the Implications for Science? *Prometheus* 15(2), 253–264.
Van de Ven, A.H., Polley, D.E., Garud, R., and Venkatamaran, S.: 1999, *The Innovation Journey*, Oxford University Press, New York.
Van Merkerk, R.O. and Van den Berg, A.: 2006, More than Technology Alone, *Lab Chip* 6, 838–839.
Verganti, R.: 1999, Planned Flexibility: Linking Anticipation and Reaction in Product Development Projects, *J Prod Innovat Manag* 16, 363–376.

# Chapter 10
# Chimeras and Other Human–Animal Mixtures in Relation to the Swiss Constitution: A Case for Regulatory Action

**Hans Peter Bernhard and Rainer J. Schweizer**

**Abstract** The progress of the biomedical sciences poses new philosophical, moral and legal questions. A transdisciplinary approach is ideal for addressing such questions and achieving a mutual understanding of the underlying biological facts. In this debate it becomes obvious that biomedical research per se does not provide a useful terminology for legislation. A transdisciplinary approach is therefore essential to achieve a common, accepted terminology and reach an agreement on the pertinent problems; both prerequisites for timely legislation.

The artificial creation of human–animal mixtures for research and therapeutic purposes touches very sensitive moral and legal issues. The novel procedures raise the question of the relevance of natural boundaries between species. The extent to which the species barrier has and will be breached by such experimental procedures which use biological entities at various developmental stages, in particular in the context of the use of human embryonic stem cells, is a strong challenge for scientists, ethicists, legislators and society. In view of pending and forthcoming legislation, it is the intention of this transdisciplinary approach to review and analyse the respective biomedical research agendas which have to be taken into account in the normative discussion.

First, we present the current relevant constitutional and statute laws of Switzerland. We then review recent developments in biological research involving human–animal mixtures, including chimeras, which are likely to become issues for regulatory action. The compilation covers a broad range of human–animal mixtures, with a focus on the use of human embryonic stem cells for basic research and therapeutic procedures. We finally address some other recent biomedical developments that could have an impact on terminology and legal definitions used in the current Swiss regulations.

**Keywords:** Chimeras · Human rights · Limits of research on humans · Parthenotes · Xenotransplantation

✉ R.J. Schweizer
Faculty of Law, University of St. Gallen, St. Gallen, Switzerland
e-mail: Rainer.Schweizer@unisg.ch

G. Hirsch Hadorn et al. (eds.), *Handbook of Transdisciplinary Research*,

## 10.1 Background

Since the outset of medically assisted procreation in Switzerland there has been an awareness of the risks and ethical questions related to the introduction of non-human germ plasm, germ cells and genetic material into the human germ line. As a consequence, the Constitutional Article of 1992 on Reproductive Medicine and Gene Technology prohibits the creation of interspecies organisms. This was further specified in the Swiss Federal Act on Medical Assistance to Procreation 1998 and in the Swiss Federal Act on Stem Cell Research 2003. During the preparation of these bills it became apparent that the ethical and legal meanings of the terms chimera and hybrid need to be critically evaluated. This task was facilitated by a Federal Executive Report – Research on Supernumerary Embryos 2002 – and a study by the Swiss Centre for Technology Assessment – Human Stem Cells 2003. The authors of this contribution are continuing their support for the Swiss legislatory activities, while participating as Swiss members of the ongoing EU-Research Project ‘Chimbrids’ (SAS6-CT-2005-0167081MGBChimerashybrids) chaired by Prof. Jochen Taupitz, Institute for German, European and International Medical Law, Public Health Law and Bioethics of the Universities of Mannheim and Heidelberg. This international research project is focused on the compilation and analysis of case studies in order to develop typologies and ethical and legal recommendations.

## 10.2 Swiss Constitution and Law

### *10.2.1 Federal Constitution*

In 1992, Switzerland introduced through a vote accepted by the people and the cantons, a special regulation in the Swiss Federal Constitution concerning medical assistance to procreation and human gene technology. This regulation has been adopted in the revised Federal Constitution of 18th April 1999 together with the different explicit fundamental rights regarding human dignity, the protection of life, health and personal freedom; and includes regulations for health protection and transplantation medicine. With respect to research with chimera and hybrids, the following regulations of the Federal Constitution are decisive:

***Federal Constitution of the Swiss Confederation***
*of April 18, 1999*
***Art.** 7 Human Dignity*
*Human dignity shall be respected and protected.*
***Art. 10** Right to Live and Personal Freedom*
*1 Every person has the right to live. The death penalty is prohibited.*
*2 Every person has the right to personal liberty, particularly to corporal and mental integrity, and to freedom of movement.*
*3 Torture and any other cruel, inhuman or degrading treatment or punishment are prohibited.*

***Art. 118*** *Protection of Health*

*1 Within the limits of its powers, the Confederation shall take measures for the protection of health.*

*2 It shall legislate on:*

*a. the use of foodstuffs and of therapeutics, drugs, organisms, chemicals, and objects which may be dangerous to health;*

*b. fighting contagious, widespread or particularly dangerous human and animal diseases.*

***Art. 119*** *Medical Assistance to Procreation and Gene Technology in the Human Field*

*1 Persons shall be protected against the abuse of medically assisted procreation and gene technology.*

*2 The Confederation shall legislate on the use of human reproductive and genetic material. It shall ensure the protection of human dignity, of personality, and of family, and in particular it shall respect the following principles:*

*a. All forms of cloning and interference with genetic material of human reproductive cells and embryos are prohibited;*

*b. Non-human reproductive and genetic material may neither be introduced into nor combined with human reproductive material;*

*c. Methods of medically assisted procreation may only be used when sterility or the danger of transmission of a serious illness cannot be avoided otherwise, but neither in order to induce certain characteristics in the child nor to conduct research. The fertilization of human ova outside a woman's body shall be permitted only under conditions determined by statute. No more human ova may be developed into embryos outside a woman's body than are capable of being immediately implanted into her;*

*d. The donation of embryos and all forms of surrogate maternity are prohibited;*

*e. No trade may be conducted with human reproductive material or with any product obtained from embryos;*

*f. A person's genetic material may only be analyzed, registered or disclosed with the consent of that person, or if a statute so provides;*

*g. Every person shall have access to the data concerning his or her ancestry.*

***Art. 119*** *a Medical Transplantation*

*1 The Confederation shall legislate in the field of transplantation of organs, tissues, and cells. It shall safeguard human dignity, personality, and health.*

*2 It shall in particular establish principles for the fair distribution of organs.*

*3 The donation of human organs, tissues and cells shall be free of charge. Trade in human organs is prohibited.*

## 10.2.2 Federal Act on Medical Assistance to Procreation

In accordance with Art. 119 of the *Federal Constitution* the *Swiss Federal Act on Medical Assistance to Procreation* was introduced on 18th December 1998. Art. 2 of this Act gives the following definitions:

| | |
|---|---|
| *Letter l.* | *to clone: to artificially create genetically identical beings* |
| *Letter m.* | *creation of chimera: the combination of totipotent cells of two or more genetically different embryos into a united cell structure. Totipotent cells are embryonic cells that are able to develop into any kind of specialised cell* |
| *Letter n.* | *creation of hybrid: to provoke the introduction of a non-human sperm cell into a human egg cell or of a human sperm cell into a non-human egg cell* |

And Art. 36 of this Act defines thereafter:

*1 Those who create a clone, a chimera or a hybrid will be sent to prison.*
*2 Those who transmit a chimera or a hybrid to a woman or an animal will be punished in the same way.*

### 10.2.3 Interpretation

The interpretation of Art. 119 §2 letter b is the following:

This regulation can be seen as complementary special norm to letter a; the interdictions sometimes overlap. The aim is to safeguard the integrity of human reproductive material as the basis for human life. The mixing of humans and animals damages the dignity of humankind, but also the dignity of a single person and the identity and integrity of the unique individual. Besides prohibiting the introduction of non-human reproductive and genetic material into human reproductive material, this regulation prohibits the creation of interspecies hybrids by combining reproductive cells of different species and of interspecies chimera (for the terms, see Art. 2, letter m and n of the Swiss Federal Act on Medical Assistance to Procreation). Art. 36 of the Swiss Federal Act on Medical Assistance to Procreation furthermore prohibits the intraspecies chimera, created from two human embryos. The result would be a creature with genetic material of at least four parents (Reusser et al., 2002).

In an Explanatory Report regarding its Draft on the prohibitions in the Swiss Federal Act on Medical Assistance to Procreation the Swiss Federal Government states:

*It would grossly offend against human dignity if someone specifically allocates their hereditary disposition to a future human being. Each person has the fundamental right not to be the copy of another individual but an original and unique personality. It cannot be allowed, that the dignity of a person be marred by an act that just brought them into existence. Those who artificially create genetically identical (human) beings will therefore be put in prison, according to Art. 2, letter m in connection with Art. 36 §1.*

*The creation of chimera (for the term see Art. 2, letter m) violates in a blatant way the dignity of man and can therefore also be prosecuted under Art. 36, §1. The object to be protected is the human embryo as well as the totipotent cell.*

*The creation of hybrids (for the term see Art. 2, letter n) is also illegal according to Art. 36 §1. The primary barrier between humans and animals is destroyed as a result of the penetration of the sperm of one species into the cytoplasma of the egg cell of another species. The generation of an impregnated egg cell using animal or human zygotes across species is unacceptable. In particular, the "Goldhamster" test to determine the penetration capacity of human sperm into an animal egg cell for diagnostic purposes, is banned. The creation of hybrids is not liable to prosecution if human genetic material is introduced into the DNA of bacteria for the production of insulin, for example.*

*Liable to prosecution according to Art. 36 §2 is also the transmission of a chimera to a woman or an animal. But if a clone is implanted into a woman, the right to life prohibits killing this human being. (Paragraph 324.208 Explanatory Report to the Draft of the Federal Act, 1996)*

Besides Art. 119 §2 letter b, attention must also be paid to letter a from Art. 119 §2 with its absolute prohibition of cloning. According to the wording under letter a, not only is cloning, by splitting an embryo into halves (twinning) prohibited; so is the transfer of a cell nucleus into an enucleated oocyte and the parthenogenesis of oocytes (cf. Art. 35 Swiss Act on Medical Assistance to Procreation).

*1 Those who interfere with the genetic material of human germ cells will be punished by imprisonment.*
*2 Those who use germ cells whose genetic material has been artificially modified, for the impregnation of an egg cell which has been modified in the same way, with the aim of creating an embryo will also be punished.*
*3 §1 is not valid if the change in the human reproductive cells is the unavoidable side effect of chemotherapy, radiotherapy or another medical treatment that a person undergoes.*

On 19th December 2003 the *Swiss Federal Act on Research on Embryonic Stem Cells* included this prohibition of creating parthenotes as well.
According to Art. 3, §1 it is prohibited:

*c) to create a clone, a chimera or a hybrid (Art. 36, § 1 of the Swiss Act on Medical Assistance to Procreation dated 18th December 1998) and to extract or use embryonic stem cells from such a creature*
*d) to develop a parthenote, to extract from it embryonic stem cells and to use them*

Finally it should be noted that Art. 119, §2, letter c of the Federal Constitution and the relevant Swiss legislation explicitly prohibit the production of embryos for research purposes as well as the ectogenesis of embryos (compare Art. 29 and 30 Swiss Federal Act on Medical Assistance to Procreation).

***Art. 29*** *Abusive extraction of embryos:*
*1 Those who impregnate an embryo with the intention to use it or to allow it to be used for a purpose other than to start a pregnancy will be punished by imprisonment.*
*2 Those who preserve an impregnated egg cell with the intention to use it or to allow it to be used for a purpose other than to start a pregnancy will also be punished.*
***Art. 30*** *Development of embryos outside a woman's body:*
*1 Those who keep an embryo outside a woman's body for longer than is necessary to lodge it in the uterus will be punished by imprisonment.*
*2 Those who transfer a human embryo into an animal will also be punished.*

The main principle of the Swiss constitutional and statutory norms, with regard to the present matter, is that human germ cells may not be used for research purposes in order to create a viable human or chimeric being. Accordingly, creating human clones and parthenotes is prohibited, even if they can not or can hardly be bred, or are not viable.

***Art. 119***, §2, letter b (and only if there is no abuse) does not cover:

- the introduction of non-human genetic material into somatic cells of human beings;
- the introduction of sequences of human genetic material into non-human genetic and reproductive material, as used in the production of proteins, like insulin or interferon (Parliamentary Report of the State Chamber [2nd Chamber of the Swiss Federal Parliament] 1990).

***Art. 119***, §2, letter a does not cover

- the creation of human germ cells by a redevelopment of human embryonic stem cells, if there is no genetic interference.

## 10.3 Experimental Procedures for the Creation of Chimeras and other Human–Animal Mixtures

Chimeras are organism composed of cells derived from at least two genetically different cell types. The cells could be from the same or separate species. Table 10.1 summarises experimental approaches that could be envisaged in the near future or have already been embarked upon. The table is not intended as a complete compilation but includes representative examples of ethically and legally problematic biological constructs that may arise through experiments performed with biological entities from various biological sources and recipient organisms at different developmental stages. The procedures cover a broad range, from gene-, chromosome- and somatic nuclear-transfer to cell-, tissue- and organ-transplantation, and gamete fusion. Human–mouse chimeras for research purposes were established almost 20 years ago within the field of haematology, e.g. a mouse strain with a deficient immune system, which was designed to tolerate a haematopoetic system of human origin became extraordinary crucial to the study of AIDS (McCune et al., 1988) while early therapeutic transplants of foetal neuronal precursor cells of animal origin into human brain cells of patients with Parkinson's disease have been performed by Borlongan and Sanberg 2002.

## 10.4 Creation of Chimeras Using Human Embryonic Stem Cells (hES cells)

The recent experimental procedures involving hES cells listed in the overview 'Chimeras and other Human–Animal mixtures' are:

- Use of non-human oocytes as recipients of human somatic nuclei in nuclear transfer with the aim of generating hES cell lines without the need for human oocytes
- Incorporation of hES cells into non-human blastocysts (Pre-implantation embryo of 50–250 cells depending on age)
- Incorporation of human embryonic stem cells or their cell derivatives into post-gastrulation stages of a non-human species
- Incorporation of human embryonic stem cells or their cell derivatives into post-natal animals of a non-human species

**Table 10.1** Overview: Chimeras and other human–animal mixtures (based on Bernhard et al., 2006)

| Procedure | Source | Recipient | Result | Applications |
|---|---|---|---|---|
| Stem cell transplantation | Human embryonic SC | Animal blastocyst or postgastrulation embryo | Chimeric organism with donor/recipient ratio depending on developmental stage and evolutionary relationship ('humanised' animals) | - Testing pluripotency, and developmental capacity of hESC (Muotri et al., 2005)<br>- Studies of cell differentiation (Flax et al., 1998)<br>- Studies of human diseases<br>- Drug testing<br>- Mouse Model with human immune system (McCune et al., 1988)<br>- Repair of motor neuronal lesions in animal model |
| | Human adult SC | Adult animal with eradicated immunsystem | | |
| | Human neural SC | Adult animal | | |
| Cell, tissue, organ transplantation | Human cells, tissues, organs | Animal foetus or postnatal organism | 'Animal' chimera containing human cells, tissues or organs | - Generation of human tissues, organs for clinical transplants (Ogle et al., 2004; Almeida-Porada et al., 2004) |
| | Animal cells, tissues, organs | Human foetus or postnatal organism | 'Human' chimera containing animal cells, tissues or organs | - Xenotransplantation for therapeutics (Borlongan and Sanberg, 2002) |

**Table 10.1** (Continued)

| Procedure | Source | Recipient | Result | Applications |
|---|---|---|---|---|
| Somatic nuclear transfer (SCNT) | Human somatic nucleus | Enucleated animal ovum | Clonal human embroid body or blastocyst with animal mitochondria | - Source for cell or tissue grafts<br>- Source for hESC (Chen et al., 2003) |
| Chromosome transfer | Animal chromosomes | Human mesenchymal SC | Human mesenchymal SC with additional animal chromosome(s) | - Studies of gene expression in animal models (Vanderbyl et al., 2004)<br>- Studies of human chromosome(s) expression in animal models (O'Doherty et al., 2005) |
| | Human chromosome(s) | Animal embryonic SC | Animal embryonic SC with additional human chromosome(s) | |
| Gene transfer | Human genes | Germ cells or somatic cells | Animals with additional human genes | - Studies of human gene expression in animal models |
| Gemete fusion | Human sperm | Anilmal ovum | Hybrid zygote | - Clinical fertility testing |

The novel quality of human embryonic stem cell transplants is due to the pluripotency of the stem cells and their developmental potential which differs from other cell transplants. The transplantation of human embryonic stem cells is expected to lead to the formation of a wide range of chimeric organisms with in a quantitative and qualitative variety of human contributions, because it is assumed that embryonic stem cells are capable of integrating within a recipient organism at prenatal as well as at postnatal stages. The degree of evolutionary kinship between the two species, the developmental stage of the recipient organism and the final localisation of the stem cells in the recipient organism are considered to be responsible for the extent and the quality of the mixing and hence the ensuing chimerism.

## 10.5 Recent Biomedical Developments Which Require Evaluation with Reference to Their Significance for Regulatory Action

### *10.5.1 Altered Nuclear Transfer*

In somatic nuclear transfer experiments, cell nuclei of somatic cells are being transferred into enucleated egg cells. The resulting construct can undergo normal embryonic development and eventually proceed to a viable offspring. Nuclear transfer is at present the preferred technique for cloning. By using multiple somatic cell nuclei from the same donor the number of obtainable clones is very high in comparison with cloning by means of embryo splitting. The technique of altered nuclear transfer (ANT) uses cell nuclei which are genetically modified in order to prevent complete embryonic development. The product of ANT is a biological embryo construct with an intentionally restricted developmental potential, which does not allow nidation and can therefore not proceed into a viable embryo (Meissner and Jaenisch, 2006).

### *10.5.2 Derivation of Gametes from Human Embryonic Stem Cell Cultures*

With the in vitro derivation of gametes from cultured human embryonic stem cells the process of gametogenesis and the procurement of gametes could become separated and independent from individuals donating oocytes and spermatocytes respectively (Hubner et al., 2003; Toyooka et al., 2003).

### *10.5.3 Derivation of Gametes from Primordial Germ Cells*

Primordial germ cells can be isolated from ovarectomies or from foetal ovaries obtained after pregnancy terminations. In the case of mammals it is possible to mature such oocytes in culture and achieve fertilisation and normal development, although the process is not efficient. In humans the success has been limited and requires an intermediate transplantation into an animal of the ovarian tissue for oocyte maturation (O'Brian et al., 2003).

### 10.5.4 *Parthenogenetic Embryonic Stem Cells*

Parthenotes are organism containing only maternal chromosomes. Preliminary experiments indicate that embryonic stem cells may be obtained by means of artificially induced parthenogenetic activation of oocytes (Vrana et al., 2003).

These recent procedures and the resulting novel biological products present a new situation for the evaluation of the moral status of artificially created, partially non-viable human embryos. In view of the protection of the human embryo in the Swiss Constitution and Swiss legislation the moral status of such ANT constructs needs to be assessed in comparison with the native embryo. In this new context the meaning of cloning may also require a critical examination.

## 10.6 Recommendations

The protection of the integrity and identity of the human genome, the human germline and the resulting embryos is an important element in Swiss, European and International law. As a consequence interventions and modifications of the human genome or germline using a combination of human and animal components are not tolerated. However, recent studies show the necessity for a more specific and differentiated regulatory approach. In particular, restrictions and limitations are also needed for the protection of animals and animal rights.

Transdisciplinary research on chimeras and other human–animal mixtures should lead to an evaluation of the current legal sanctions as well as to a review of priorities, and eventually to further specific restrictions for biomedical research. Research modules such as ELSI should provide a most useful basis for introducing the transdisciplinary approach at the very beginning of the research activities. In our country this could be achieved best, if for example the Swiss National Science Foundation (SNFS) would, on principle, finance such accompanying research activities as part of the current granting practice for selected biomedical research applications.

The ongoing biological and medical research on germ and stem cells and on the early development of human life includes various experiments which mix human and animal cells and rearrange their respective genomes. Involved researchers from natural sciences and humanities are therefore challenged to explain their scientific activities and results to each other and to participate in the public ethical and legal debates. With our contribution we are following an interdisciplinary and iterative process between natural and legal scientists who are directly involved in the formulation of legislation. This iterative process, of clarifying definitions and terminology between biological and normative sciences, needs to start as early as possible and is imperative for timely legislation.

## References

Almeida-Porada, G., Porada, C., and Zanjani, E.D.: 2004, Plasticity of human stem cells in the fetal sheep model of human stem cell transplantation, *Int J Hematol* 79, 1–6.

Bernhard, H.P., Rubin-Lucht, B., and Schweizer, R.J.: 2006, *Chimeras and Hybrids in Comparative European and International Research*, CHIMBRIDS Opening Conference, Mannheim.

Borlongan, C.V. and Sanberg, P.R.: 2002, Neural transplantation for treatment of Parkinson's disease, *Drug Discov Today* 7, 674–82.

Chen, Y., He, Z.X., Liu, A., Wang, K., Mao, W.W., Chu, J.X., Lu, Y., Fang, Z.F., Shi, Y.T., Yang, Q.Z., Chen, D.Y., Wang, M.K., Li, J.S., Huang, S.L., Kong, X.Y., Shi, Y.Z., Wang, Z.Q., Xia, J.H., Long, Z.G., Xue, Z.G., Ding, W.X., and Sheng, H.Z.: 2003, Embryonic stem cells generated by nuclear transfer of human somatic nuclei into rabbit oocytes, *Cell Res* 13, 251–263.

Federal Constitution of the Swiss Confederation of April 18, 1999, Retrieved June 5, 2007, from http://www.admin.ch/ch/itl/rs/1/c101ENG.pdf.

Flax, J.D., Aurora, S., Yang, C., Simonin, C., Wills, A.M., Billinghurst, L.L., Jendoubi, M., Sidman, R.L., Wolfe, J.H., Kim, S.U., and Snyder, E.Y. : 1998, Engraftable human neural stem cells respond to development cues, replace neurons and express foreign genes, *Biotechnology* 16, 1033–1039.

Hubner, K., Fuhrmann, G., Christenson, L.K., Kehler, J., Reinbold, R., De La Fuente, R., Wood, J., Strauss, J.F., Boiani, M., and Scholer, H.R : 2003, Derivation of oocytes from mouse embryonic stem cells, *Science* 300, 1251–1256.

McCune, J.M., Namikawa, R., Kaneshima, H., Shulz, L.D., Liebermann, M., and Weissmann, I.L.: 1988, The SCID-hu mouse: murine model for the analysis of human hematolymphoid differentiation and function, *Science* 241, 1632–1639.

Meissner, A. and Jaenisch, R.: 2006, Generation of nuclear transfer-derived pluripotent mouse embryonic stem cells from cloned Cdx2-deficient blastocysts, *Nature* 439, doi.10.1038, 1–4.

Muotri, A.R., Nakashima, K., Toni, N., Sandler, V.M., and Gage, F.H.: 2005, Development of functional human embryonic stem cell-derived neurons in mouse brain, *Proc Natl Acad Sci USA* 102, 18644–18648.

Ogle, B.M., Butters, K.A., Plummer, T.B., Ring, K.R., Knudsen, B.E., Litzow, M.R., Cascalho, M., and Platt, J.L.: 2004, Spontaneous fusion of cells between species yields transdifferentiation and retroviral transfer in vivo, *Faseb J* 18, 548–550.

O'Brian, M.L., Pendola, J.K., and Eppig, J.K.: 2003, A revised protocol for in vitro development of mouse oocytes from primordial follicles dramatically improves their developmental competence, *Biol Reprod* 68, 682–1686.

O'Doherty, A., Ruf, S., Mulligan, C., Hildreth, V., Errington, M.L., Cooke, S., Sesay, A., Modino, S., Vanes, L., Hernandet, D., Linehan, M., Sharpe, P.T., Brandner, S., Bliss, T.V.P., Henderson, D.J., Nizetic, D., Tybulewicz, L.J., and Fisher, E.M.C.: 2005, An aneuploid mouse strain carrying human chromosome 21 with Down syndrome phenotypes, *Science* 309, 2033–2037.

Reusser, R. and Schweizer, R.J.: 2002. In: B. Ehrenzeller, Ph. Mastronardi, R.J. Schweizer, and K. Vallender (eds), *Commentary to the Swiss Federal Constitution*, Zürich, Commentary to Art. 119, 1211–1229.

Toyooka, Y., Tsunekawa, N., Akasu, R., and Noce, T.: 2003, Embryonic stem cells can form germ cells in vitro, *Proc Natl Acad Sci USA* 100, 11457–11462.

Vanderbyl, G.N., MacDonald, G.M., Sidhu, S., Gung, L., Telenius, A., Perez, C., and Perkins, E.: 2004, Transfer and stable transgene expression of mammalian artificial chromosomes into bone marrow-derived human mesenchymal stem cells, *Stem Cells* 22, 324–333.

Vrana, K.E., Hipp, J.D., Goss, A.M., McCool, B.A., Riddle, D.R., Walker, S.J., Wrrstein, P.J., Studer, L.P., Tabar, V., Cunniff, K., Chapman, K., Vilner, L., West, M.D., Grant, K.A., and Cibelli, J.B.: 2003, Non human primate parthenogenetic stem cells, *Proc Natl Acad Sci USA*, 100 Suppl 1, 11911–6.

# Part III
# Problem Analysis

# Chapter 11
# The Development of Multilateral Environmental Agreements on Toxic Chemicals: Integrating the Work of Scientists and Policy Makers

**Nuria Castells*** **and Ramon Guardans***

**Abstract** In this paper we present, first, a brief summary of the historical development of the main components of contemporary international environmental agreements. Next, we argue, based on the examples of a few recent international agreements concerning environment and health, that the procedures established in the framework of these agreements provide a solid international base for stable and effective scientific, industrial and political cooperation. Later, we identify some of the present and future problems and difficulties that should be addressed. In the final section we underline several points that have been shown to be relevant to ensure fair and effective action through international agreements on global issues of environment and public health. In conclusion, the paper argues for strengthening the dialogue, and the extent of transdisciplinary research, between the scientific community and policy makers in order to develop synergies from various areas of knowledge to address complex environmental issues and to coordinate effective tools to mitigate the impact of anthropocentric activities.

**Keywords:** Integrated assessment models · Multilateral environmental agreement (MEA) · Science and policy · Long-range transboundary air pollution · Persistent organic pollutants

## 11.1 Background

Multilateral Environmental Agreements (MEAs) are in general the result of a bottom–up process in societies and countries where environmental problems are of deep concern to the population and public officials due to serious health or natural resource problems, such as the impact of the acidification of rivers and lakes on

✉ N. Castells
United Nations Conference on Trade and Development, Geneva, Switzerland
e-mail: Nuria.Castells@unctad.org

*The views expressed in this article are those of the authors in their personal capacity and do not necessarily reflect those of the United Nations.

G. Hirsch Hadorn et al. (eds.), *Handbook of Transdisciplinary Research*, 

fisheries in Scandinavia during the 1960s (Bölin, 1972). Since the legal framework of national sovereignty over national resources is often too narrow to resolve the environmental problem, national authorities bring their concern to the international arena and raise awareness in neighbouring countries so that, together, they can coordinate efforts in reducing transboundary pollution.

Over the past centuries science, industry and society have had very dynamic interactions. The interplay between these actors has shaped the development of local, regional and international rules and agreements that deal with the main components of the process: innovation, development, and commerce. Innovation patterns in policy making about environment, health and risk prevention follow patterns that can be compared to innovation patterns in technology (Nijkamp and Castells, 2001).

During the course of history the activity of human societies has always resulted in the intentional and unintentional diffusion of substances into the environment: air, water and soil (Martinez-Cortizas et al., 1997). By the end of the 18th century the environmental and health impact of urban and industrial combustion was well established and early regulations and norms were proposed (Newell, 1997). In the early years of the 20th century the work on ionising radiation had started and in 1928 the precursor to the International Commission for Radiation Protection was established (Lindel and Dunster, 1998) and has produced, over the years, many of the components of environmental and medical monitoring, global scale assessments and regulations that have been applied to other health and environmental hazards.

The early nuclear explosions in 1945, the multiple tests and a few accidents in the following years, showed clearly how long-range atmospheric transport could occur in short time periods and ultimately lead to the global dispersion of substances in the atmosphere (Chamberlain, 1953; Pasquill, 1961). The atmospheric models and operational procedures such as global monitoring networks developed in the nuclear technology environment were applied to other forms of pollution in the mid 1960s and established the framework on which the problem of acid rain was identified (Eliassen and Saltbones, 1975). The theoretical and empirical evidence gathered between the 1950s and the 1970s demonstrated the importance of long-range transport of air pollution.

From a theoretical point of view and for practical purposes, atmospheric transport is divided into: (1) local transport including distances perhaps up to 50 km within seconds or hours. The main phenomena are mixing within the lower atmosphere or boundary layer, prevailing winds and deposition, (2) mesoscale transport up to a few hundred kilometres within hours up to a few days, for which large scale synoptic weather patterns and chemical processes in the air masses are central, and (3) hemispheric and global transport where the identification of the balance of sources and sinks and the transport pathways are important. These distinctions are part of the academic division of labour and are relevant, together with information about the length of time pollutants stay in the atmosphere (residence time), to practical matters such as model building and deciding the priorities for monitoring.

The continuity of diffusion and transport processes at multiple scales, from local to global, and the long residence time of some pollutants made it apparent that the large scale (air) pollution problems could only be tackled in international, multilateral agreements. When the United Nations was established in 1945 the globe was organised in a limited number of geo-political regions. One of them, the Economic Commission for Europe (ECE) of the United Nations, included the entire European continent, including the whole area of what was then the USSR, the United States and Canada. UN-ECE was established to facilitate concerted action for the economic reconstruction of Europe, for raising the level of economic activity and strengthening the cooperation of European countries. Other analogous regional commissions were established all over the world. The ECE region included in 1945, and includes today, many of the wealthiest and most powerful countries in the world. It was in this region, during the second half of the 20th century, that many of the practical problems of very large scale industrialisation were addressed for the first time.

The UN-ECE Convention on Long-range Transboundary Air Pollution (www.unece.org/env/LRTAP), established in 1979, has provided a framework in which MEAs (e.g. several protocols to the Convention) have been negotiated, agreed upon, implemented and shown to work (De Vries and Posch, 2003). The work under LRTAP has benefited over the years from synergies with other processes in the ECE region. Working groups, laboratories and funding involved in other important institutional frameworks such as OSPAR (www.ospar.org), MAP (www.unepmap.org) and AMAP (www.amap.no) have historically played important roles and still do; but it can be argued that the binding reporting obligations and the assessment methods developed in the ECE/LRTAP Convention have been particularly innovative in linking policy and science in a long-term dynamic integrated framework. The increased awareness within some social groups and the political concerns of scientists and policy makers, led to a unique configuration of transdiciplinary design embedded in the definition of the agreement itself. Converging concerns focused on issues of health, environment, social impact (e.g. of loss of traditional means of subsistence due to acidification of lakes in Scandinavia) and policy definition: leading to the creation of the framework for the LRTAP Convention.

The political process in the UN-ECE framework (see Figure 11.1), led by the annual meetings of the Executive Body, provides a unique and pioneering framework for linking science and policy, to mutually enhance the effectiveness of the research undertaken to fulfil a political mandate based on sound scientific data and processes. It can be said that the experience gained under the umbrella agreement of the LRTAP Convention made a seminal contribution to the patterns of making environmental policy in Europe, and later, on a larger scale, to address complex environmental problems where solutions would require the collaboration between science and policy. Network building, scientific work mandated by policy requirements, monitoring networks, and research networks beyond political boundaries (at the time of East and West confrontation) provided an innovative model for other agreements to follow. The Center for Integrated Assessment Modelling (CIAM) at the International Institute for Applied Systems Analysis (IIASA) in Laxenburg, Austria, added the necessary scientific contribution to balance the components between science and policy and ensured the means and the objectivity required to provide countries with robust modelling scenarios, while the EMEP network facilitated the series of meteorological and air quality data on which to base future abatement strategies for resolving the environmental problem at stake.

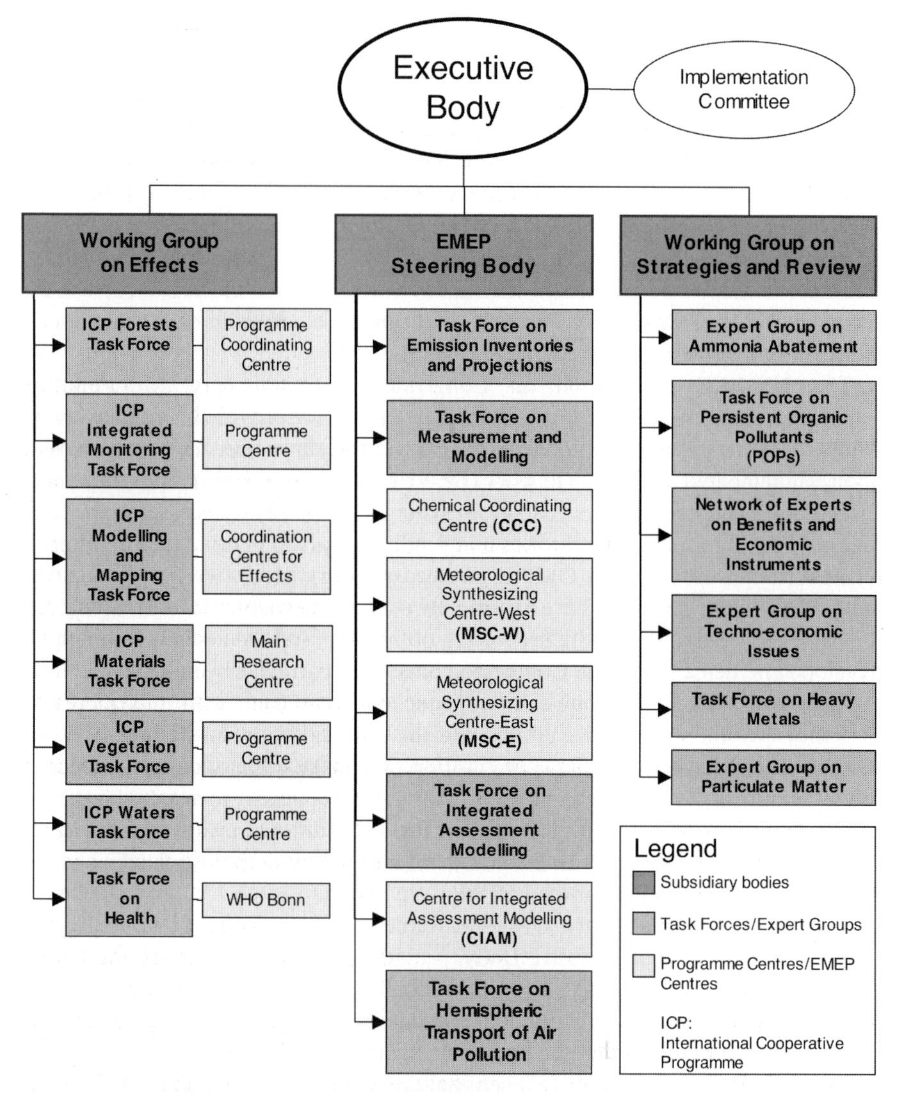


**Fig. 11.1** The structure and organisation of the LRTAP convention and its subsidiary bodies in 2006 (Source: http://www.unece.org/env/lrtap/13 January 2006)

## 11.2 Multilateral Environmental Agreements (MEAs) and Sustainable Development

Many developing countries have found themselves constrained to adhere to a process aimed at international environmental governance in the name of global welfare and shared responsibilities on global commons, while they have not directly benefited from the economic welfare associated with the industrialisation process of the pioneering countries. Indeed, environmental policy compliance is often used as one of the parameters allowing them to benefit from technical assistance in terms of technology transfer and investment. Even when compliance is not formally binding, if they are not part of the agreements, they can find themselves in economic difficulty since the non-participation in these MEAs can lead to economic consequences in terms of trade and development goals. For instance, exporting countries increasingly need to comply with environmental requirements of importing countries, be these private standards or not, and this can have significant consequences. Countries exporting products not complying with specific environmental requirements can experience real difficulties in accessing export markets (UNCTAD, 2006). Although the norms might not be legally binding, in practice the market mechanisms will exclude them from the international market. This situation can lead, de facto, to a scenario in which polluting productive processes and products with lower requirements in terms of health and environment will be authorised only for the local market of developing countries, while products for the export markets will have to adapt to international environmental and health related requirements.

The preceding analysis raises the question of 'which environment for whom' and how the trade offs between sustainable development, and environmental protection and public health, should be addressed. In economic terms, developing countries cannot afford to allocate the necessary means to regulate, implement and enforce environmental legislation to comply with international environmental regulations. Although the global environment can be perceived as an issue of global concern, developing countries do require more decisive and significant support in economic and technological terms to enable them to meet the targets and ambition levels in the agreements. The Rio Declaration took into account the fact that countries have common but varying responsibilities since they have different reference starting points. The means to implement this common responsibility has to be given to all countries, taking into account the role they play in the global environment, and their own needs, while respecting their sovereignty and policy space. Many MEAs do attempt measures to provide technical and financial assistance to facilitate compliance by developing countries, and while successfully applied in some cases, they most often prove difficult to implement due to attached conditions, lack of transparency, non-user-friendly allocating procedures, and absence of early warning systems to anticipate and adapt to forthcoming regulations.

The LRTAP Convention, provides an example of effective collaboration allowing countries to deal with a transboundary environmental problem, despite different perceptions of its magnitude and importance, its causes, and its distributional impact in terms of harm. The burden sharing of its implementation costs was discussed within an international framework and addressed by means of integrated assessment models (IAMs), which attempted to evaluate all these different parameters and to provide different policy scenarios that enabled abatement strategies to be agreed and implemented.

Several innovations introduced by the LRTAP Convention have proven to be effective. These include (i) mandatory reporting and the engagement by parties to provide the best available information through a network of cooperative research efforts; (ii) cooperative and consistent monitoring and modelling of atmospheric transport and effects on ecosystems and health on a continental scale; and (iii) an iterative process of assessment and simulation performed by a network of technical groups including representatives of all countries working with agreed data, on emissions, transport, effects and abatement costs and using a simulation framework to integrate the information. This process can identify strategies that, according to the best available knowledge, achieve maximum environmental effects at minimum costs.

The LRTAP framework displayed in Figure 11.1 was set-up in a stepwise process. The Convention in Article 9 'Implementation and further development of the cooperative programme for the monitoring and evaluation of the long-range transmission of air pollutants in Europe' gave the specifications of what would be formally established, five years later, by the 1984 'Protocol on Long-Term Financing of the Cooperative Programme for Monitoring and Evaluation of the Long-Range Transmission of Air Pollutants in Europe EMEP' (www.emep.int). This agreement established a stable budget to run the programme shared by a list of mandatory contributions from all the participating countries specified in the agreement and updated every year. The coordination and intercalibration of chemical air quality and precipitation measurements are carried out at the Chemical Coordinating Centre (CCC) in Oslo. The storage and distribution of reliable information on emissions and emissions projections is the task of the Meteorological Synthesizing Centre–West (MSC-W), in Oslo. The MSC-W is also responsible for the modelling assessment of sulphur, nitrogen photo-oxidant pollutants and atmospheric particles. The modelling development for heavy metals and persistent organic pollutants (POPs) is the responsibility of the Meteorological Synthesizing Centre–East (MSC-E) in Moscow. In 1999 the Executive Body of the Convention decided to include integrated assessment into the core activities of EMEP and to establish a Center for Integrated Assessment Modelling (CIAM) in Laxenburg, Austria, building on past modelling work, in particular the RAINS model.

EMEP (May 2004) includes 186 monitoring stations in 38 countries working with the same standards and performing intercallibration exercises. It has developed and maintains continental size transport models that calculate, based on the best available information on atmospheric processes and atmospheric conditions, the movements and transformations in six hour steps of a range of chemical substances across the European continent, using a grid of cells of 50×50 km and 30 vertical atmospheric layers. These models include extensive official databases of activities and estimated pollutant emissions, compiled from the reports submitted by each country. EMEP provides sound scientific support in the areas of emission inventories and projections, atmospheric monitoring and modelling, and integrated assessment modelling. Results from the monitoring network, together with emission data and atmospheric transport modelling, have shown how air pollution moves through the atmosphere, and have made it possible to quantify the source–receptor relationships that attribute different fractions of atmospheric pollutant concentration in a region to sources in different countries. This science is used by policy makers to design actions across wider regions such as Europe and North America, as well as to address emissions in their local areas.

The evidence provided by EMEP was central to the negotiation that produced the 1985 'Protocol on the Reduction of Sulphur Emissions or their Transboundary

Fluxes by at least 30%'. This agreement underlined the problem with the choice of 30% (25% may have a similar effect at lower cost; 40% may have a slightly higher cost but have a much more significant effect) and the necessity to develop effects oriented strategies that integrated all available information and to try to find ways of maximising the environmental impact of abatement strategies in the most cost-effective way.

The effects based approach was established in the 1988 'Protocol concerning the Control of Nitrogen Oxides or their Transboundary Fluxes' that says, inter alia, in Art. 6

> *the Parties shall ... seek to ...*
> *(a) Identify and quantify effects of emissions of nitrogen on humans, plant and animal life, waters, soils and materials, ...;*
> *(b) Determine the distribution of sensitive areas;*
> *(e) Develop, in the context of an approach based on critical loads, methods to integrate scientific, technical and economic data in order to determine appropriate control strategies.*

The objective of developing effects oriented abatement strategies was described and formally established in the 1994 'Protocol on Further Reductions of Sulphur Emissions'. This agreement was clearly effect oriented and relied on EMEP and the Working Group on Effects (www.unece.org/env/wge) to provide scientific background and to assess the impact of alternative pollution abatement scenarios. The Working Group on Effects was established to provide the Convention with information on the degree, geographic extent and dynamism of the impacts of major air pollutants on the environment and human health. There are six International Cooperative Programmes (ICPs). These deal with: (1) the effects of air pollution on fresh waters – (it runs a continental network of over 30 stations and works in the language of limnologists), (2) the effects on forests – monitoring is carried out at thousands of plots, and intensive monitoring at hundreds of sites across Europe and speaks the language of forestry science, (3) effects on crops (using the terminology of agricultural science), (4) effects on materials (using engineering terminology ), (5) integrated monitoring – focusing on measuring mass balances, and ecosystems science in heavily instrumented research sites such as watersheds, and (6) modelling and mapping – integrating the main findings of these groups in coherent and consistent geographical databases, graphic representations and modelling procedures. The Joint Task Force of the World Health Organization (WHO) and the Executive Body for the Convention on the Health Aspects of Long-range Transboundary Air Pollution assesses the significance of LRTAP in public health strategies.

The Center for Integrated Assessment Modelling under EMEP is located at the International Institute for Applied Systems Analysis. IIASA is a non-government research organisation. It conducts interdisciplinary scientific studies on environmental, economic, technological and social issues within the context of human dimensions of global change. IIASA is located in Austria near Vienna and is sponsored by its National Member Organizations in Africa, Asia, Europe, and North America (www.iiasa.ac.at). It has a central role in the development of effects oriented abatement strategies in LRTAP. Over 20 years the Transboundary Air Pollution (TAP) Group in IIASA has developed and maintained an integrated assessment modelling framework linking detailed quantitative scenarios, emissions, atmospheric transport and effects, for several pollutants and types of effects.

The Secretariat of the LRTAP Convention at the UN in Geneva has played a very relevant role in the timely preparation of the meetings and documents and is responsible in a significant measure for the overall success of the effort.

The work carried out under the 1994 Protocol was the basis for the development of the 1999 'Protocol to Abate Acidification, Eutrophication and Ground-level Ozone' where, for the first time, several pollutants and several effects are addressed in a consistent effects oriented strategy, which was negotiated through multiple simulations and assessments of different scenarios.

The ground work laid out by EMEP concerning methods for building and maintaining emission inventories and air pollution monitoring networks, has been instrumental in the development of other initiatives such the Framework Convention on Climate Change and its Kyoto Protocol.

By 1990, the evidence gathered in the Arctic by AMAP and others concerning the long-range atmospheric transport of persistent toxic chemicals confronted the LRTAP Convention with a new challenge: these chemicals, by their mode of propagation and their effects, could not be dealt with using the existing effects oriented methods and conceptual frameworks. At the same time it was apparent that there was a solid argument to include these pollutants in the work of the LRTAP Convention.

The 'Protocol on Persistent Organic Pollutants' under LRTAP, signed in Aarhus in 1998, is not effects oriented and did not make any use of integrated assessment models due to the difficulty in establishing source–receptor relations of substances that can be deposited from the atmosphere and revolatilised multiple times. The Protocol prepared the ground for the global agreement on persistent organic pollutants under the United Nations Environmental Programme, signed in Stockholm in 2001, that established reporting obligations and cooperation procedures on a global scale.

The description of this process clearly highlights the benefits of transdiciplinarity, which enabled the definition of multi-effects oriented strategies based on information coming from different scientific sources and different policy interests. The latest LRTAP Protocols are clear examples of the gains that can derive from cooperation between the different branches of science, and between the social sciences and the hard sciences. Complexity cannot be addressed from a narrow knowledge framework. Global problems require cooperation from different actors: from policy-makers to scientists and the actors in society. Transdiciplinarity in this case is not an epistemological question but a result of a real need to ensure an effective design and an effective implementation of the MEAs.

## 11.3 Integrated Assessment Models at the Interface of Science and Policy Making. The Use of RAINS Model in the LRTAP Convention

The role of integrated assessment models (IAMs) in this process has been instrumental in enabling scientists and policy-makers to find common tools for working together on environmental issues. Relying on the EMEP Programme and the

Working Group on Effects to provide the required scientific knowledge and necessary data, CIAM has developed methods and procedures that enable scientists to address an environmental problem while producing an outcome that can be understood by policy makers, in terms of policy scenarios, without the need to understand the complexity in the IAM. Furthermore, by incorporating the necessary variables and indicators into the design the socio-economic impact of the environmental strategies is addressed and evaluated in the different policy scenarios. The social impact can then be calculated thanks to the powerful modelling tools that are available nowadays.

The implementation of the LRTAP Convention and the EMEP programme, as well as the activities of ICPs and their Task Forces under the Working Group on Effects, led over some 25 years to the development of a stable network of hundreds of laboratories and thousands of scientists across the European continent and North America. These regularly exchange information concerning new knowledge, methods and data, thus providing robust, consistent and comparable information to modellers and policy makers across a sizable part of the northern hemisphere (Bull et al., 2001).

It is important to underline the strong synergy between environmental objectives and foreign policy objectives that has been central to the development and implementation of the LRTAP Convention, as well as to other successful, pioneering examples of integrated environmental strategies, such as the Arctic Monitoring and Assessment Programme, OSPAR or MAP. The shared environmental and health targets described in agreements, and the trust building scientific cooperation efforts resulting from them, have played a significant role in controlling and mitigating environmental risks and in establishing procedures aimed at more effective, transparent and accountable governance. In addition to scientific knowledge from the hard sciences, a key instrument for linking this knowledge to the policy making process has been the development of powerful integrated computer models. These simulate policy scenarios, transport and diffusion pathways, environmental and health exposures and effects, and are based on the best available data compiled through the cooperation under the agreements.

Among the different protocols developed during the implementation of the overarching mandate of the LRTAP Convention, the 1994 Sulphur protocol marked a turning point in the history of policy making on environmental issues. It was the first time that negotiators agreed to use an Integrated Assessment Model (IAM), the RAINS model, as a key tool to elaborate policy scenarios on air pollution abatement strategies. IAMs, such as the RAINS model, have become essential technical tools used to simulate *ex ante* the potential effects of alternative air pollution abatement strategies (Alcamo et al., 1990; Sørensen, 1994; Mayerhofer et al., 2002). Since their results facilitate the negotiations by providing objective data about the possible consequences, the performance of such tools has facilitated the development of more detailed international agreements in terms of the effects, on public health and the environment, of alternative strategies. They provide alternate scenarios for several time horizons in each country, and different emissions abatement hypotheses and their costs. Nonetheless, these IAMs are not able to, and they are not supposed to, take account of the reflexive properties of the systemic issues at stake when defining environmental policies. A scientific tool will not replace a process of decision taking, which by definition ends up being a political agreement. The mutual supportiveness of science and policy will pave the way for policy options to be identified on a sound scientific basis and for negotiations to be shortened by the availability of better information on which to base the final scenario.

The RAINS model became a pioneering model for supporting the decision making process towards an international agreement. In fact, the literature refers to this model as the first one that integrated the use of computer tools into the bargaining process on an international environmental issue. Futhermore, it has been a tool for learning about the dynamic interplay between scientists and policy makers: several seminars, including representatives of both communities, have been organised around its application, in order to improve the communication. Scientists learned more about what the policy makers expected from their assessment, while the policy makers better understood what RAINS allowed them to ask of the scientists. This is reported by Alcamo (Alcamo et al., 1996):

> One lesson learned by the RAINS modelling team was that the interaction between computer modellers and policy-makers was a key ingredient in having RAINS used in negotiations. For example, the modelling team met many times directly with negotiators or their advisors and responded to specific requests for further analysis.

It can be said that the use of the RAINS model corresponded to the application of the so-named regulatory science. This implies using scientific knowledge for regulatory purposes. It diverges from pure science mainly due to the fact that the final values, based on the updated scientific knowledge and 'weighted' by some specific policy requirements, do not result purely from the application of scientific methods. There is an interplay between these scientific methods and the policy purposes to which they are supposed to provide assessment.

The analysis of IA approaches and the related use of IAM has led the discussion towards fundamental questions that shape the building of environmental policies, related to the role that scientific knowledge plays in the decision making process. This is a central point in transdisciplinary research, which discusses – as an important characteristic of the institutional innovation process in building MEAs – the role of scientific expertise to support policy decisions on salient environmental issues.

In the case of the LRTAP Convention, there was an internal debate about which IAM to use as the basis for defining policy scenarios in terms of abatement strategies. Although the overarching goals and the input parameters would be similar, the different computer models could provide a variety of abatement options. These differences could lead to changes in burden sharing among the countries involved in the overall pollution abatement exercise. This is not the place to discuss in detail the technicalities, but this discussion provides an example of the subjective dimension of scientific tools, which must be taken into account when using them for defining political options. Policy makers are not necessarily aware of the existence of alternative tools and the implications of selecting one among the possible set of options. It is necessary that scientists and policy makers, hard sciences and social sciences, open a channel for dialogue in order to consider all these *soft* elements in the definition process of strategies. (Castells and Funtowicz, 1997)

## 11.4 Evolution of Tools and Models in Environmental and Public Health Policy Making

In recent decades, risk assessment methods have evolved significantly in two directions that are relevant to this chapter. First, the vast improvement in the available analytical methods, e.g. Mass Spectroscopy (MS), Gas Chromatography (GC) and the Electron Capture Detection (ECD), have made it possible to identify precisely very small amounts of complex molecules in small samples of the environment and organisms (including people). These improved tools were applied in extensive monitoring exercises around the world. In the 1970s these tools demonstrated the presence of thousands of different artificial chemicals in samples from even the most remote locations in the planet (Jensen et al., 1969). Second, the research into these artificial chemicals and their effects on organisms has benefited from the development of molecular biology, genetics and, of course, computer science. These changes have made it possible to characterise environmental and health effects of toxic substances taking into account multiple endpoints: not only tumours, but also more subtle negative effects such as immune system disruption, endocrine disruption and developmental impacts, all of which can be identified in cellular and molecular processes and in their ecological consequences.

A classic example of the changes in the perception of the pathways and endpoints of toxic chemicals is Rachel Carson's *Silent Spring* (Carson, 1962). Carson included forceful and well documented descriptions of the environmental effects of DDT and other pesticides and the book significantly raised the profile of this issue. Today, the detailed understanding of cellular and genetic processes makes it possible to identify the effects and modes of action of toxic molecules interfering with cellular and developmental pathways with enhanced precision. This means that potential effects on immune or hormonal processes can be inferred from activity at the cellular receptors, such as the *Ah* receptor. Furthermore, there is now ample evidence of chronic effects associated with low doses of POPs and mixtures of POPs. Exposure over long periods, even at relatively low concentrations, can have significant impacts at particularly sensitive periods in the development of neural, hormonal and inmune systems.

Risk assessments carried out in the 1990s (De March et al., 1998) identified the threats associated with volatile, persistent, liposoluble and toxic substances. Because they are persistent, these substances can remain in the environment and in organisms for long periods; because they are volatile and persistent, they can travel long distances in the atmosphere; because they are liposoluble, and thus dissolve better in fats than in water, they accumulate in organisms and concentrate along food webs. Some of these substances, identified as persistent organic pollutants

(POPs) or persistent toxic substances (PTS), can be toxic at very low concentrations, thus potentially interfering with genetic, immune, endocrine and developmental processes.

The mounting evidence about this very relevant component of long range atmospheric transport in the global distribution of these toxic substances, led the countries of the Arctic to identify the UN-ECE/LRTAP Convention as the most effective forum to initiate wider international action to control the risk of harm posed by POPs, to the environment and health.

The recent work on persistent organic pollutants has had several important consequences in the conceptual framework to address the assessment of risk from air pollution and formulate risk reduction strategies. The way POPs move in the environment and in organisms is different from the way pollutants such as sulphur, nitrogen oxides or ozone move or intervene in the environment and in organisms. As a consequence the models of emission, atmospheric transport and effects for POPs are different in several aspects from the IAM models developed for acidification and ground level ozone. POPs are volatile and can thus go back to the atmosphere from the soil, water or vegetation for hours days, weeks or years following deposition if the temperature or concentration conditions change. This means that the relatively simple source–receptor relations that can be used for sulphur, nitrogen and ozone are of no help for POPs and models based on fugacity or non-steady state assumptions have been developed. Due to their tendency to accumulate in lipids and to their mode of action POPs can be a risk to human health and the environment at very low environmental concentrations. Consequently the lengthy pathways and routes that lead to exposure and the resulting dose needs to be included into the monitoring and modelling efforts to identify effective and practicable ways to decrease exposure and environmental levels.

## 11.5 International Action on Persistent Organic Pollutants

By the last decade of the 20th century the LRTAP Convention had developed an elaborate process of consultation and analysis that had been shown to be effective in decreasing the intensity of the flows of transboundary air pollution. The initial steps of the process led to the agreement on POPs technical groups. Policy makers had to develop new approaches to deal, within the LRTAP Convention system, with a problem that was clearly relevant but could not be dealt with using the existing assessment framework and expertise. At the same time, however, experts working on POPs identified the LRTAP Convention as the best way to achieve effective international action on POPs. Interestingly the LRTAP process was not only able to deal with that challenge, and with the help of interested parties, to produce in 1998 an agreement on POPs but, in doing so, it prepared the ground for a global agreement under UNEP and for relevant developments in the European Union legislative framework concerning toxic chemicals.

The risk assessment process that substantiated the case for an internationally binding agreement on POPs under LRTAP Convention was taken up by UNEP.

In 1998, its Governing Council initiated the process to establish a global agreement that launched the construction of an international framework to provide legal, political and technical tools to decrease the risk from certain POPs to health and the environment. The May 1995 meeting of the UNEP Governing Council (GC) adopted Decision 18/32 on POPs, inviting several UN programmes dealing with toxic chemicals, to initiate an expeditious assessment process – Inter-Organization Programme for the Sound Management of Chemicals (IOMC), working with the International Programme on Chemicals Safety (IPCS) and the Intergovernmental Forum on Chemicals Safety (IFCS). It initially began with a short-list of twelve POPs: aldrin, chlordane, DDT, dieldrin, endrin, heptachlor, hexachlorobenzene, mirex, polychlorinated biphenyls, polychlorinated dibenzo-*p*-dioxins, polychlorinated dibenzofurans and toxaphene.

In February 1997 UNEP Governing Council agreed on Decision 19/13C, which called for the establishment of an Intergovernmental Negotiating Committee (INC) to develop a global instrument to address POPs and to initiate a number of immediate actions pertaining to exchanging POPs information, identifying alternatives to POPs, developing guidance for identifying sources, and managing and disposing of wastes and materials containing POPs.

The initial list of substances considered by UNEP represented a compromise aimed at establishing the initial steps of a process within a relatively low budget, and thus included only substances for which scientific research in the preceding decade had produced ample evidence of long range transport and toxicity. Importantly, these substances did not require much further research and were already subject to stringent restrictions and regulations in many countries, thus facilitating compliance in the initial stages. The Convention includes a procedure to add other substances in the future.

This process led by UNEP and the Global Environment Facility (GEF), organised a series of meetings, convening experts and policy makers from all over the world and elaborated a global survey published in 2002 (www.chem.unep.ch/pts/regreports). The Convention on Persistent Organic Pollutants was signed on 22 May of 2001 at the Conference of Plenipotentiaries in Stockholm (www.pop.int). The agreement came into force on the 14 May 2004 with 151 countries signing and 110 ratifying and thus being parties to it (September 2005), and initating a large international transdisciplinary and cooperative effort on this issue.

In summary, the methods developed under the LRTAP Convention have established a solid base for long-term international action on air pollution relying on cooperation of scientists and policy makers through a network of working groups and models. The evolution of measurement and monitoring techniques and the resulting changes in risk assessment and perception has brought POPs to the attention of scientists, policy-makers and society at large. These then, relied on the work of the LRTAP Convention, which addressed the problem effectively and had an instrumental role in developing a global approach to POPs. By doing so, the LRTAP Convention enlarged its scope and conceptual framework. In addition to the real success of many of the protocols born under the LRTAP umbrella agreement, its main achievement at a more structural level has been to stimulate and drive an interactive process between the scientific community and the political community, translating into policy making language the results and simulations emanating from scientific models and creating a stable and well-coordinated research network on key air pollution issues. This dynamic learning process of interaction among policy makers, scientists and stakeholders concerned by the problems – government representatives, industry, NGOs, research institutes, academy – has provided an operative model to address complex environmental issues of a transboundary nature in an effective and collaborative way. This has increased the possibility of mitigating impacts and minimising risks associated with anthropocentric activities on the environment and public health (Guardans, 2002).

## 11.6 Current Challenges

Undoubtedly, the successful implementation of international environmental agreements has resulted in better trends in some specific areas of transboundary environmental pollution. The experience has led to an increased acceptance of the need for international collaboration to avoid unsustainable trends and damage to the environment and public health. It has shown the need to improve dialogue and develop trust between stakeholders and to find common languages to discuss complex issues and to define policy scenarios in effective terms.

Nevertheless, this success story might read differently when looked at from the perspective of stakeholders who might feel they are being requested to contribute to solving a problem and therefore to sharing the costs of implementation: be they real or simply the costs of not being allowed to follow certain development paths that they would have followed had they not been obliged to be part of these international environmental agreements. Transboundary air pollution is a global issue to which different countries have contributed in substantially different proportions across the world. The emissions per capita of air pollutants over time and by regions have been very unevenly distributed. Indeed, industrialised countries accelerated the air pollution process and launched environmental policies once their industrialisation process was already consolidated. Being the richest countries in the world, they were able to shift their priorities and devote part of their economic resources to mitigating the damage they had been inflicting on the environment for decades, in order to address the increased concern of their populations. The situation faced by developing countries now cannot be compared to the one faced by pioneering countries in the industrialisation process. New frameworks are required to address their needs while preserving their development potential and not putting global sustainability at stake.

The methods and experience gathered through these international cooperation processes in heavily industrialised and polluted countries can help, to some extent, avoid the repetition of costly and avoidable mistakes, in the contemporary process of industrialisation of other areas of the world.

Models can tell us about an environmental problem and the alternative scenarios for policy strategies. They do not shed light on what all the political options for the countries involved should be. A model is a powerful policy tool, but it cannot be better than the data provided or the analytical structure embedded in it. The weight given to the different parameters in terms of policy options can to some extent reflect priorities of those designing the policy tool. More often than not, these tools have been developed in countries with well-advanced knowledge about technology and environmental problems. Unless the parameters for these modelling tools are discussed in a transparent process among the stakeholders in the different regions and social sectors concerned by the final policy scenarios, and there is a transparent consultation process for the selection of values in the model, the resulting set of scenarios could embed a set of values that might not reflect the priorities of developing countries (Castells, 1999).

In the future, the process of constructing international policy tools to protect the environment and public health will have to develop further, in line with the integrated models described here, to identify common ambition levels and implementation strategies that take into account the very different baselines or starting points reinforcing the *ex-ante* policy process of consultations among the different countries and sectors involved.

Several features of the POPs issue – such as the multiplicity of sources, including products, their persistence and long range transport and the multiple effects end points – make it clear that it is urgent to enhance the transdisciplinarity of the approach to identify the main aspects of the problem and find effective ways to deal with it from local to global. This leads to a new generation of integrated models and other innovations that will have to consider the factors that modulate exposure (Sexton et al., 1995; UNEP/ILO/WHO, 2000), the dynamic filter that intervenes between environmental concentrations and the body burdens in wildlife and people.

## 11.7 Recommendations

Many substances can be harmful at high concentrations but are innocuous or required at lower concentrations (e.g. metals, sulphur etc.). Radiation and persistent organic pollutants are potentially harmful at all doses. The probability of negative effects decreases with the intensity of the aggression but there is no lower threshold of effect. Below some level the response is not proportional to the dose, it is stochastic, the probability of effects can be established but no deterministic prediction is possible about precise effects on individuals, locations or times. These differences are obviously relevant for the design of models, monitoring strategies and policy instruments and should be taken into account during the interactive process of building integrated knowledge for policy making.

Practically, in terms of political and social debate, there has always been a tension between the identification of pollution problems and the resulting risk to environment and health, and the implementation of effective action at scales relevant to confronting the problem. In particular in the Atlantic societies of the northern hemisphere, over the last century, the means of production and the sectors of activity in society that produce wealth were lacking a dialogue framework for building consensus on how to prevent risks without jeopardising the differing interests of the parties. Democratic procedures are not equally developed and fully efficient among the parties. A prosperous democratic society has, in theory, the means to prevent risk of predictable harm to the environment and public health but the practical balance of power is favourable to short term benefits. Consequently, in the domain of environmental and public health legislation there is a time lag between the identification of a problem with enough factual evidence and the implementation of effective action that is in the range of 20–30 years.

It has been shown that a plural and political process to develop effective strategies to avoid harm to environment and health is possible but heavy and slow because it has to harmonise the communication between activity sectors, social groups, geographical regions and multiple scales of action, from a household to the United Nations.

The LRTAP Convention has played a central role in creating awareness and concern regarding large scale air pollution and the necessity for fair and effective policies to handle the problem, involving multiple actors and multiple scales of action. Factors that made this agreement particularly effective include:

- *Involving relevant stakeholders*: country representatives – including both polluters and polluted-, scientists, policy makers, international organisations, academic institutions, NGOs.
- *Creating a 'transdisciplinary framework for dialogue'*: The establishment of mandatory reporting obligations, and the engagement by parties to provide the best available information made it possible to develop and use Integrated Assessment Models, adopted for the first time in international law as a means to design policy scenarios on which to base political decisions.
- *Institutional innovation*: The process by which ambition levels and scenarios where specified and agreed involved the settling of a micro–macro coordination, involving national resources; local monitoring and collecting data networks; international resources; mediation of international–regional organisations such as the UN, and the European Commission (Clean Air for Europe).
- *Bringing together scientists and policy makers* and translating complex issues into real world policy scenarios through stable networks of working groups and mandatory reporting procedures.
- *Ex-ante estimation of the burden sharing*: The use of IAMs enabled policy makers to take decisions while being aware of estimated costs of implementing the decisions they would be endorsing. Previous experiences had resulted in policy commitments that were unbearable in practice by the countries, but the lack of tools and scientific assessment prior to the policy commitment impeded the estimations being known prior to the implementation phase.

The history of the LRTAP Convention and the definition of some of its protocols shows that the interaction of scientist and policy makers in the framework of a stable long-term process can result in win–win scenarios in terms of effective environmental policy being designed, agreed and successfully implemented in the most effective and efficient way. For such a pattern to be reproduced, the participatory process involving stakeholders and the dialogue between science and policy seems to be a prerequisite, which does not ensure success but paves the way for it.

The views expressed in this article are those of the authors in their personal capacity and do not necessarily reflect those of the United Nations.

## References


Alcamo, J., Shaw, R., and Hordijk, L. (eds): 1990, *The RAINS Model of Acidification. Science and Strategies in Europe*, Kluwer Academic Publishers, Dordrecht, XIX+402pp.

Alcamo, J., Kreileman, E., and Leemans, R.: 1996, Global models meet global policy, *Global Environ Change* 6(4), 255–259

Bölin, B.: 1972, *Sweden's case study for the United Nations Conference on the Human Environment: Air Pollution Across National Boundaries,* Norstadt and Sons, Stockholm.

Bull, K.R., Achermann, B., Bashkin, V., Chrast, R., Fenech, G., Forsius, M., Gregor H.-D., Guardans, R., Haussmann, T., Hayes, F., Hettelingh, J.-P., Johannessen, T., Kryzanowski, M., Kucera, V., Kvaeven, B., Lorenz, M., Lundin, L., Mills, G., Posch, M., Skjelkvåle, B.L., and Ulstein, M.J.: 2001, Coordinated effects monitoring and modelling for developing and supporting international air pollution control agreements, *Water Air Soil Pollut* 130, 119–130.

Carson, R.: 1962, *Silent Spring*, Mariner Books, Houghton Mifflin, Boston.

Castells, N.: 1999, *International Environmental Agreements. Institutional Innovation in European Transboundary Air Pollution Policies*, Ph.D. Thesis, Free University Amsterdam, 287pp. Retrieved June 11, 2007, from http://mypage.bluewin.ch/N_Castells/LRTAP_CastellsThesis1999.pdf.

Castells, N. and Funtowicz, S.: 1997, The use of scientific inputs for environmental policy-making: The RAINS model and the sulphur protocols. *Int J Environ Pollut* 7(3), 512–525.

Chamberlain, A.C.: 1953, *Aspects of Travel and Deposition of Aerosol and Vapour Clouds*. AERE Report H.P.l1261. Atomic Energy Research Establishment, Harwell, Berkshire

De March, B.G.E., de Wit, C.A., and Muir, D.C.G.: 1998, Persistent organic pollutants. In: *AMAP Assessment report*: *Arctic Pollution Issues*, Oslo, Norway, pp. 183–372.

De Vries, W. and Posch, M.: 2003, Critical levels and critical loads as a tool for air quality management. In: C.N. Hewitt and A.V.Jackson, *Handbook of Atmospheric Science*, *Principles and Applications*, Blackwell Publishing, London, pp. 562–602.

Eliassen, A. and Saltbones, J.: 1975, Decay and transformation rates for S02 as estimated from emission data, trajectories and measured air concentrations, *Atmos Environ* 9, 425–429.

Guardans, R.: 2002, Estimation of climate change influence on the sensitivity of trees in Europe to air pollution concentrations, *Environ Sci Pol* 5, 319–333.

Jensen, S., Johnels, A.G., Olsson, M., and Otterlind, G.: 1969, DDT and PCB in marine animals from Swedish waters, *Nature* 224, 247–250.

Lindel, B. and Dunster, H.J.: 1998, *International Commission on Radiological Protection: History, Policies and Procedures*, Swedish Radiation Protection Institute (SSI), SE-171 16 Stockholm, Sweden.

Martinez-Cortizas, A., Pontevedra-Pombal, X., Munos, J.C.N., and Garcia-Rodeja, E.: 1997, Four thousand years of atmospheric Pb, Cd, and Zn deposition recorded by the ombrotrophic peat bog of Penido Vello (northwestern Spain), *Water, Air, Soil Pollut* 100, 387–403.

Mayerhofer, P., de Vries, B., den Elzen, M., van Vuuren, D., Onigkeit, J., Posch, M., and Guardans, R.: 2002, Long-term, consistent scenarios of emissions, deposition, and climate change in Europe, *Environ Sci Pol* 5, 273–305.

Newell, E.: 1997, Atmospheric Pollution and the British Copper Industry, 1690–1920, *Tech Cult* 38, 655–689.

Nijkamp, P. and Castells, N.: 2001, Transboundary environmental problems in the European Union: Lessons from air pollution policies, *J Environ Law Pol* 4, 501–517.

Pasquill, F.: 1961, The estimation of the dispersion of wind-borne materials, *Met Mag* 90, 33–49.
Sexton, K., Callahan, M.A., Bryan, E.F., Saint, C.G. and Wood, W.P.: 1995, Informed decisions about protecting and promoting public health: rationale for a national human exposure assessment survey, *J Expo Anal Environ Epidemiol* 5(3), 233–256.
Sørensen, L.: 1994, Environmental Planning under Uncertainty, Ph.D. Thesis, Lyngby: Technical University of Denmark.
United Nations Conference on Trade and Development: 2006, *Trade and Environment Review*, Retrieved June 11, 2007, from www.unctad.org/en/docs/ditcted200512_en.pdf.
United Nations Environmental Programme/International Labour Organization/World Health Organization: 2000, *Environmental Health Criteria 214 Human Exposure Assessment*, Retrieved June 11, 2007, from http://www.inchem.org/documents/ehc/ehc/ehc214.htm.

# Chapter 12
# Climate Protection vs. Economic Growth as a False Trade Off: Restructuring Global Warming Mitigation

**Hermann Held and Ottmar Edenhofer**

**Abstract** The project asks how to derive and then promote a balanced portfolio of climate policies and technologies currently under discussion as efficient means to avoid ('mitigate'), or at least limit, global warming. Thereby, the agenda responds to the emerging consensus within the climate research community that anthropogenic greenhouse gas emissions – such as generated during combustion – would induce global warming with serious potential consequences. The ongoing climate debate about how much global warming should be mitigated serves as a starting point. While 'climate environmentalists' opt for strict emission reductions, influential economists proposed – at least at the beginning of the project – that the requested reductions would severely hamper the world economy. The 'metamethod' of the project disentangles the deterministic and normative arguments of the debate. Within that setting, the methodology of the project displays two main characteristics. Firstly, robust deterministic knowledge about the climate system and the energy technology sector is captured by mechanistic models. This 'deterministic branch' would allow us to represent the effects of greenhouse gas emissions on global warming as well as the effects of investment decisions on the competitive advantage of renewable sources over greenhouse gas emitting technologies. Secondly, the project distills the present line of normative settings involved in the climate debate – what impacts of global warming on the one hand, and of strict emission reduction targets on the other hand, are acceptable. A further normative issue is how to decide, under present-day uncertainties that modulate our knowledge, about the causal links from potential political actions to their impacts. The approach of the project is to search for climate policies that would observe the minimum requests of each of the two major disputing 'camps' and to thereby maximise the chance for societal consensus. The minimum request of the 'environmentalists' is to guarantee that global warming shall not transgress 2°C. The minimum request of economists is that welfare loss due to climate protection should be somewhat below 1%. As a primary result, the project has qualified the systems dynamics, enabling the identification of

✉ H. Held,
Potsdam Institute for Climate Impact Research (PIK), Potsdam, Germany
e-mail: held@pik-potsdam.de

G. Hirsch Hadorn et al. (eds.), *Handbook of Transdisciplinary Research*,

investment paths that are likely to observe the minimum requests of both parties. Key transdisciplinary challenges of the ongoing project are as follows: distilling the major epistemic (knowledge on systems dynamics) and major normative arguments, and alleged disagreements, as the two are intimately entangled within economic theory; finally, keeping track of assumptions and desires of major players to ensure that the project's stylised solutions will in fact catalyse a societal consensus.

**Keywords:** Climate change · Global warming mitigation · Economic growth · Costs of mitigation · Consensus driven approach

## 12.1 Background

The project attempts to tackle the alleged trade off between global warming mitigation and economic growth that has continually been proclaimed in the scientific as well as the public debate over the last years. While environmental NGOs tend to focus on the former aspect, most large energy suppliers are more concerned with the latter. Furthermore, both points of view are strongly supported, if not shaped, by members of the climate and economic research communities, while the engineering community is split depending on whether they work on renewable or non-renewable sources. Due to the importance of fossil fuels in discussions of climate change, geologists also have a strong impact on the debate about the costs of climate protection, and are themselves divided into two camps. One camp predicts that the extraction of conventional fossil fuels will reach a maximum within a decade; the other camp claims that fossil fuels will remain cheap and abundant for the rest of this century. The latter view is strongly supported by the majority of economists who argue that price signals will induce substitution processes so that potential bottlenecks in the extraction of conventional oil can be overcome. If fossil fuels are relatively cheap and abundant, then the costs of mitigation are relatively high because any limits set on $CO_2$ emissions effectively devalue assets invested in the fossil fuel sector. The specific role that researchers play within the debate on climate policies is characterised by a delicate entanglement of target knowledge and systems knowledge arguments. Hence, a significant proportion of resources within our project is devoted to clarifying not only the validity, but initially, the category of those arguments.

Furthermore, our assessment requires the integration of various disciplinary paradigms: while the natural sciences strive to extract the key dynamic mechanisms of the system under study, and attempt to be as neutral as possible regarding normative settings, economic growth theory traditionally proclaims to illuminate which future paths (in our case, which investment paths) should be chosen to maximise welfare. As we will outline in the following sub-section, this seeming clash of an 'objective' versus a 'normative' discipline is easier to tackle than is achieving a balanced view of the uncertainties involved in the individual disciplines together with the further complications caused by looking at issues from the engineering perspective. We have to clearly state that any attempt to comprehensively compare the validity of results derived by various disciplines goes beyond the scope of any individual project and rather represents a long-term enterprise. However, we strive to make the assumptions upon which our assessment is based as explicit as possible, while at the same time trying to develop schemes that either bypass pertinent uncertainties or shift the burden of reducing those uncertainties from the research community to powerful market based institutions. We emphasise that these market based institutions have to be embedded in a broader regulatory framework enabling not only economic but also social and ecological sustainability.

In the course of our project we have found it necessary to massively invest in several of the disciplines involved before we could use their results for integration: each community had defined its own priorities, which were not optimised for later integration.

Finally, we attempt to integrate social attitudes by striving for new solutions that satisfy the minimum claims of the mitigation side while leaving global welfare roughly untouched, thereby satisfying requirements for a conservative welfare function as proposed by opponents of strong mitigation measures. We represent the minimum mitigation conditions by the requirement that global warming remains within the 'climate window' defined in WBGU (1997): the increase of global mean surface temperature shall not transgress 2°C (compared to its pre-industrial value) and the rate of change shall stay below 0.2°C per decade if large-scale ecosystem or climate system disruptions are to be avoided. The significance of this climate window has been renewed in WBGU (2003). We frequently check the potential of our iteratively substantiated solutions for consensus by informally contacting various opposing stakeholders.

On 16th February 2005, the Kyoto protocol, that for the first time implements an emission cap for carbon dioxide, was put in operation. However, the emission reduction achieved by this framework represents only a small fraction – markedly below 10% – of the reductions necessary to achieve the climate window. Therefore, at present, a 'post-Kyoto regime' is under heavy negotiation. Our project shall contribute to the development of a powerful 'post-Kyoto regime' resulting in significant, yet politically realistic mitigation measures.

## 12.2 The Climate Debate

The overwhelming majority of climate scientists expect that the Earth's mean surface temperature will increase in response to anthropogenic emissions of greenhouse gases in ways unprecedented over tens of thousands to millions of years. More specifically, in 2001 the Intergovernmental Panel on Climatic Change (IPCC) established as a scientific consensus that the temperature would rise between 1.5 and 6.2°C (Houghton et al., 2001) by the year 2100. This rather large temperature spread is due in roughly equal parts to the considerable uncertainty in intrinsic properties of the climate system and to the span of future greenhouse gas emission scenarios. The latter implies that climate policies would in fact have a strong influence on the extent of global warming. In fact, several global players such as governmental bodies,

NGOs and energy suppliers (European Commission, 2000; Browne, 2004) have stated that global warming must be constrained to an upper limit of 2°C to avoid potential massive, or even irreversible, impacts. Although an increasing number of statements have been made in the past few years asking that significant measures be taken to mitigate global warming, a consensus on which options may facilitate an ambitious climate goal has not been established. Typically, individual options such as massive implementation of solar power or life-style changes are considered as either economically inefficient (Nordhaus and Boyer, 2000; Lomborg, 2004), impractical in engineering terms on a large scale, or politically implausible (Lomborg, 2004). At the same time, supporters of these measures disseminate numbers suggesting an enormous potential for mitigating global warming.

Our project aims at clarifying these somewhat contradictory and non-integrated assessments of mitigation options. We expect that an intelligent mix of options rather than one optimal solution will facilitate mitigation (of massive global warming) and analyse the following major options accordingly:

- Transformation of the worldwide energy system to renewable energy sources
- Enhancement of energy efficiency
- Carbon capture and sequestration (CCS)

We aim to determine for the economically optimal mix of investment streams for these options. The optimisation is performed under the boundary condition that the WBGU climate window as defined above (Section 12.1) is observed. For our optimisation, *business as usual* (in particular usage of fossil fuels) is treated as a fourth option. While the first two options have some tradition within the political debate on a decadal time-scale, carbon capture and sequestration is only now entering the public discourse. The latter technology suggests leaving the infrastructure based on fossil fuels intact, while extracting carbon dioxide from the stream of effluent gases emitted from large point sources such as power stations, and compressing, transporting and injecting it into geological formations (Edenhofer et al., 2005a; Held et al., 2006).

Such an assessment is typically called *integrated* as several options are analysed in a joint manner, and several disciplines are involved: climate science (for the impact of emissions on global mean temperature), engineering and economics.

The procedure outlined so far would qualify as an integrated assessment, according to the state of the art. However, we would like to take seriously the fact that major deterministic properties involved in the analysis (such as the climate dynamics) are uncertain, and ask for policy recommendations that are *optimal under uncertainty*. In a formalised setting, the latter is the topic of decision theory. Furthermore, as various mitigation options are characterised by significant potential for undesirable side-effects typically not represented in economic terms, we plan to nest our economic investigations into a qualitative risk assessment using the terminology of risk classes (WBGU, 1998).

Finally, we would like to extend the analysis further towards the actual needs of climate policy. First, our recommendations shall take the view of present-day global actors (such as the EU, the US, India, China, NGOs, corporations) rather than the view of a single global planner (such as a fictitious world government). Second, the social optima derived from such economic analyses still do not tell much about how policies shall be implemented in the existing economic system, and in particular, which financial instruments would induce the desired development at the private enterprise level. Our project aims at developing such financial instruments that shall bridge the gap between abstract policy recommendations and the room for maneuvering that actual policy makers have.

For these reasons, our project involves representatives from climate science, economic growth theory, systems theory and statistics. We import engineering expertise from external partners. Research progress is informally communicated to the German Federal Ministry of Economy, to the Ministry of the Environment, to German energy suppliers and to the NGO Greenpeace. We attempt to shape the scientific information such that a societal consensus on a mitigation policy can be catalysed on rational grounds, and attempt to distill the deterministic from the normative issues involved.

## 12.3 The First Iteration of an Integrated Assessment

How do we position ourselves as interdisciplinary researchers within the fields spanned by the climate–economy debate? As a key motivation for our project we comply with the above mentioned statements by the climate community on anthropogenic emission induced global warming. While we accept their systems knowledge on the expected climate dynamics in the sense of a most plausible range of behaviours, we note that the last IPCC report suppresses major sources of uncertainty. We will come back to this issue in the following sub-section. Furthermore, the goal of limiting global warming to 2°C arises from a composite of both systems and target knowledge: the former refers to the impacts of global warming while the latter states that those impacts ought to be avoided. Although the relevant systems knowledge is less certain than the link between emissions and global warming, we still regard the accumulated evidence as sufficient to continue on that basis. In fact, the 2°C target has been further supported in a symposium held in 2004 (European Climate Forum, 2004). Finally, we find the normative position that aims at avoiding those impacts to be highly plausible.

For these reasons, to design a first iteration of assessment we restrict our exploration to those paths that are compatible with the 2°C climate window, noting that our analysis may adjust to lower or higher targets once more comprehensive knowledge on impacts of global warming are produced in the future. We even comply with the stricter climate window that also restricts the *rate* of warming to 0.2°C per decade (WBGU, 1997). By observing such a strict climate window we expect to win over the stakeholders in the climate and environmental communities to the stylised compromise we attempt to derive.

Bruckner et al. (1999) have argued that on the basis of the precautionary principle one must simply derive the funnel of emission paths that fall within the climate window (as they did), and that analysing the costs of corresponding climate policies should be a secondary priority. However, commercial use of coal, oil and gas since the end of the 18th century has tied industrial society so closely to increases in anthropogenic greenhouse gas emissions, that wealth and carbon dioxide emissions due to the use of fossil energy have become almost synonymous terms. Therefore, anyone demanding a reduction of greenhouse gas emissions in the name of the precautionary principle with regards to the climate system provokes fears about whether, and how soon, a zero-emissions capitalist economy is possible. This is exactly the reasoning used by some economists in the past, calling upon an 'economic precautionary principle' and warning against 'excessive climate protection' (Nordhaus and Boyer, 2000).

In order to also win over the stakeholders within the economic community for our stylised compromise we adopt their main tool of evaluation: cost-benefit analysis (CBA). Before we make explicit the way in which we comply with their tradition, and the way in which we deviate, we would like to briefly sketch CBA in its standard form.

As Fig. 12.1 shows, three information bases are fed into the CBA: a first module describes the dynamics of the climate system, a second module represents the dynamics of the economy and the third module includes instruments such as taxes and certificates used to achieve control paths. The economy influences the climate system via emissions, which in turn, through an increase in global mean temperature, induces damage to the economic system. The aim is now to determine future emissions in a way that yields an optimum balance between costs of $CO_2$ avoidance and of climate damages. The amount of emission reduction is therefore determined simultaneously with the policy instruments (taxes, certificates) that should be applied for achieving the calculated emission target. CBA attempts to integrate complex knowledge (meaning here: 'not easy to see through at the negotiation table') on climate and, economic dynamics by a method of weighting goods –

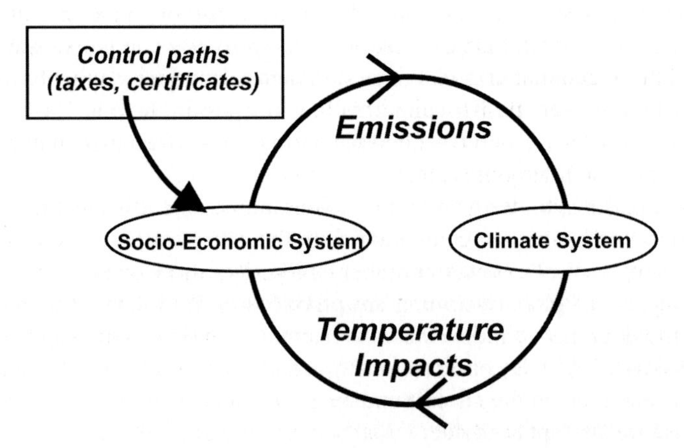


**Fig. 12.1** Cost-benefit analysis (CBA) as the standard tool within environmental economics that is supposed to deliver optimal control paths over the coupled economic and climate dynamics

based on an intergenerational order of preferences, along with ancillary dynamic conditions – so that a consistent 'social optimum' is derived. This optimum, however, can only be determined with the help of an evaluation function – it is usually assumed that the utility function of present and future generations can be derived exclusively from per-capita consumption. Thus, standard CBA presumes the predominant liberal economic paradigm preference order: absolute (monetary) calculability. This ensures CBA a high degree of compliance with the economic model dynamics it tries to incorporate, and has eased its access to the political level so far.

When asking about integration of disciplinary paradigms one should note that CBA accomplishes such integration by a subtle cut in the way it represents the socio-economic sphere. At the level of aggregation that is required by the problem-setting, it defines some economic system dynamics that are the emergent, quasi non-normative behaviours of a system made up from multiple actors, each potentially optimising according to their individual welfare functions. However, by interpreting historical developments on decadal to secular scales, economic growth theory extracts such an economic deterministic kernel (left ellipse in Fig. 12.1) for the purposes of CBA. That kernel can then be coupled to further deterministic dynamics such as the climate dynamics (right ellipse in Fig. 12.1). The joint systems dynamics is then treated as a boundary condition for intertemporal optimisation via control paths (rectangle in Fig. 12.1) according to the economic welfare function.

We follow this standard approach in the sense that we adopt the 'predominant liberal economic' welfare function; one that does not respect non-monetary categories such as moral or aesthetic qualities. That way we maximise support from the present day economics community. However, in contrast to our analysis, standard CBA does not utilise the concept of a climate window. Instead, it monetarises global warming damages and optimises without being restricted by a climate window. We deviate from standard applications of CBA because of our impression that the necessary systems knowledge (coupled economic/climate) as to future impacts of global warming beyond the climate window, is inadequate for such an analysis at present. Also, the auxiliary assumptions in climate models (in particular for the older, more established ones) are generally oriented towards *current* climatology; we therefore expect that they, in general, underestimate the potential for abrupt climate changes on the continental or even global scale (Schellnhuber and Held, 2002) when operated beyond the climate window. Protagonists applying CBA often use damage functions based on even smoother input–output relations and hence overestimate the relative costs of mitigation. By restricting our optimisation to the climate window ('cost-effectiveness analysis') and conservatively (from the point of view of mitigation sceptics) disregarding warming damage costs, we bypass those conceptual traps.

We involved a further deviation from standard CBA of the climate problem that refers to the economic systems knowledge: in standard analyses technological change is modelled 'exogenously', i.e. by a fixed rate (typically of 1% per year). That way, newer technologies (renewable sources) will by definition always lag behind their well-established counterparts (fossil fuels) in terms of cost-efficiency, no matter how much one would invest in the former. In our model we cured this apparent defect by explicitly 'endogenising' technological change to be driven by

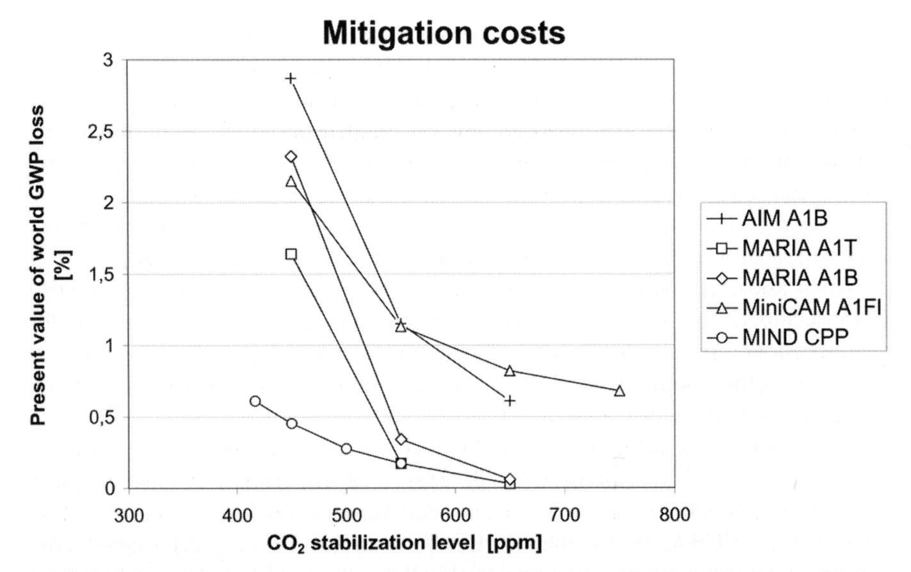


**Fig. 12.2** Discounted economic costs of climate protection until 2100 for a climate protection path (CPP) in the model MIND (developed in this project), compared to other macroeconomic models (such as AIM, MARIA, MiniCAM for emission scenarios A1B, A1T, A1Fl that observe, among other items, stabilisation levels as indicated; see Edenhofer et al., 2004; Morita et al., 2000). A stabilisation level of 400 ppm – as against higher levels – allows the system to stay within the climate window, assuming the most plausible of the uncertain climate system parameters. Our model suggests drastically lower costs of significant mitigation than other models do, due to the endogenisation of technological change

investment decisions. First results (Edenhofer et al., 2005a) show that the conclusions of traditional CBA need to be revised quite fundamentally: induced technological change drastically reduces the costs of climate protection (Fig. 12.2) to a level that seems acceptable within economic circles. Subsequently, the climate window requirement can be met by a change in investment strategy. The results of this first iteration indicate that the often proclaimed trade off between global warming mitigation and economic welfare is a mere construct of deficient modelling and that, quite to the contrary, policies are possible that would satisfy both mitigation protagonists as well as sceptics.

## 12.4 Towards a Practical and Robust Assessment

In the following we would like to outline why these results must be further substantiated as well as extended, and which preliminary, although not yet integrated, results we have obtained along those lines so far.

- While we eliminated a major source of uncertainty in climate systems dynamics by using the climate-window approach, significant further uncertainties were not tackled in our first iteration. Among those is the sensitivity of the climate system to greenhouse gases. Not only do different state-of-the-art climate models predict diverging warming trends when driven by the identical emission path (Section 12.2), but any climate model also contains several uncertain tuning parameters. The effect of the latter was not tackled in the last IPCC report. To close that gap, several research groups undertook *Bayesian learning* on *climate sensitivity*. Bayesian learning represents a branch of statistics that synthesises a posterior belief on an uncertain entity out of subjective prior knowledge and new objective information. Within the paradigm of standard decision theory, the so derived posterior belief represents an optimised basis for taking rational decisions (Berger, 1985). Climate sensitivity encapsulates most of the uncertainty in climate predictions that are conditioned on a prescribed path of carbon dioxide concentration. It is defined as temperature response after increasing the atmospheric content of carbon dioxide from 280–560 ppm. Bayesian learning on climate sensitivity then resulted in roughly twice as large an upper limit for climate sensitivity per model (Stainforth et al., 2005) than that stated in the IPCC report for the whole group of models. This result would be important since it implies great difficulty assessing climate policies as one can no longer define a 'safe' upper limit of emissions. We performed such a Bayesian analysis ourselves, and included paleoclimate information from the period of the last glacial maximum, which displays a much better signal-to-noise ratio than the weak warming signal of the 20th century. Preliminary results (Schneider von Deimling et al., 2006) show that the range for climate sensitivity published in the last IPCC report can be re-established, this time on much more objective grounds. Uncertainty about climate sensitivity could be further reduced if fluctuation properties of the climate system are also taken into account (Held and Kleinen, 2004).

- On the economic side, in future work major uncertain parameters will be subjected to Bayesian learning, replacing the current best guess values. As a final goal, stylised decisions on optimal paths under joint climate/economic uncertainty can be derived. In this context, another innovation appeared as necessary: Bayesian methods imply a somewhat automated weighting of potential futures that we regard as helpful, since a purely subjective assessment of the 'high-dimensional space of futures' would be too demanding. However, Bayesian methods require the inclusion of subjective, prior, expert knowledge on the uncertain entities. As the modelling effort itself constitutes a form of expert knowledge, we basically consider that drawback as acceptable. Nevertheless, the Bayesian approach will meet its limitations if an expert has few a priori preferences for concurring parameter constellations due to a lack of experience. In this case, the Bayesian method subtly suggests to the analyst, apparently firm information, which, however, underestimates and plays down (from the point of view of a pessimist) the probability of generically critical regimes. Modern approaches to imprecise probabilities (Kriegler and Held, 2005; Held and Schneider von Deimling, 2006) offer a valuable compromise between the Bayesian approach and overly conservative, hence mostly non-informative interval methods.
- An Integrated Assessment that aims to take into account the whole portfolio of avoidance options, including carbon capturing and sequestration (CCS), requires

taking one more step back. We consider it impossible that any modelling effort will ever be able to take into account all possible impacts of mitigation measures. The diverse options should therefore be pre-evaluated by applying qualitative metacriteria. Environmental chemistry has developed an interesting proposal for avoiding the problem of a highly complex chain of impacts (Scheringer, 1996): one proposal for the pre-evaluation of environmental chemicals is to draw upon their spatial range (which can be measured rather easily) as a metacriterion. The existing minimal statistical-deterministic knowledge – that is, that substances diffuse at certain rates – is optimally used and transformed into practically relevant knowledge. First considerations indicate that this concept could constitute an important building block for a semi-qualitative analysis of CCS options. For a complementary approach that attempts to delegate risk assessment to the captial market, see Edenhofer et al. (2005b).

- Regionalisation should be conducted in a way that identifies potential conflicts of interest and possible compromises between important emitters: main actors like Europe, China, India, USA, corporations, and NGOs should be included in semi-qualitative modelling. First attempts to systematically incorporate issues of distribution and justice systematically into CBA were undertaken by Uzawa (1995, 2003). In this context it will become necessary to assess whether standard economic modelling requires fundamental changes in order to maintain coherence between differing orders of preferences in the optimisation of CBA, within the optimisation and preference order based economic modules. A main issue will be to explicate and relate historical and future orders of preference. Practically, such preference orders will have to be established iteratively because actors are likely to become aware of their own preferences only in the context of calculated 'optima' with an asserted value.
- Finally, one cannot be content with taking all possible uncertainties into account; one also has to aim at reducing them, at least in part. Here 'one' refers to the view of one of the actors named above; for example, Europe as represented by the EU Commission. This reduction can be achieved by granting research funds, but we would also like to urge that market forces, the most powerful actors in the current world theatre, be considered in integrated assessment efforts: Edenhofer et al. (2004) make proposals to transfer parts of the – regionally manageable – risks onto the beneficiary of a certain energy technology. By so doing, liability does not remain only with the state, but is shared substantially by the company that is actually making profits. Market instruments can contribute to taking into account the public's desire for environmental safety in manifold ways – maybe in a more democratic and efficient way than an indispensable environmental agency alone could ever achieve.

## 12.5 Implementation

Our project has been defined as an enterprise of analysis rather than implementation. Moreover, it would be premature to report to what extent our project has been responsible for the implementation of climate policy, which is in a very early state. Nevertheless, environmental diplomats, policy makers and NGOs are very aware of the fact that the future of the Kyoto process depends on setting up a reasonable accompanying architecture ('KyotoPlus'). Members of our project are involved in these discussions and we can report the kind of debates our project is already influencing:

- Together with other research institutes like FEEM (Fondazion Eni Enrico Mattei, Milan), DIW (Deutsches Institut für Wirtschaftsforschung, Berlin) and the Tyndall Centre, Norwich, we have founded the Innovation Modelling Comparison Project (IMCP). Within this project we launched a modelling inter-comparison exercise to improve the understanding of how technological change determines mitigation costs and mitigation strategies in different modelling frameworks. It turns out that induced technological change reduces mitigation costs substantially (Edenhofer et al., 2006; Giles, 2006). We are now exploring the implications of these results for a 'KyotoPlus Architecture'. The German Environmental Protection Authority and the EU Commission are at present supporting follow-up projects assessing new ideas for a climate policy regime beyond 2012. The proposition that, costs of climate protection are overestimated by models without taking into account technological change has attracted the interest of different stakeholders, such as the Duma, the CEO of BP, large electricity suppliers in Germany, NGOs and journalists.
- Members of our project have been involved in high level debates about the regulatory framework of CCS and about the promotion of renewable energies. In our project we are now starting an in-depth analysis of climate and energy policy instruments.
- The research agenda of a major proportion of the climate research community will reflect an intensifying of attempts to assimilate paleoclimate information, as we do (PMIP, 2006). This could further substantiate the desirability of avoiding most risky engineering-type mitigation options (Keith, 2000).
- We promote imprecise probabilities within decision theoretical analyses of the climate problem; this will result in improved climate policy recommendations, as the current downplay of extreme events will be replaced by more adequate representations of uncertainty.

## 12.6 Recommendations

This chapter sketched our transdisciplinary project on the integrated assessment of global warming mitigation options. The project began by observing that the debate about climate policies seemed to be somewhat stuck between opposing views, which were substantiated by a misleading admixture of normative settings and systems science arguments, with the latter often implemented below disciplinary standards. For those reasons the pool of potential climate protection policy paths to which most opposing views could subscribe was yet to be scanned in an appropriate way. We showed that this could be achieved by a series of improvements at the disciplinary level, combined with some serious disentangling of systems versus normative arguments. After two years, we then identified policy paths with the desired property. At present, we modify these paths to make them robust under uncertainty and potential side-effects of mitigation measures. In follow-up projects we will incorporate a frequently observed demand by stakeholders into our analysis and replace the one single fictitious global optimiser perspective with that of interdependent actual global actors. We regard this as a way to address, in a stylised manner, what incentive schemes could attract the main emitters to join a KyotoPlus policy regime observing binding warming mitigation targets.

- One could ask in what sense is our approach transdisciplinary? The answer lies in the very fact that we develop stylised solutions to the climate problem that have a fair chance of catalysing political decisions in favor of the very decision paths we derive. Our impression is that such catalysis is not unlikely for the following reasons: we integrate paradigms of opposing actors in such a way that any of those actors views the boundary conditions of the field in which he or she is a stakeholder (economy or climate) as being represented in a scientifically sound way. At the beginning of our project, the two authors, in some sense represented the first iteration of a stakeholder dialogue: one of us was in touch with the – often implicit – normative background of climate scientists, while the other understood the normative assumptions underlying standard economic theory and influential economists. Therefore, within our framework, we have been able to identify pathways that at least satisfy the minimum requests of those opposing groups of actors. Our iterative informal stakeholder dialogues ensure and successively qualify these aspects.
- To what extent could our experiences be transferred to other transdisciplinary projects? Starting with a theoretically trivial, however practically most demanding item, we addressed a publicly debated science–policy issue whose solution had been hampered by approaches too much oriented along disciplinary lines – and from that we distilled one single researchable question to be answered within three years. We tried to work with the simplest rather than the most advanced representations of the involved fields whenever possible, e.g. worked with the most reduced climate model in the first iteration of our analysis.
- At the same time we advanced disciplinary fronts when necessary, such as endogenous economic growth or upper limits of climate sensitivity. For that, a profound disciplinary background or at least a supportive environment proved crucial.
- Before bringing the disciplines together the team leaders had to develop a joint language across disciplines in order to decide which categories and terminologies would enable the disciplines to interact. A deep understanding of the conceptual foundations and limitations of the involved disciplines, proved far more crucial than is usually the case when tackling standard disciplinary research questions.
- Finally, iterative informal stakeholder dialogues provided a key to whether we adequately represented the preferences of conflicting actors in our analysis and

whether we in fact made significant statements about decision options – and not on a too stylised quantity – a decision maker really had at their disposal. An evocative example of the latter point is that it seems much more influential in the climate policy debate to formulate our results in terms of investment strategies rather than desirable emission paths.

**Acknowledgments** We would like to thank R. Brecha and K. Lessmann for helpful discussions. This work has been supported by the Volkswagen Foundation within their programme 'Promotion of junior scientists on transdisciplinary environmental research' under the grant II/78470.

## References

Berger, J.O.: 1985, *Statistical Decision Theory and Bayesian Analysis*, Springer, New York.

Browne, J.: 2004, Beyond Kyoto, *Foreign Aff*, 83(4), 20–32.

Bruckner, T., Petschel-Held, G., Toth, F.L., Füssel, H.M., Helm, C., Leimbach, M., and Schellnhuber, H.-J.: 1999, Climate change decision support and the tolerable windows approach, *Environ Model Assess* 4, 217–234.

Edenhofer, O., Bauer, N., and Kriegler, E.: 2005a, The Impact of Technological Change on Climate Protection and Welfare: Insights from the Model MIND, *Ecol Econ* 54(2–3), 277–292.

Edenhofer, O., Held, H., and Bauer, N.: 2005b, A regulatory framework for carbon capturing and sequestration within the post-Kyoto process. In: E.S. Rubin, D.W. Keith, and C.F. Gilboy (eds), *Proceedings of the 7th International Conference on Greenhouse Gas Control Technologies*, Elsevier, Amsterdam, pp. 989–997.

Edenhofer, O., Carraro, C., Köhler, J., and Grubb, M. (Guest eds): 2006, Endogenous Technological Change and the Economics of Atmospheric Stabilization, *Energ J Special Issue* (27).

Edenhofer, O., Schellnhuber, H.-J., and Bauer, N.: 2004, Risks and Opportunities of Planet Protection, *Internationale Politik, Transatlantic Edition* 4/2004, 5, 61–68.

European Climate Forum: 2004, *What is Dangerous Climate Change? Initial Results of a Symposium on Key Vulnerable Regions Climate Change and Article 2 of the UNFCCC*, 35pp., Retrieved June 11, 2007, from http://european-climate-forum.net/.

European Commission: 2000, *Communication of the Commission on the Precautionary Principle, COM (2000)* 1 (renewed in its minutes 20 December 2004; supported by the European Parliament 2005).

Giles, J.: 2006, Economists claim carbon cuts won't break the world's bank, *Nature* 441, 264–265.

Held, H., Edenhofer, O., and Bauer, N.: 2006, How to deal with risks of carbon sequestration within an international emission trading scheme. In: *Proceedings of the GHGT-8 conference*, 19–22 June 2006, Trondheim, Norway.

Held, H. and Kleinen, T.: 2004, Detection of climate system bifurcations by degenerate fingerprinting, *Geophys Res Lett* 31, L23207.

Held, H. and Schneider von Deimling, T.: 2006, Transformation of possibility functions in a climate model of intermediate complexity, *Adv Soft Comput* 6, 337–345.

Houghton, J.T., Ding, Y., Griggs, D.J., Noguer, M., van der Linden, P.J., Dai, X., Maskell, K., and Johnson, C.A. (eds): 2001, *Climate Change 2001: The Scientific Basis, IPCC Third Assessment Report*, Cambridge University Press, Cambridge.

Keith, D.: 2000, Geoengineering the climate: history and prospects, *Annu Rev Energ Environ* 25, 245–284.

Kriegler, E. and Held, H.: 2005, Utilizing random sets for the estimation of future climate change, *Int J Approx Reason* 39, 185–209.

Lomborg, B.: 2004, *Global Crisis, Global Solutions*, Cambridge University Press, Cambridge.

Morita, T., Nakicenovic, N., and Robinson, J.: 2000, Overview of Mitigation Scenarios for Global climate Stabilisation based on the New IPCC Emissions Scenarios (SRES), *Environ Econ Pol Stud* 3(2), 65–88.

Nordhaus, W.D. and Boyer, J.: 2000, *Warming the World. Models of Global Warming*, MIT Press, Cambridge.

PMIP: 2006, *Paleoclimate Modelling Intercomparison Project Phase II*, Retrieved June 11, 2007, from http://www-lsce.cea.fr/pmip2/.

Schellnhuber, H.-J. and Held, H.: 2002, How fragile is the earth system?. In: J. Briden and T. Downing (eds), *Managing the Earth: The Eleventh Linacre Lectures*, Oxford University Press, Oxford, pp. 5–34.

Scheringer, M.: 1996, Persistence and spatial range as endpoints of an exposure-based assessment of organic chemicals, *Environ Sci Tech* 30, 1652–1659.

Schneider von Deimling, T., Held, H., Ganopolski, A., and Rahmstorf, S.: 2006, Climate sensitivity estimated from ensemble simulations of glacial climates, *Clim Dynam* 27, 149–163, DOI 10.1007/s00382-006-0126-8.

Stainforth, D.A., Aina, T., Christensen, C., Collins, M., Faul, N., Frame1, D.J., Kettleborough, J.A., Knight, S., Martin, A., Murphy, J,M., Piani, C., Sexton, D., Smith, L.A., Spicer, R.A., Thorpe, A.J., and Allen, M.R.: 2005, Uncertainty in predictions of the climate response to rising levels of greenhouse gases, *Nature* 433, 403–406.

Uzawa, H.: 1995, Global Warming and the International Fund for Atmospheric Stabilisation. In: *Workshop on Equity and Social Consideration Related to Climate Change. Proceedings of IPCC*, WG III, 1995, pp. 49–54.

Uzawa, H.: 2003, *Economic Theory and Global Warming*, Cambridge University Press, Cambridge.

WBGU: 1997, *Ziele für den Klimaschutz*, German Advisory Council on Global Change (WBGU), Retrieved June 11, 2007, from http://www.wbgu.de/wbgu_sn1997.html.

WBGU: 1998, *Welt im Wandel – Strategien zur Bewältigung globaler Umweltkrisen*, annual report by the German Advisory Council on Global Change (WBGU).

WBGU: 2003, *Climate protection strategies for the 21st century: Kyoto and beyond*, Special report by the German Advisory Council on Global Change (WBGU).

# Chapter 13
# Policy Analysis and Design in Local Public Management: A System Dynamics Approach

**Markus Schwaninger, Silvia Ulli-Beer and Ruth Kaufmann-Hayoz**

**Abstract** The purpose of this article is to draw lessons for transdisciplinary modelling endeavours in social contexts. The chosen approach is to explore a research project that was part of the Swiss National Science Foundation's 'Swiss Priority Programme Environment' (SPPE). The project focused on ecological issues and was realised in collaboration with local actors from a Swiss municipality. The case study design is used to analyse methodological issues related to the study of complex dynamic challenges. For example, for the purpose of policy design and learning, pertinent examples of the roles of heuristic frameworks and of simulation models are examined. Another challenge is to involve problem owners in the inquiry process. Furthermore, the transdisciplinary approach, in order to gain a better understanding, is addressed. These issues will be illustrated by means of system dynamics modelling and analysis of current challenges in the realm of solid waste management. Finally, the chapter makes several recommendations to researchers who want to investigate multifaceted, dynamically complex issues together with practitioners.

**Keywords:** Policy design · Policy analysis · Recycling management · Model-based theory-building · System dynamics

## 13.1 Background

This chapter refers to a research project, which is presented as a case study. The project originated within the 'Swiss Priority Programme Environment' (SPPE) of the Swiss National Science Foundation (SNSF).

The work we are referring to is called '*Citizens' Choice and Public Policy*'. In this research project, the general issue of citizens' choice is related to the study of environment related behaviour of citizens and exemplified by a modelling and simulation study to identify effective public policies. The concrete study object chosen

---

✉ M. Schwaninger
Institute of Management, University of St. Gallen, St. Gallen, Switzerland
e-mail: markus.schwaninger@unisg.ch

was the *management of solid waste at the local level.* The objective of the study was to explain the dynamic interaction between public policies and environmentally relevant behaviour of citizens, and to provide solutions for environmental management in communities.

In the following, we will elaborate on the research design of the project under study. A report of the different phases of the project with an overview of the heuristics and instruments used will ensue. Finally, several lessons for researchers will be drawn from the analysis of the project.

## 13.2 Research Design

In view of the purpose of the research project, the main questions to be answered were:

- What are the factors triggering citizens' environmentally relevant behaviour?
- How can actors be influenced towards an environmentally friendly behaviour pattern?

To answer these questions, the specific objective was to build and explore a simulation model, which should lead to better decision making.

Given the complexity of the issue under study, the basic orientation of the project had to be a *transdisciplinary* one. It was clear from the outset that the knowledge necessary to accomplish the project reached beyond one single field of study. The relevant knowledge had to come from different disciplines and included system experts, i.e. persons who were knowledgeable about the system under study. Also, several constituents had stakes in the inquiry, i.e. they were affected by or interested in the inquiry. Therefore, we needed more than a merely multidisciplinary procedure, by which a number of specialists would be consulted: we required mutual communication that would cross boundaries not only between different disciplines but also between science and practice.

Transdiciplinary processes of inquiry or design are characterised by one essential feature: they unite experts from different fields – usually disciplinary – to cooperate, the interaction being facilitated by a common framework or code enabling effective communication across different disciplines and perspectives. The common code is, in principle, a formal language provided by sciences such as system theory, cybernetics or mathematics. In our case, the *System Dynamics* methodology was the core device for integrating concepts from different disciplines in a simulation model and analysing the issues.

System Dynamics (SD) is essentially a system theory based methodology for the modelling and simulation of complex systems, which provides an effective platform for a transdisciplinary dialogue in policy design and learning. It is a widely used methodology developed by Prof. Jay Forrester at MIT-Massachusetts Institute of Technology (Forrester, 1961). SD models are made up of closed feedback loops and the systems modelled are simulated as continuous processes; the mathematics of the models are based on differential equations and integration (Sterman, 2000; Barlas, 2006). SD is particularly useful for discerning a system's dynamic patterns of behaviour, which may be 'counterintuitive' (Forrester, 1971).

The research strategy chosen is one of *action science*. Argyris et al. (1985) define action science as a research approach that combines the study of practical problems with research that contributes to theory building and testing. The authors start from two premises. Firstly, it is necessary to understand the world in order to manage or transform it in a desired direction. Secondly, in order to understand the world, an attempt should be made to change it. Consequently, the protagonists of action science advocate issue- or problem orientation and favour the participation of exponents of different disciplines, and practitioners. Hence, this kind of research implies a transdisciplinary approach to strengthen the inquiry process.

Action science aims at improving social and human systems. Many applications have been documented, including cases where researchers and decision makers cooperated to address issues of organisational learning (Argyris and Schön, 1996). As this literature reveals, a crucial concern of action science is the development of strategies for sustainable change. Efforts to reveal defensive forces that try to maintain the status quo play an important role, as does understanding the consequences of actual and potential interventions, in the light of context dependent action of human beings.

The research project was composed of two parts:

*Part I*: A research methodology was developed, and a conceptualisation of the issues under study – environmentally relevant behaviour of citizens and effective public policies – was accomplished. The method followed in this part was essentially a study of literature and the deductive development of a heuristic framework.

*Part II*: The development of a dynamic theory. The object of the study was the interaction between policies and environment related human behaviour. The management of solid waste was the concrete setting for studying these issues. The process of theory development was guided by the System Dynamics (SD) methodology (Schwaninger and Hamann, 2005). An initial understanding of the system in question was developed applying the Group Model Building (GMB) methodology (Vennix, 1996; Andersen and Richardson, 1997). It delineates the model building procedures involving both researchers and problem owners – people affected by and experts knowledgeable about the issues under study.

The vehicle for the policy analyses, which was developed, was a differential equation model for Solid Waste Management (SWM model). The quality of the model, and the policy analyses realised with the help of that model, were of crucial importance to reaching the objective of the project.

We expected to gain valuable insights from this research design, enabling the formulation of a dynamic theory about citizens' choice and public policy, in the context of environmental management.

## 13.3 Part I: Problem Structuring and Conceptualisation

In the following, we will describe the activities undertaken and the results obtained during the phases of the project. To begin, the researchers conceptualised a design for a guiding framework. In parallel, and throughout Part I, a literature review and three Group Model Building (GMB) workshops were undertaken. These events brought together researchers, experts in environmental management, and the client group, i.e. a set of 'local system experts' who were knowledgeable representatives of the system under study. They were familiar with the system, and therefore in a position to judge the utility of whatever was developed in the project. Workshop participants were chosen according to two criteria. The first was a willingness to participate and cooperate in transdisciplinary research dealing with issues of sustainable development at the local level. The second was experience with projects aimed at promoting sustainable development. The workshops were facilitated by one of the authors (S. U.) and technical supporters. The objective of the workshop was the elaboration of a conceptual framework which would orientate further research. The main inputs to that event were the results of a higher level project of the SPPE, the Integrated Project 'Strategies and Instruments' (Full title: 'Strategies and instruments for sustainable development: Bases and evaluation of applications, with special regard to the municipality level with nine subprojects' SNSF no. 5001–48826.

The main outcome of these efforts was a 'Policy Framework for Solid Waste Management' (Fig. 13.1). It focuses on the determinants of environmentally

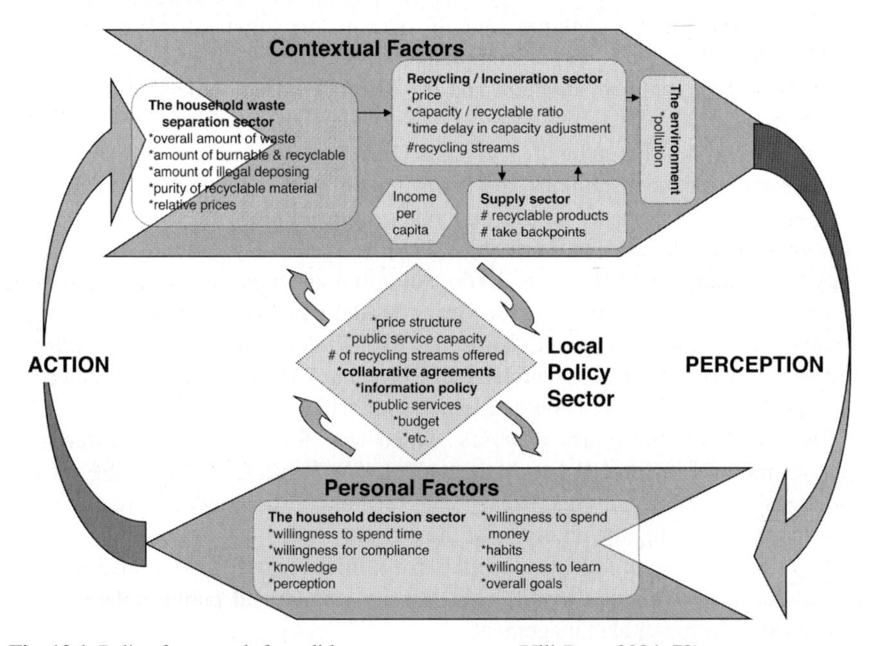


**Fig. 13.1** Policy framework for solid waste management (Ulli-Beer, 2004, 79)

responsible behaviour: the interaction between contextual constraints and opportunities on the one hand and personal, motivational and cognitive factors on the other (Kaufmann-Hayoz, 2006). This framework reflects the main assumptions underlying the ensuing work of model building and use in Part II.

According this framework, policy makers can alter either the internal or the external structure in order to induce changes in behaviour (Kaufmann-Hayoz et al., 2001). Different types of policies, such as economic interventions, service and infrastructural measures, as well as budget constraints, can alter the external structure. On the other hand, communication policies and collaborative agreements can modify the individual factors. A more comprehensive description of this framework is given by Kaufmann-Hayoz (2006).

## 13.4 Part II: Developing a Dynamic Theory

Part II involved developing a dynamic theory. It made use of a specific case study, the subject of analysis being the management of solid waste in a Swiss community. The interaction between policies and environment related human behaviour was to be examined. The core of the process of theory building was the construction of a System Dynamics model and its use for policy analysis.

This part of the project was designed along the lines of four phases: Model – Test – Design – Change, and the respective steps. These phases and steps correspond to Integrative System Methodolgy (ISM), a heuristic methodology developed for management in complex settings (Schwaninger, 1997). The researchers made use of ISM as a metamethodology in their project because it allows the combination of different methodologies and focuses on dealing with complexity. In Table 13.1 the four phases are translated into the language of the System Dynamics (SD) methodology (Richardson and Pugh, 1981; Sterman, 2000).

**Table 13.1** Phases and steps of Part II

| Phases | Tasks/steps | Corresponding activities in the SD modelling process |
|---|---|---|
| Model | Ascertaining relevant perspectives, the goals and factors critical for attaining them | Problem recognition |
| | Surfacing issues | System conceptualisation |
| | Elaborating models | Building model |
| Test | Apprehending the dynamics of the system | Exploration and evaluation of model behaviour |
| | Simulating and exploring scenarios | |
| | Interpreting and evaluating simulation outcomes | |
| Design | Ascertaining control levers | Policy analysis |
| | Designing strategies/action programmes | |
| Change | Realising strategies | Model use |

The municipality chosen for the action research was Ittigen. This is a community located in the agglomeration of Berne (Switzerland). It has about 11,000 inhabitants and is recognised for taking an innovative approach to its environmental policies. For example, Ittigen participated in the *light electro mobiles* (LEM) pilot and demonstration programme and in the Global Action Plan (GAP), an international environmental initiative. It also was the first municipality in Switzerland to implement the environmental norm ISO 14001. The authorities of Ittigen were willing to gain a deeper understanding of factors that determine the environmentally relevant behaviour of their citizens. They provided abundant information about their municipality for the investigation. The authors greatly appreciate the commitment and the support of the authorities of Ittigen in this research project.

### *13.4.1 Model*

As reported above, the modelling phase started with a literature review on environmentally relevant behaviour, and environmental policy. In addition, one of the two 'gatekeepers' of the client group, the environmental manager of Ittigen (the other gatekeeper was the mayor of Ittigen), helped to determine the relevant actors that should participate in the model building process. Gatekeepers came from the client organisation and were responsible for the project. They acted as 'liaison' between the model builders and other people involved in the project. An important role of the gatekeepers was to motivate people in the client organisation, enhancing their commitment and participation (Vennix, 1996; Andersen and Richardson, 1997).

Three Group Model Building sessions were held in order to map the causal chains and feedback loops of interest, as observed in the municipality. At that point the modelling was of a qualitative type, i.e. no quantification took place yet. This kind of modelling is often called 'Qualitative System Dynamics'.

Ten experts, representatives of the local government, the local administration and consulting firms responsible for local management and policies in the realms of energy, transport, water/waste water, solid waste and consumption were involved. They shared their mental models about problems of environmental policy making to elicit patterns of policy resistance and compliance. The objective was to explain and reach consensus about causes and consequences of environmentally harmful behaviour of citizens, as well as the factors conducive to favourable behaviour. Although the interactions and processes still remained unclear it became obvious that policy intervention would influence different variables and subsequently change the various preconditions.

It was at that point that the 'Policy Framework' (Fig. 13.1) developed in Part I, showed its value in guiding the model building process. The driving forces specified in the framework (e.g. relative prices, number of recycling streams) could now be linked with the phenomena observed in the system (e.g. change in the amount of waste, purity of recyclable material).

To sum up, the framework enhanced the efficiency of the model construction and facilitated theory building. It was also crucial to the policy design phase, helping to widen the problem focus to the extent that preconditions for action choices were systematically considered.

Early on in the modelling phase, discussions were held with the authorities responsible for local solid waste management. The focus was on specific aspects of solid waste management. These discussions helped identify and clarify the main issues concerning solid waste management in the community, as seen by the authorities. Subsequently, the objectives of the modelling study were further refined. Hence in the system inquiry, both kinds of environmentally significant behaviour could be addressed: waste separation behaviour of citizens, i.e. behaviour that directly causes environmental changes; and solid waste management policies that shape the context in which citizens' waste separation behaviour takes place; and their interaction. Reference modes of the crucial variables of the local solid waste management system were mapped. A reference mode is a set of graphs and descriptive data that shows the development of the issue or problem under study, over time (Sterman, 2000). Figure 13.2 gives an example of such a reference mode showing both the past development path and a hypothesis about future developments – of the fraction of waste separated for recycling and the number of streams of the different recycled materials. As shown, a continued stepwise increase of the number of recycling streams, could lead to an inefficient logistic (excess capacity), a pattern known as overshoot behaviour.

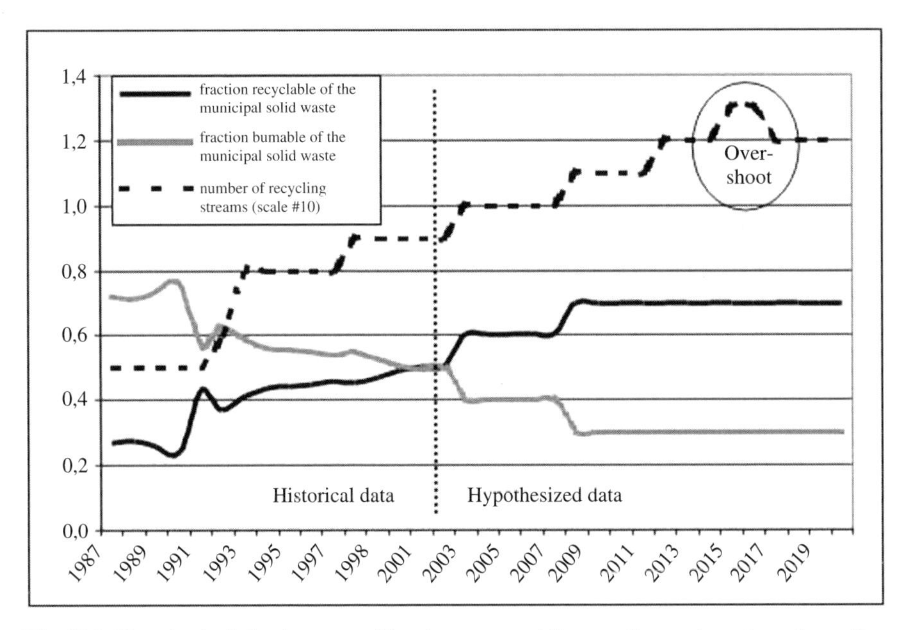


**Fig. 13.2** Hypothesised development of fraction separated for recycling and number of recycling streams (Ulli-Beer, 2004, 91)

Based on the heuristic framework and the reference modes, a dynamic hypothesis was stated (see the description in the following box and the causal loop diagram in Fig. 13.3). To show how the dynamic hypothesis and the reference modes link up, let us, for example, take the time series of the 'number of recyling streams' in Fig. 13.2. It is then hypothesised, as shown in Fig. 13.3, that these time series are controlled by the two loops: 'limiting propensity from time cost' and 'deficit limits investment'.

## 13.5 Problem Description and Dynamic Hypothesis

A dynamic hypothesis is a problem description made transparent by fleshing out causes and consequences of system behaviour. As a starting point for improvements the focus is mostly on a diagnosis of dysfunctional behaviour. In a nutshell, the picture was as follows (see also Fig. 13.3).

Since citizens in the case community were failing to separate various streams of solid waste for recycling, the local authorities gave price incentives in the form of a garbage bag charge (implemented in 1991). The intended effect was to promote the desired separation behaviour. As a consequence the fraction of separated waste increased and the relative amount of solid waste for burning (variable 'fraction to be burned') decreased. The unintended effect was that not only the relative amount of waste disposed for burning decreased, but also the revenue generated from the garbage bag charges ('revenue from burnable material') declined. Therefore, the municipal budget deficit started to grow (via declining 'profit from solid waste management'). A further increase in the price for burnable material had almost no effect

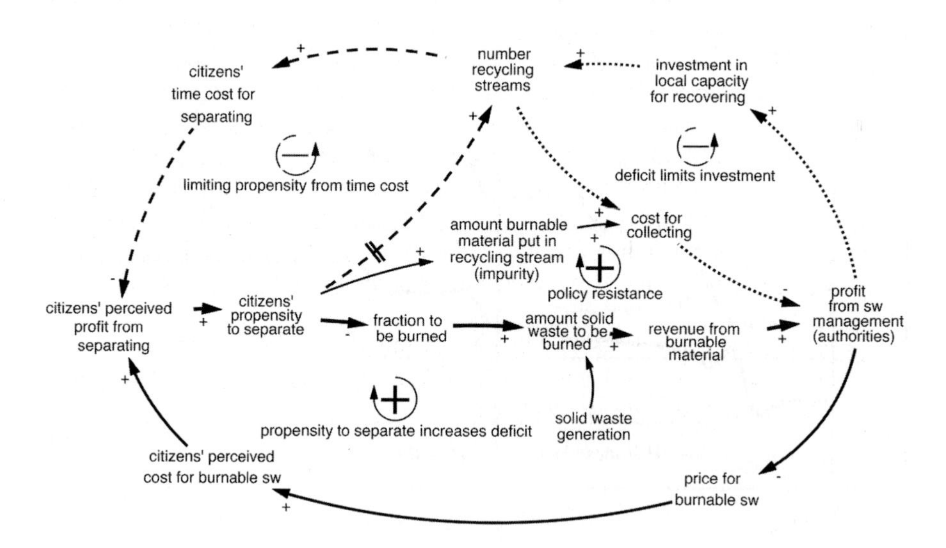


**Fig. 13.3** Dynamic hypothesis for local solid waste management (Ulli-Beer, 2004, 88)

on the separation behaviour, since the number of recycling streams remained nearly constant. The citizens had no real legal option to avoid higher costs for disposing of burnable material. Another unintended consequence was that the quality of the separated material decreased as citizens started to put burnable material into the recycling streams. However, this effect was only occasionally observed and could not be exactly quantified. The model calibration was substantiated by the data available for the 1987–2001 time period, which revealed the historical pattern of the development of budgets and the amount of separated material.

The illustration of this dynamic hypothesis in Fig. 13.3 makes the features of feedback, delays, closed loops and multiple causes visible.

Feedback is symbolised by the arrows. Arrow polarity between two variables (say X $\rightarrow$ Y), which is denoted as '+' means: all else being equal, if X increases (decreases), then Y increases (decreases) above (below) what it would have been. Arrow polarity which is denoted as '−' means: all else being equal, if X increases (decreases), then Y decreases (increases) below (above) what it would have been (Sterman, 2000). Furthermore, referring to Fig. 13.3, a '+' at the centre of a loop signifies self-reinforcement (reinforcing loop), while a '−' denotes self-attenuation (balancing loop). A delay, i.e. a retardation of consequences, is visualised in this diagram by the bar on the arrow from 'citizens' propensity...' to 'number recycling streams'.

The next step specified the concrete concept of the model, including its boundary. Thereupon, one of the authors (S.U.) developed the quantitative SWM model with the support of experienced modellers. The model building process started with a simple first order 'concept model', which was a crude representation of the dynamic hypothesis (Richardson, 2006). The modelling process was not linear but rather a winding search for the right focus and approach – culminating in a promising structural representation of the system, which properly described the observed phenomena. From that point on, the model development process was guided by the dynamic hypothesis and the sector diagram of the solid waste management system. The model was made up of the following sectors: household waste separation, household decision making and local solid waste management. Sector by sector, the modules of the SD model were formulated and iteratively tested (Ulli-Beer, 2003). For the quantitative modelling process the Vensim software was used. Vensim is a software program and a registered trademark of Ventana System, Inc. Co., Harvard, Massachusetts.

The hypotheses were developed in conjunction with the gathering, coding and analysing of data. The resulting dynamic theory, therefore, is grounded in both theoretical frameworks and empirical data.

### *13.5.1 Test*

In order to ensure the quality, strength and utility of the model for reliable policy analysis, the validation procedures adopted were a crucial aspect of the testing.

Validity, in this context, is the level of confidence one can have in a model. The quest for proper validation resulted in a number of steps (Forrester and Senge, 1980; Barlas, 1996; Sterman, 2000). These take into account direct structural tests (reviews of model structure), indirect structural tests (computer-based tests of model structure) and behavioural tests (in the sense of behaviour replication, i.e. the fit of data resulting from the simulation model and the empirical data from the real system). The boundaries of this chapter would be overstretched, if all of these validation tests were described. We shall limit ourselves to an overview.

The broad theoretical background and the intensive exchange with experts who understood the system helped to build a dynamic theory about the mechanics of the solid waste management systems. Specifically, it helped to build the hypotheses about causal relationships that are captured in the model. Besides knowledge about the real system, theoretical frameworks also guided the modelling process, resulting in a model structure that was grounded in the relevant knowledge of the system. As model building progressed, each new structural module was evaluated with respect to the plausibility of its behaviour (reality check): Does it produce the right behaviour for the right reasons? Is it consistent?, etc. The model building and testing processes were helped by the reference modes as patterns of model behaviour could be checked in relation to these modes, during the different phases of model construction. By applying the behaviour reproduction test, flaws in the structure and mathematical functions were discovered and corrected. The final model showed a good fit with the reference modes. Testing the model was an iterative process by which an increase in confidence in the model was achieved in a stepwise manner. Once the basic model structure had been elaborated, different test procedures were conducted in order to fully understand the dynamics of the system and to test its robustness, consistency and correspondence with real data. In the middle of this phase an expert validation was held, in the form of a further meeting with the core audience – the knowlegeable representatives of the system. Their feedback was quite encouraging since the model structure, its assumption, its dynamics and the derived insights seemed plausible to them.

This group pleaded that the audience for the next meeting be expanded to include representatives from other municipalities as well as state and national agents. Like quality management, validation is a never ending process. Therefore, such an expansion would be a beneficial exercise in the next round.

### *13.5.2 Design*

In this phase, two distinct classes of policy questions were addressed (Zagonel et al., 2004). Firstly, policy analysis experiments addressing 'What If' questions were conducted. A typical question was 'What might happen if we were to make such and such policy change?' With the model as a virtual laboratory, policy experiments were set up that looked back at previous policy initiatives (backcasting) or forward to untried future policies (forecasting). Secondly, scenario analysis experiments were set up addressing questions such as 'What might happen if some parameters or variables not under our control were to change dramatically?' The behaviour patterns resulting from the different policy experiments, using similar and different scenarios, were compared and explained. Furthermore, sensitivity analysis was used to explore the effects of uncertainty in the system. Figure 13.4 shows a simplified version of the model structure. The diagram is in the original SD notation with stocks (symbolised by boxes) and flows, which change the stocks (denoted by the valves). The picture represents a core aspect of the model – the diffusion process by which people become willing to separate their garbage into different streams of waste. The policy intervention points for influencing ('steering') the diffusion process are depicted by hexagons.

Detailed results of the different policy and scenario analysis experiments can be found in Ulli-Beer (2004, 2006) and Ulli-Beer et al. (2007).

During the policy design phase we reverted to a 'Practical Guide for Facilitating Environmental Policy Compliance', which was one of the by-products of the framework building phase (Ulli-Beer, 2006) This was essentially a checklist of possible measures for facilitating environmentally sound action, which helped us to enrich the policies indicated on the grounds of the quantitative computer simulations. All in

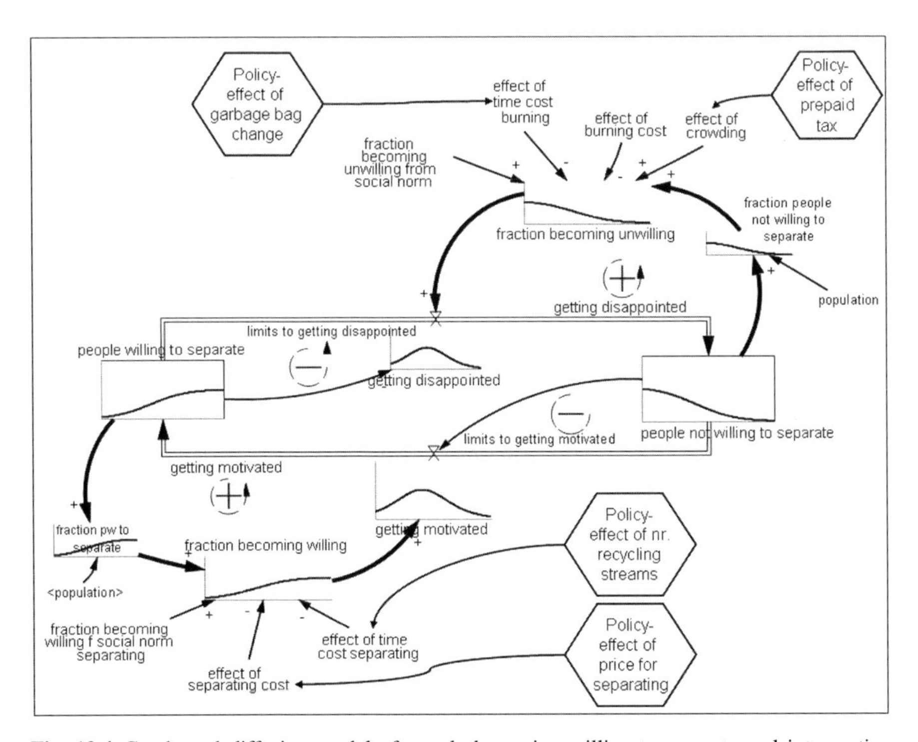


Fig. 13.4 Condensed diffusion model of people becoming willing to separate and intervention points for steering the diffusion process (adapted from Ulli-Beer, 2004, 120)

all, the policy programmes were designed so that not only the obvious problems were addressed. Rather, the approach aimed at strengthening the preconditions for desirable environment related behaviour of citizens and to overcome policy resistance in the system, i.e. increase policy compliance.

### *13.5.3 Change*

In line with action science, this project has been executed together with two groups of collaborators. Firstly, those interest groups that would be most affected if there was a change in the environmental policies applied in the community under study. Secondly, the involvement of those decision makers who would be most effective in changing structures and rules in order to improve the situation. The task of the system inquiry was to understand how real world behaviour emerges out of real world structure. Understanding the crucial processes that explain undesirable patterns are important preconditions for influencing actors' motivation and knowledge and for implementing the most promising policy measures. Hence, the commitment of these collaborators was crucial for the acceptance of the policy recommendations and ultimately for the success of the project.

However, it had not been planned that the project include the implementation of the policy measures being developed. So where is the change?

The intended immediate change was not directly observable, i.e. a change of the collaborators' mental models due to a better understanding of the problem structure.

Furthermore, the logic of the System Dynamics methodology suggests a paradigmatic change. As the modelling and design progresses, the mindsets of the participants will change. At some point, instead of thinking in terms of simple cause and effect schemes, they will change to a more sophisticated rationale – thinking in feedback loops. They will assimilate a new understanding of the environmental issues, learn to look at the system from a feedback perspective, focusing on closed loops, delays and multiple causalities. This is seen as an important step towards a new management approach aimed at enforcing desirable developmental trends and avoiding undesirable ones.

Although this paradigmatic change is hard to substantiate we could observe some delayed change in the community indicating the probability that the mental model of the collaborators evolved. The mayor of Ittigen reported that an educational programme about municipal solid waste and recycling management was started at the local primary schools. Although we do not know definitely if this initiative is a direct consequence of our study, it directly reflects one of the main findings of the project: economic instruments are not sufficient to yield an efficient and robust policy outcome; further educational and service interventions are indicated.

However, in order to enhance the transformation of knowledge into action in other communities, further modelling workshops could have been rolled out. That way, the usefulness and adequacy of the developed model and theory could be further tested and extended. The action science approach would call for such a continuation.

## 13.6 Recommendations

The objective of this article was to draw lessons for transdisciplinary modelling endeavors in social contexts, based on an in-depth study of a complex process of inquiry. The case was profoundly characterised by an involvement of 'problem owners' and other stakeholders into the research process.

The described approach resulted in a dynamic theory that explains observed behaviour patterns of a local solid waste management system and provides policy recommendations for its improvement. Model interfaces and policy experiments can also be used, in the future, as a scientifically grounded decision support system that informs the debate on policy design in solid waste management.

Having reviewed the phases of the whole project, a puzzling phenomenon emerges, which can be coded in one polarity, the polarity between plurality and unity. It leads to questions of the following kind:

- 'How could a coherent body of knowledge come about under these circumstances?'
- 'How could inputs from different stakeholders with distinct disciplinary backgrounds and interests result in a consistent model?'
- 'How could a process of change in a dynamic social context be induced?'

The key to these questions is transdisciplinarity. The different views of all the people involved were integrated and condensed using the SD model. Such a model is essentially a language based on system theory. It allows for variables originating from different domains, e.g. economic, social, ecological and technological, to be connected to form a consistent picture. The system theory language was the key to transdisciplinarity. Initially, it was owned by the facilitators. As the dialogue between the different constituencies of the project evolved, the systemic thinking and the command of the new language spread among participants, enabling genuinely transdisciplinary conversations. To conclude, without such a transdisciplinary approach, it would not have been possible to develop a model such as the one outlined here or induce a policy learning process.

We would like to conclude with a set of recommendations for researchers who want to explore and investigate multifaceted, dynamically complex issues together with stakeholders. We underline that this is pragmatic advice rather than scientific teaching. Here are seven points, which appear to be very important.

According to these assumptions, the flow of human action is regarded as part of the continuous interaction – through the processes of perception of action – between the internal structure of the actor (personal factors and processes) and the external structure (contextual factors and processes with cultural, socio-economic, institutional, and physical aspects). The external factors offer options but they also constrain human behaviour. They are the result of a multistage decision making process in the political/administrative, technological and economic domains. As far as they are perceived by actors, they influence actors' actions, but actors can also change the external structures through their actions.

*Design the project for a transdisciplinary inquiry.* The traditional disciplinary and even multidisciplinary approaches are in principle insufficient for adequately dealing with complex issues. Therefore, a transdisciplinary setting for the inquiry must be created. The interaction between the holders of different kinds of expertise must be facilitated, and it should rely on a common, integrative language.

*Use frameworks to manage the relevant body of knowledge.* Identifying relevant knowledge, e.g. helpful disciplinary concepts, is an iterative research process. In this process, scientifically grounded frameworks can serve as heuristic devices of great importance. They help to integrate different disciplinary perspectives and to organise blocks of insights, as well as to structure the issue in a transdisciplinary mode. Most notably, they are apt to trigger a systematic process of modelling and learning.

*Apply a unifying perspective when structuring the issues under study.* The systems view is a powerful conceptual approach to complex projects such as the one reported here. In this case, it proved most helpful in determining the field of investigation and then in integrating the different perspectives. It was also crucial for the exploration of relationships between the different system components, i.e. the essential factors that had to be represented in the model. For example, the 'Policy Framework for Solid Waste Management', is an essentially systemic model. It visualises the basic controlling loop between actors and their context. Equally, the SD methodology of modelling and simulation embodies a systems theory, which enables unifying distinct perspectives in one and the same model.

*Choose a proper methodology for dealing with dynamically complex issues.* A diagnosis, a system design and its implementation can only be as good as the underlying methodology. The Integrative Systems Methodology applied in this case offers a reference methodology that allows for combining complementary research approaches with a focus on dealing with complexity. Group Model Building and the qualitative System Dynamics approach proved to be useful for the development of an insightful dynamic theory on recycling dynamics. The quantitative model building process helped identify and integrate useful disciplinary concepts and findings so that the dynamic theory could be consolidated and a useful decision support tool could be developed. Another significant result was that microprocesses that explain macrobehaviour could be modelled and analysed.

*Customise the policy mix.* A model such as the one developed here gives evidence that the interaction and impact of different policies are situation-dependent, i.e. they vary with the contextual and personal factors of the concrete behavioural setting. When sensitive policy parameters of the system are identified, knowledge of the mode of functioning of different policy instruments is extremely important so that the most effective policy package will be chosen. Adequate policy making tools such as a System Dynamics model can allow the policy mix to be customised close to an optimum.

*Master the crux of policy implementation and policy resistance.* The representation of both internal (personal) variables and external contextual variables in the decision rules of the model provides a broader basis for policy analysis. For local authorities, it opens up richer options for intervention and implementation. This is fundamental for coping with complexity, in the sense of Ashby's Law of Requisite Variety (Ashby, 1956). Furthermore, focusing on the interplay between contextual and personal influences is crucial in order to understand the phenomenon of policy resistance. In addition, it points out important system requirements vital to achieving high policy compliance. There may be a lot of 'sand' in the 'mechanism'; however, in order to be effective, it is important to understand the dominant processes guiding policy compliance.

*Effective system change requires a participative research design.* One must add that the set of participants in the process of inquiry should include different stakeholders, and that these should be included throughout all research phases. The colloquial term 'audience' is often used for the participants who are not researchers or moderators. Figure 13.5 suggests one template of a research design that systematically includes practitioners in the inquiry process. It is built on the ISM and SD modelling phases and is inspired by Beer's methodology of topological maps and scientific modelling (Beer, 1984). Also, the modelling steps and Group Model Building scripts offer crucial stimuli for improved learning in complex decision making situations. Finally, the perceptions of participants should be monitored carefully so that the interventions can be improved and collective learning be achieved.

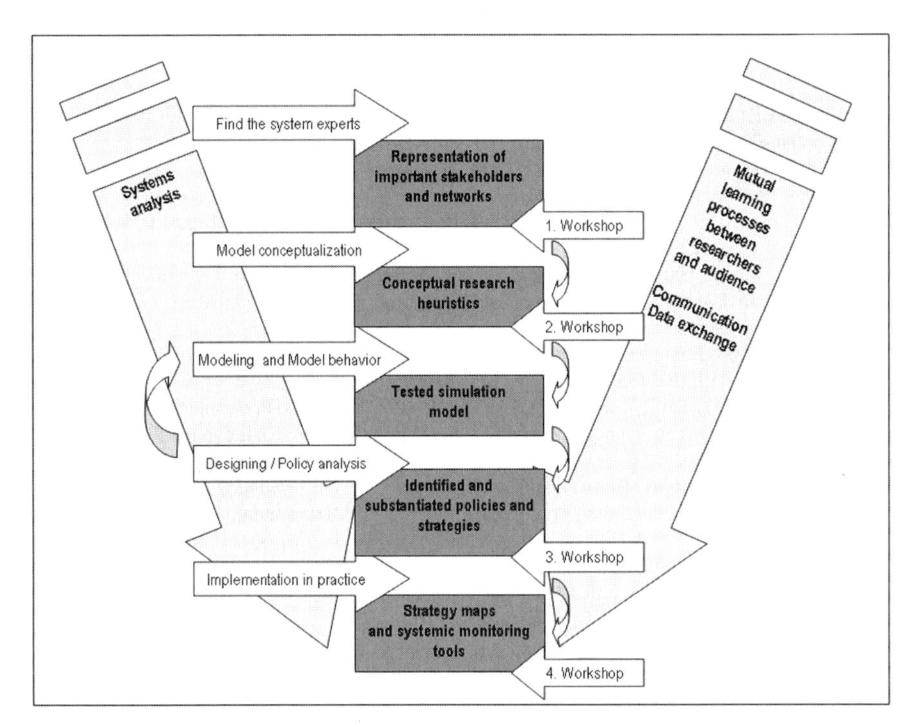


**Fig. 13.5** Template of a transdisciplinary approach to modelling and policy learning (Ulli-Beer et al., 2006b)

## References


Andersen, D.F. and Richardson, G.P.: 1997, Scripts for Group Model Building, *Syst Dynam Rev* 13, 2, 107–129.

Argyris, C., Putnam, R., and Lain-Smitz, D.M.: 1985, *Action Science: Concepts, Methods, and Skills for Research and Intervention*, Jossey-Bass Inc., San Francisco.

Argyris, C. and Schön, S.A.: 1996, *Organizational Learning II: Theory, Method, and Practice*, Addison-Wesley publishing company, Reading.

Ashby, W.R.: 1956, *An Introduction to Cybernetics*, Chapman & Hall, London.

Barlas, Y.: 1996, Formal Aspects of Model Validity and Validation, *Syst Dynam Rev* 12, 3, 183–210.

Barlas, Y.: 2006, System Dynamics, in Encyclopedia of Life Support Systems (EOLSS). In: *Developed Under the Auspices of the UNESCO*, EOLSS Publisher, Oxford.

Beer, S.: 1984, The Viable System Model: Its Provenance, Development, Methodology and Pathology, *J Oper Res Soc* 35, 1, Jan 7–25.

Forrester, J.W.:1961, *Industrial Dynamics*, MIT Press, Cambridge.

Forrester, J.W.: 1971, Counterintuitive Behavior of Social Systems, *Technol Rev* 73, 3, 52–68.

Forrester, J.W. and Senge, P.M.: 1980, Tests for Building Confidence in System Dynamics Model. In: A.A. Legasto, J.W. Forrester, and L.M. Lynes (eds), *System Dynamics*, North-Holland Publishing Company, New York, pp. 209–228.

Kaufmann-Hayoz, R.: 2006, Human Action in Context: A Model Framework for Interdisciplinary Studies in View of Sustainable Development, *Umweltpsychologie* 10, 1, 154–177.

Kaufmann-Hayoz, R., Bättig, C., Bruppacher, S., Defila, R., Giulio, A.D., Flury-Kleubler, P., Friederich, U., Garbely, M., Gutscher, H., Jäggi, C., Jegen, M., Mosler, H.-J., Müller, A., North, N., Ulli-Beer, S., and Wichtermann, J.: 2001, A Typology of Tools for Building Sustainability Strategies. In: R. Kaufmann-Hayoz and H. Gutscher (eds), *Changing Things- Moving People: Strategies for Promoting Sustainable Development at the Local Level*, Birkhäuser, Basel.

Richardson, G.P.: 2006, Concept Models. In: *Proceedings of the 24th International Conference of the System Dynamics Society*, Nijmegen, NL.

Richardson, G.P. and Pugh, A.L.: 1981, *Introduction to System Dynamics – Modeling with Dynamo*, Productivity Press, Cambridge.

Schwaninger, M.: 1997, Integrative Systems Methodology: Heuristic for Requisite Variety, *Int Trans Oper Res* 4, 2, 109–123.

Schwaninger, M. and Hamann, T.K.: 2005, Theory-Building with System Dynamics. Computer Aided System Theory. In: R. Moreno-Diaz, F. Pichler, and A. Quesada-Arencibia (eds), *EUROCAST*, Springer, Berlin, pp. 56–62.

Sterman, J.D.: 2000, *Business Dynamics. Systems Thinking and Modeling for a Complex World*, Irwin McGraw-Hill, Boston.

Ulli-Beer, S.: 2003, Dynamic Interactions Between Citizen Choice and Preferences and Public Policy Initiatives: A System Dynamics Model of Recycling Dynamics in a Typical Swiss Locality. In: *21st International Conference of the System Dynamics Society*, New York City, USA.

Ulli-Beer, S.: 2004, *Citizens' Choice and Public Policy. A System Dynamics Model for Recycling Management at the Local Level*, PhD-Thesis, Institute of Management, St. Gallen, University of St. Gallen.

Ulli-Beer, S.: 2006, *Citizens' Choice and Public Policy*, Shaker Verlag, Aachen.

Ulli-Beer, S., Andersen, D.F., and Richardson, G.P.: 2007, Financing a Competitive Recycling Initiative in Switzerland, *Ecol Econ* 62, 727–739.

Ulli-Beer, S., Bruppacher, S., Grösser, S., Geisshüsler, S., Müller M., Mojtahedzadeh, M., Schwaninger, M., Ackermann, F., Andersen, D., Richardson, G., Stulz, R., and Kaufmann-Hayoz, R.: 2006b, Introducing an Action Science Venture: Understanding and Accelerating the Diffusion Process of Energy-efficient Buildings. In: *Proceedings of the 24th International Conference of the System Dynamics Society*, Nijmegen, NL.

Vennix, J.A.M.: 1996, *Group Model Building. Facilitating Team Learning Using System Dynamics*, John Wiley & Sons, Chichester.

Zagonel, A.A., Rohrbaugh, J., Richardson, G.P., and Andersen, D.F.: 2004, Using Simulation Models to Address 'What If'-questions about Welfare Reform, *J Pol Anal Manage* 23, 4, 890–899.

# Chapter 14
# Constructing Regional Development Strategies: A Case Study Approach for Integrated Planning and Synthesis

**Alexander I. Walter, Arnim Wiek and Roland W. Scholz**

**Abstract** This article presents a transdisciplinary research approach illustrated by a case study on sustainable regional development in the Swiss canton Appenzell Ausserrhoden (AR). The canton is a typical central European rural landscape, struggling with problems of structural change and migration. Different industry sectors (timber, dairy farming and textile production), as well as different aspects of regional development (land use, mobility, landscape protection and tourism) were analysed in a two year transdisciplinary research project. Its goal was to generate cross-sectoral long-term development strategies based on different scenarios for future land use. The study shows that current development strategies will not lead to a desirable future state. There are several desirable developments, but clear political decisions are necessary for targeting a specific scenario.

The project was jointly led by a professor of ETH Zurich and the governing chairman of AR. The study involved numerous officials and inhabitants of AR, as well as scientists and advanced students from different research institutes (Scholz, 2002; Scholz et al., 2003). The study combined teaching, research and application.

The case study is presented as an example of the Transdisciplinary Integrated Planning and Synthesis (TIPS) approach (Scholz et al., 2006). TIPS is an approach to embed a formal, scientifically based, integrated planning approach into a real world setting, allowing for mutual learning among scientists and practitioners. It starts from a faceting of the case, then uses system analysis and scenario construction for the problem investigation procedure. In the problem transformation (or 'implementation') phase, the preferences of the stakeholders are evaluated through multi-criteria procedures and area development negotiation. The results from the different facets are integrated, and cross-sectoral development strategies for the case are formulated.

The chapter gives an overview of the organisation, methodology and epistemology of the TIPS approach. The following three sections describe the transdisciplinary TIPS approach: problem identification, problem investigation and problem transformation. Each step of the TIPS approach is explained and illustrated by examples from the case study. Special emphasis is put on the integrated project architecture and on the

✉ A.I. Walter
Institute for Environmental Decisions (IED), Natural and Social Science Interface, ETH Zurich, Zurich, Switzerland
e.mail: alexander@wackazong.com

G. Hirsch Hadorn et al. (eds.), *Handbook of Transdisciplinary Research*, 

linkage between problem investigation and problem transformation, these being the main challenges from a transdisciplinary point of view. The last section draws some conclusions on the applicability of the TIPS approach and presents recommendations for similar projects.

**Keywords:** Mutual learning · Integrated project architecture · Methodology · Multi-criteria assessment · System analysis

## 14.1 Background

The TIPS approach presented in this paper strives for a planning process suitable to initiate and foster sustainable development (Ravetz, 2000). The approach can be applied from urban and rural cases to business cases, all dealing with complex, ill-defined problems concerning human–environment interactions. It combines a scientific analysis of the problem with an extensive transdisciplinary process of knowledge integration, mutual learning and joint research. It relies on an elaborated ontology, epistemology, methodology, and project management (Scholz and Tietje, 2002; Scholz et al., 2006). From a broader perspective, the TIPS approach contributes to the emerging field of 'Sustainability Science' (Clark and Dickson, 2003; Bammer, 2005).

Since 1994, ETH has conducted one large-scale transdisciplinary case study each year. The TIPS approach has also been applied in other universities internationally (Scholz et al., 2006; Steiner and Posch, 2006). The result of TIPS is the formulation of development strategies based on a common system understanding, a spectrum of possible future developments, transparent preference structures and shared future visions of all key agents. TIPS is expected to lead to societally robust decisions as opposed to 'top–down' planning approaches, because it explicitly combines and integrates developments of the different sectors and layers of the system (Meppem and Gill, 1998).

At ETH, the TIPS approach is used in different projects, including project-based teaching applications. It has proven to be an efficient instrument to teach graduate students the art of complex problem-solving. Direct contact with a case greatly enhances a student's learning experience and motivation. Nevertheless, case study learning is a demanding task, especially in a transdisciplinary context, since the goals of teachers, stakeholders, and students have to be balanced (Stauffacher et al., 2006).

### *14.1.1 The Societal Context: Regional Development in Appenzell Ausserrhoden (AR)*

With an area of 24,000 hectares, AR is one of the four smallest cantons in Switzerland. A Swiss canton has its own constitution, law-making, and government, and corresponds to a state. AR lies in the north-eastern part of Switzerland. It is a sub-alpine area with the altitude ranging from 400 m (shore of the Lake Constance) to 2,500 m above sea level (peak of the Säntis mountain). In 1999, AR had a population of 53,800 inhabitants. Land use was 56% agriculture, 33% forestry, and 11% settlements and unproductive area. About 8% of the employees worked in the first sector. AR consists of 20 municipalities structured into three administrative parts. The *Mittelland* is located close to St. Gallen, the *Hinterland* is the most remote part of the canton and the *Vorderland* faces the Bodensee and the Rhine valley. Only the capital Herisau has a semi-urban structure. AR is an example of a complex rural area located in the immediate vicinity of an urban sub-centre (St. Gallen), and at the remote periphery of a major urban centre (Zurich). These rural areas are distant from inter-regional traffic infrastructure and major centres and have to cope with a decrease in population, labor opportunities, and wealth.

The canton suffers from structural economic problems caused by changes in the local economy. There has been some early industrial development in the canton, especially in the textile sector. In the 18th century, AR had one of the highest proportions of industrial workers in Switzerland. But there has been a constant, high fluctuation in economic performance and, consequently, in the number of inhabitants. Today, the urban sub-centre of St. Gallen plays a major economic role, which has a positive impact on the number of inhabitants, especially in the municipalities close to it. There is an ongoing loss of inhabitants in the other municipalities.

### 14.1.2 The Scientific Context: Transdisciplinary Integrated Planning and Synthesis (TIPS)

The TIPS approach is based on a strong understanding of transdisciplinarity, which goes beyond classical interdisciplinary or participatory approaches. In our understanding, transdisciplinarity is based on a sociocultural–constructivist paradigm (Stauffacher et al., 2006), which emphasises that individuals and organisations actively construct knowledge in a specific social setting (Resnick, 1987), according to existing knowledge and experiences. This understanding of science differs strongly from a positivist and rationalist perspective. We follow the position that science can no longer prefabricate solutions to upcoming complex problems, but must set ‘aside the idealised context of science in order to produce practically relevant knowledge’ (Pohl, 2005). This leads to the insight that mutual learning between science and society (Thompson-Klein et al., 2001) and between stakeholders (Meppem and Gill, 1998) is necessary to produce relevant knowledge. The most advanced learning paradigm which respects these insights is the sociocultural-constructivist paradigm, the core idea of which was developed by Berger and Luckmann (1966), claiming that ‘reality is socially constructed’. This has some important consequences if science tries to tackle real world problems in multi-actor contexts:

- The situational context becomes important: knowledge of concrete real world problems has to be produced in exactly the same or comparable contexts in order to be meaningful. This is a strong argument for a case based approach. It also implies that transdisciplinary research becomes a goal oriented process rather than a knowledge production process per se.
- The active involvement of the concerned individuals and institutions (stakeholders) is important. Only by letting them participate in the knowledge production process is it possible to include their specific expertise, and view of the problem, in the process. In the TIPS approach, stakeholders are co-leaders of the case studies, contributing to it with financial and human resources, and sharing the responsibility for the results of the study. In this way they can develop a sense of 'problem ownership', which is crucial for their motivation.
- Interaction between the participants plays a crucial role: one of the major goals of such a process has to be to develop a shared view of the situation, the problem, and the utility of possible solutions/developments of the case: 'construction of meaning is tied to specific contexts and purposes. People develop shared ways of responding to patterns and features in particular contexts.' (Wilson and Myers, 2000). The TIPS approach is designed to support this interaction process.
- An important differentiation has to be made regarding the role of scientists in transdisciplinary research. In order to meet scientific standards, there have to be two kinds of scientists involved in transdisciplinary projects: disciplinary experts, who are responsible for contributing their disciplinary knowledge and methodology, and transdisciplinary scientists, whose expertise lies in methods of knowledge integration. The latter are responsible for the overall research design, the embedding of the project into the real world context, and the successful integration of disciplinary and real world knowledge through a scientific methodology. At the same time, their research focuses on evaluating and improving these methods (Bammer, 2005; Tress et al., 2005).

Starting from these insights, we derive the following principles for transdisciplinary research on complex human–environment problems. Firstly, transdisciplinary research is the scientific treatment of a real world problem. Therefore the major challenge is the 'embedding' of the research design into the real world context. Secondly, integration of knowledge from different actors is crucial for the success of transdisciplinary research. Insights from science have to be combined with values and knowledge from society in a mutual learning process to ensure validity, objectivity and reliability of the results on the one hand, as well as utility, applicability and relevance on the other.

In order to meet the stated requirements, the TIPS approach is comprised of an epistemology, an organisational structure and a detailed methodology. The organisational structure and the methodology provide the framework for the knowledge integration process, while the epistemic model allows for a de-composition and re-composition of the problem in an analytic process. This also defines the interface between scientific and societal spheres (Scholz et al., 2006).

Figure 14.1 illustrates the epistemic model with the case of AR: on the one hand, the systemic sphere of the case is embedded into the normative sphere of the preference structures of the stakeholders. This embedding has to be taken into account in the research design. On the other hand, in the systemic sphere, the case can be described on different cognitive levels. The level of 'understanding' is the most encompassing. This is where the case study starts and ends. During the process of problem investigation, the case is conceptualised and analysed with disciplinary scientific methods, reducing the complexity of the case but increasing the concreteness (Scholz et al., 2006). The achieved results have to be integrated up to the level of understanding, in order to be of value for the stakeholders. A successful implementation process largely depends upon this synthesis process.

Regarding the project methodology and organisation, transdisciplinary projects usually differentiate between three phases: problem identification and structuring, problem investigation and problem transformation (or 'implementation', see Abstract). The project architecture of the TIPS approach is further elaborated, but these three phases can be distinguished as the basic framework (Table 14.1). Special

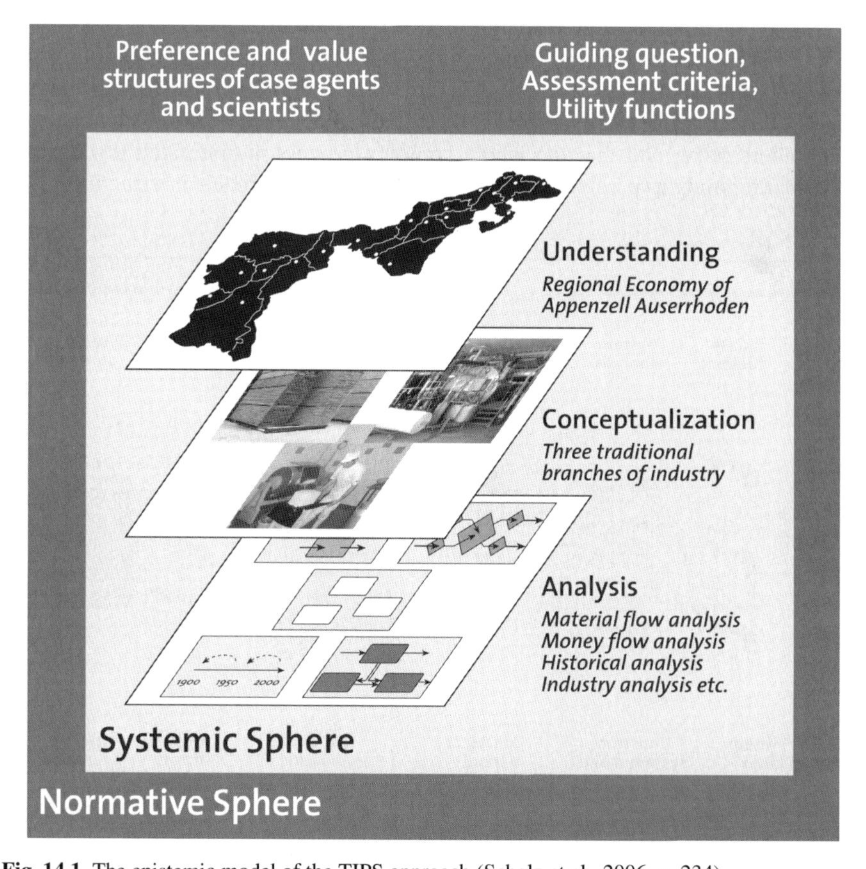


**Fig. 14.1** The epistemic model of the TIPS approach (Scholz et al., 2006, p. 234)

**Table 14.1** The different steps of the TIPS approach

| General project framework of transdisciplinary projects | Related steps in the TIPS approach | Knowledge types mainly involved in the specific step |
|---|---|---|
| Problem identification and problem structuring | Case definition<br>Case faceting | Target knowledge<br>System knowledge |
| Problem investigation | System analysis<br>Scenario construction | System knowledge<br>System knowledge |
| Implementation (referred to as problem transformation in this chapter) | Area development negotiations<br>Facet integration<br>Development strategies | Target knowledge<br>System knowledge<br>Target knowledge<br>Transformation knowledge |

attention is paid to the integration of different types of knowledge. System knowledge, target knowledge, and transformation knowledge are required to support sustainability transitions (Hirsch Hadorn et al., 2006; Chapter 1). Therefore, each step of the TIPS methodological framework uses specific methods to integrate these different types of knowledge with appropriate methods (Fig. 14.2).

To ensure that the different planning steps of TIPS lead to a meaningful result, a sound methodology, which relies on a functional perspective, is necessary (Wiek et al., 2006). In order to ensure that TIPS meets the expectations of scientists and stakeholders alike, and that the whole complex process is orientated towards the goals of the study, we rely on a backward planning approach in structuring the

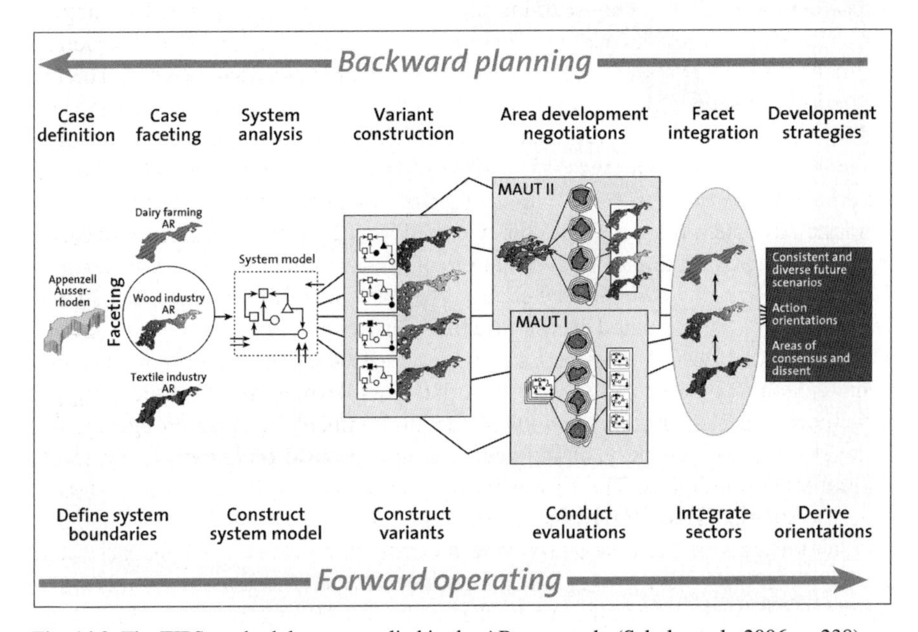


**Fig. 14.2** The TIPS methodology as applied in the AR case study (Scholz et al., 2006, p. 238)

process. All steps and procedures are functionally defined in relation to the goal of deriving operational strategies for the case (Holmberg and Robèrt, 2000; Scholz and Tietje, 2002). This aids in keeping the process aligned to the goals of the study.

Figure 14.2 illustrates the TIPS methodological framework. Starting with the intended goals and products of the study, each methodical step is defined in the planning phase according to the functional, formal and content related requirements of its successor. During the operational phase, the steps are completed in the other direction. As shown in Fig. 14.2, the case is first divided into a number of separate thematic facets. The following steps are undertaken for each one of the facets. After a systems analysis of the case facet, scenarios are constructed. The scenarios are assessed in two different multi-criteria assessment procedures, combining an expert and a layperson based approach. After that, the results and models of the different facets are reintegrated to subsequently derive development strategies for the case as a whole. The following sections give a more detailed insight into this process.

### *14.1.3 The Institutional Context: Organisation of the Transdisciplinary Case Study*

As the case study is organised as a transdisciplinary process, actors and institutions from the case are involved in establishing, leading, and conducting the case study. Transdisciplinary research has to rely on a multitude of interfaces between science and society to enable mutual learning processes and facilitate knowledge integration. Therefore, the TIPS approach comprises a complex organisational structure in order to bring together adequate representatives of society and science. The organisation of the case study reflects the position of the case study between science and society (Fig. 14.3). For each board on the science side there is a corresponding group on the case side, and vice versa. All responsibilities, rights and obligations of the groups have to be agreed upon in the preparation phase of the case study.

In the AR study, the two co-leaders from ETH and the canton presided over the *steering group* consisting of representatives of the scientific disciplines involved in the case (*scientific committee*) as well as representatives of the canton. The makeup of this group was determined during case faceting process. The *case working group*, which is a subgroup of the steering group, consisted of the governing chairman of the canton, the director of the cantonal business development, the secretary for agriculture and forestry, the director of the environmental protection agency, the regional secretary, private planning agencies, the cantonal planning agency, farmers and local politicians. The case working group met roughly every four weeks, while the steering group came together only roughly every two to three months. The operative leaders of the case study were the case study office (TdLab, ETH) and the Pivot (AR). The Pivot is a person that has to liaise between the scientists and the stakeholders. The scientific work takes place in *project groups*, of which there are as many as there are case facets. The project groups are counterbalanced on

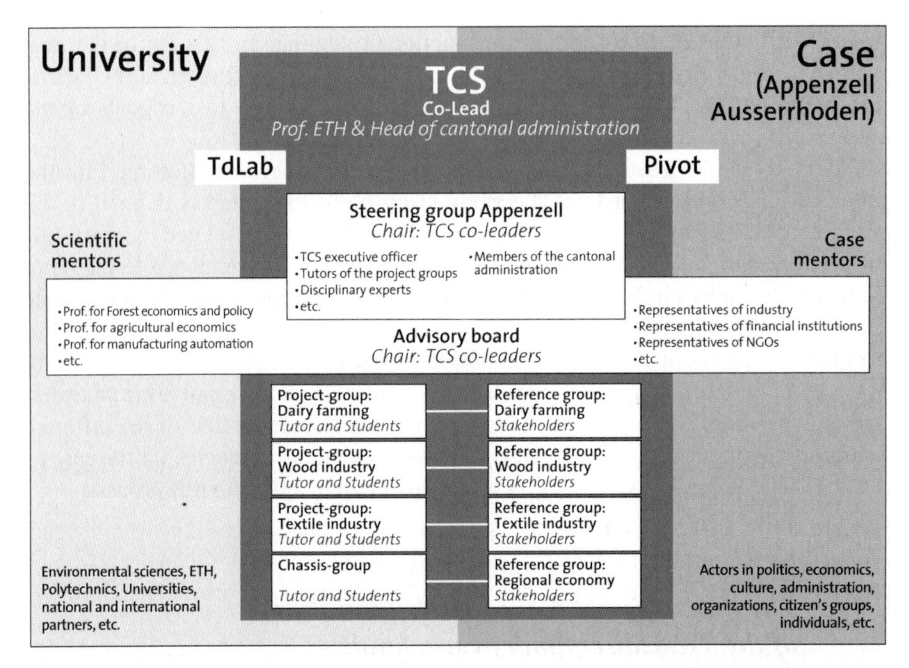


**Fig. 14.3** The organisational structure of the second year of the AR case study (Scholz et al., 2006, p. 243)

the case side with so called *reference groups*, i.e. committees of stakeholders, who are involved in the analysis of the case facets. The reference group meets regularly (every three weeks) with its corresponding project group to discuss the results and subsequent steps of the work. In terms of numbers, the different categories of agents involved in each project year are listed in Table 14.2.

One of the transdisciplinary requirements of the TIPS approach is the joint funding of the projects (materials and personnel). The canton agreed to bear a substantial proportion of the costs of the case study (ca. 30%), while the rest was funded from the university teaching and research budget. The canton also agreed to supply the

**Table 14.2** The agents involved in the Appenzell Ausserrhoden case study

| | Case study 01: Landscape development | Case study 02: Economic development |
|---|---|---|
| Students | 38 | 36 |
| Tutors | 14 | 7 |
| Headship | 3 | 3 |
| Administration | 2 | 2 |
| Steering Board | 12 | 11 |
| Reference Groups | 30 | 40 |
| Experts | 11 | 30 |
| Reviewers | 24 | 29 |

case study with the necessary data and expertise. These agreements were held down in a contract between ETH and the canton.

## 14.2 Problem Identification and Problem Structuring

As explained before, the TIPS approach deals with this phase in two steps. The next two sub-sections will deal in detail with the phases of case definition and case faceting.

### *14.2.1 Case Definition*

A case has to be selected from the viewpoint of science, which aims at deriving generally valid insights from a number of representative cases (Yin, 1984). At the same time, the case also has to be relevant from a societal perspective (Scholz and Tietje, 2002). Therefore, a case represents a general problem, but in a specific and unique shape. The question of the transferability of the results has to be part of the scientific research.

At the beginning, there is no 'case' and no interrelation between practitioners and scientists. Most often, an informal contact initiates a 'case'. Based on this, the next step in establishing a transdisciplinary research project is to determine whether transdisciplinarity is the appropriate method. Ill-defined problems (Scholz and Tietje, 2002), which touch upon societal norms and values but which are still at least partly treatable with scientific methods, are predestined for a transdisciplinary approach.

At this stage it is important that the rights and obligations of scientists and stakeholders are clarified. Both sides have to profit in some way from the project, otherwise collaboration will not occur in a goal oriented way. The kinds of 'products', which the case study will produce, has to be agreed upon. The result of this phase is a 'willingness to cooperate' between the stakeholders and the scientists, which should be formally recorded in letters of intent from both sides. It has to include a first definition of the problem to be worked upon. It is important, in this first phase, that all relevant decision makers are included in the process, to foster a development of problem ownership.

Subsequently, the main thematic areas of the case have to be identified, in order to prestructure and allow for the scientific handling of the case. Departing from the issue formulated in the letters of intent, important aspects of the problem are analysed applying problem structuring methods (Mingers and Rosenhead, 2004) to form a 'problem landscape'. The main direction of the case study is set in this step, which should not be subject to change in later stages of the process. Based on the problem structuring, the goals of the study should be identified and agreed upon. This procedure also leads to a clearer perception of the problem itself and forms a first common knowledge base of the problem that is supported by all stakeholders. The result of this phase is a first draft of the *guiding question*, which defines the issue and goal of the case study in a clear and generally understandable way. The guiding question connects a complex, real world problem to a societal goal system and can be seen as the basis for transdisciplinary cooperation.

In the case of AR, the team from ETH was looking for a case to use for its yearly transdisciplinary case study, while the canton of AR was struggling with the typical problems of peri-urban and rural regions: decrease of population and workplaces, level of service and quality of life. ETH had preselected a number of regions in Switzerland as being suitable for a case concerning sustainable regional development, and AR was one of them. In several meetings between the government of AR and ETH, a cooperation contract as well as a general project outline were defined. Both sides agreed that the TIPS approach would serve their needs. The canton agreed to take over the co-leadership and to co-finance the study. The village of Urnäsch was selected as a focal case area due to its representativeness for the *Hinterland* of AR and for other Swiss regions in a similar situation. There is no clear methodological procedure for this process, because the form of cooperation between the project partners is too informal to allow for structured methods. The operative procedure is most similar to 'setting the scene' in consensus building approaches (Susskind, 1999).

Two guiding questions were formulated during intensive problem structuring processes. For the first case study year, the guiding question was: 'How can the landscape of the canton Appenzell Ausserrhoden be preserved or improved in its ecological quality, while at the same time increasing the economic added value? For the second year, the guiding question was: What are the prerequisites for the regional economy in Appenzell Ausserrhoden in order to operate in a sustainable way and in harmony with the environment and the socio-economic needs?

Based on these questions, the goal of the case study was defined: to provide strategic orientations for the future development of the canton regarding the issues addressed in the case study. In the following preparatory phase, a cluster of theses and small explorative projects prepared the case study. Relevant sources of data were identified or newly acquired, e.g. through material flux analysis or psychological surveys. To model the influence of external factors on the canton, special context scenarios were constructed through a formative scenario analysis. They served as a joint reference point for all working groups.

### 14.2.2 Case Faceting

In order to treat the problem in question in a scientific way, the case and its guiding question have to be scientifically framed in addition to being agreed upon by the stakeholders. The case has to be embedded into a theoretical model of the situation, which draws from existing (and mostly disciplinary) scientific knowledge. Therefore, in cooperation with the stakeholders, the scientists create a general model of the problem situation, which allows for the application of relevant fields and their theories. This structure is then translated into different *facets*. These facets are used to divide the case on a conceptual level. The different facets have to be designed in a way that makes them scientifically treatable, but still allows for later synthesis of the results. It is important to plan the re-integration of these facets at the time of creating them. Moreover, there are always dependencies among the facets, which have to be carefully considered throughout the process by a cross-sectoral integration group. The goal of the faceting is the formulation of a rough *research concept*, which is written by both scientists and practitioners.

In the AR case, using a functional approach to landscape and regional research, a facet model was developed on the basis of the guiding questions. This facet model (for the first year) differentiated the three facets 'Settlements', 'Nature and landscape' and 'Leisure and tourism' as main aspects of the case according to the guiding question. The cross-sectoral integration group as a fourth group had the task of accompanying the process and ensuring an orientation towards the final integration of the results. The actual methodical procedure of re-integration was also a task of the cross-sectoral integration group (Wiek and Walter, submitted).

By following this procedure, at the beginning of the main project phase, the orientation of the project is already defined. A guiding question for the whole project exists as well as a facet model, dividing the subject area into interconnected facets, each of which is analysed by one facet group. A case area has been selected which is considered representative for the region and also offers the structural and organisational possibilities for conducting a case- study. Most importantly, a research concept clearly sets out the methods to be used and how these methods feed into each other to create a final result.

In the AR case study, it was decided to conduct a system analysis of the problem situation, to identify the most important impact factors. The identified impact variables were then used to extrapolate the current state into the future, with the help of a scenario construction method. The resulting scenarios were then to be rated by the stakeholders as well as by the scientists to deliver results in the form of preference structures for different developments of the case. This methodology was to be conducted in each facet group. In the end, the results were integrated between the different groups, to form the base of a strategy derivation process for the whole canton (Fig. 14.1).

## 14.3 Problem Investigation

During the problem investigation phase, knowledge of the case is gathered from scientists and stakeholders and integrated into a general system model, which is based on the guiding question and on the facets. In the TIPS approach, a system analysis is used, based on soft operations research methods. The resulting system model is used to construct scenarios, which describe the possible future states the case could develop into. These scenarios are then assessed in a multi-attributive evaluation process, which already forms a part of the problem transformation phase (Section 14.5).

### *14.3.1 System Analysis*

The goal of this step is to come to a representation of the problem situation via a system model (Checkland, 1999). The system model integrates the heterogeneous knowledge of the stakeholders into one abstract representation of the case. In transdisciplinary workshops with stakeholders, the main impact variables of the model are identified and described, using moderation techniques and problem structuring methods. These workshops are moderated by transdisciplinary scientists. In a second step, the relations between these impact variables are classified in an impact matrix, leading to a system model of the case. This process is organised in parallel groups, in order to estimate the validity of the representation and to identify areas which have to be discussed in detail. The system model is constructed according to the principles of sufficiency and adequacy (Scholz and Tietje, 2002). The goal is to construct a functional system model, which serves as a basis for the construction of case scenarios in the next steps and represents the case well enough to draw necessary conclusions from it. Moreover, the system modelling is an important learning process for the stakeholders, which results in a shared representation of the jointly constructed case situation. This constructive aspect greatly enhances the learning process (Stauffacher et al., 2006).

In a second step, the system model is analysed, applying software to identify classes of variables and important feedback loops. The results are discussed with representatives from the case, in order to test the results for validity and achieve new insights into the case. This process leads to a new conceptualisation of the case: important but so far neglected feedback loops and interrelations are integrated into the stakeholders' and scientists' cognitive representation of the case. The functional result of this step is a set of impact variables that can be used for the construction of case scenarios.

The system analyses for the AR case study were separately conducted by each facet group including students and a reference group. In the second project year, focusing on regional economy, this step encompassed analyses of the historical industrial development, agent networks, production chains, creation of added value, and economic, social, and environmental performance indicators of exemplary companies. Based on the data gathered, relevant system variables were derived and selected. Finally, these variables and the evaluated data were integrated into a semi-quantitative system model of the specific regional business sector.

### *14.3.2 Scenario Construction*

Based on the system models, future scenarios of the case are constructed in each facet group (Scholz and Tietje, 2002; Scholz et al., 2006; Wiek et al., 2006). In an initial workshop, possible future developments of each impact variable are identified. In a second workshop, the possible future developments are combined and rated according to their pair-wise consistency, i.e. the possibility of being realised simultaneously.

The identified possible developments of the impact variables and their corresponding consistencies are then combined, again with the help of software (Tietje, 2005), to form all possible case scenarios. This process leads to a great number of case scenarios, which are then ordered by their total consistency, derived from the pair-wise consistency values. The highly consistent case scenarios are then compared with intuitive case scenarios, which were developed by the stakeholders beforehand, on the basis of the system analysis. The results are fed back into the transdisciplinary process. For the assessment step, the case scenarios are visualised in order to give the assessing groups (which are also stakeholders) an adequate basis for their cognitive approach.

In the second year of the AR case study, based on the system model, four to six scenarios were constructed for each industry, which described possible future states of the entire sector and its companies. For instance, the five scenarios for the timber industry were: (i) 'Business as usual', (ii) 'Active marketing', (iii) 'Large firms', (iv) 'Special products', and (v) 'Timber cooperation' (Table 14.3). Because these scenarios strongly depend on developments of external factors, their robustness under previously constructed context scenarios was additionally estimated.

## 14.4 Problem Transformation

The problem investigation phase has already generated a number of products, which are built on a transdisciplinary knowledge base and constructed on a basis of scientific methods combined with the stakeholder's expertise. One part of the requirements for transdisciplinary processes, the mutual learning process between science and society, has therefore been fulfilled. Because of this process, the stakeholders already have an understanding of the situation and possible future developments. The classical implementation step, which starts with the communication of the study results to the stakeholders, is therefore replaced by a new process, which we refer to as problem transformation.

To actually start a problem transformation process based on the newly acquired knowledge, two things are necessary (Bammer, 2005): firstly, the knowledge generated needs to be linked to the values and preferences of the stakeholders. Otherwise, no prioritisation of actions is possible, and the result of the project will be a long list of desirable measures, without any implementation plan. This is why in the TIPS approach, the next step of the project architecture (Fig. 14.2) is a multi-attributive assessment of the scenarios, both by stakeholders and by scientists. Secondly, the results generated need to be synthesised to a level that is understandable by stakeholders and decision makers (Fig. 14.1). The language and framing has to be changed from a scientific perspective to fit into the specific setting of the case. The facets defined in the case faceting step have to be re-integrated again (Fig. 14.2). Also, disciplinary fragmentation has to be overcome to deliver a full picture to societal decision makers, who are not used to thinking in terms of scientific disciplines. Otherwise, only incoherent results will be produced, which are not usable in decision

**Table 14.3** The five scenarios derived for the wood industry in AR as derived by the formal scenario construction process (Scholz et al., 2003, translated by the authors)

| Scenario Impact Var. | Business as usual | Active marketing | Large firms | Special products | Timber cooperation |
|---|---|---|---|---|---|
| Training and development | Little choice | Strongly enhanced | Little choice | Strongly enhanced | Strongly enhanced |
| Operating costs (CHF per year) | Small firms: >100,000 Medium-sized firms: >450,000 Large firms: >1 Mio. | Small firms: >100,000 Medium-sized firms: >450,000 Large firms: >1 Mio. | Small firms: >100,000 Medium-sized firms: >450,000 Large firms: >1 Mio. | Small firms: >100,000 Medium-sized firms: >450,000 Large firms: >1 Mio. | Small firms: >100,000 Medium-sized firms: >450,000 Large firms: >1 Mio. |
| Relations to other firms | Strong competition | Competition of products | Strong competition | Strong competition | Cooperation |
| Cash-flow (CHF per year) | Less than 30,000 | More than 110,000 | More than 110,000 | More than 110,000 | More than 110,000 |
| Energy balance (kWh per $m^3$ roundwood) | Purchase of more than 40 | Purchase of more than 40 | Purchase of more than 40 | Purchase of more than 40 | More or less balanced |
| Investment | Not possible | Possible | Possible | Possible | Possible |
| Know-how | Low | High | Low | High | High |
| Customer-oriented production | Not available | Available | Not available | Available | Available |
| Relations to supply chain | Bad | Good | Good | Bad | Good |
| Marketing | No marketing | Strong marketing | Strong marketing | Strong marketing | No marketing |
| Machines and infrastructure | Old | Old | State of the art | Old | State of the art |
| Product type | Mainstream | Special products | Mainstream | Special products | Special products |
| Certification | None | All | All | None | Some |

support. Only after this process has taken place, can the decision makers together with the scientists work on actual decision alternatives, i.e. concrete measures or strategies (transformation knowledge, Table 14.1). This procedure also avoids an early lock-in on specific action plans.

### 14.4.1 Area Development Negotiations

The method of area development negotiations (ADN) was developed by Scholz et al., 1996 to combine a participatory process of land use planning with knowledge from psychological decision research on negotiation and bargaining (Bazerman and Neale, 1993; Pruitt and Carnevale, 1993). It can be considered an 'analytic variant of mediation' (Loukopoulos and Scholz, 2004). Loukopoulos and Scholz (2004) describe the procedure in detail. The goal of ADN is to reach a consensus among stakeholders, but in a systematic and transparent fashion. Decisions are made only after the diversity of values, interests and knowledge bases of the stakeholders has been mapped out. In a TIPS case study, ADN can rely on the cooperation that already has been initiated and consolidated in the preceding phases of the project and is therefore relatively easy to implement (Scholz and Stauffacher, 2007). The function of ADN in this context is to map out the hidden preference structures of the stakeholders and compare their analytical and intuitive judgments.

In order to assess the scenarios developed during the problem investigation phase on a common basis, a set of criteria is developed by the facet groups, based on the concept of sustainable development, defining different sustainability domains. Different aspects of these domains are deduced, translated into criteria and operationalised by indicators, leading to one hierarchical goal system of sustainability indicators. This process is coordinated closely between the different facet groups. To define the relative importance of the criteria, weightings are assigned according to the preferences of the agents. The combination of the interests, preference structures and utility functions of the stakeholders into a goal system can again be considered a integration process (Wiek and Walter submitted).

In two separate assessment processes the case scenarios are evaluated by the stakeholders as well as by the scientists in a so called 'exploration parcours' (Loukopoulos and Scholz, 2004). The stakeholders rate the case scenarios intuitively as well as on the basis of the aforementioned indicators, while the scientists make a data-based assessment of the indicators. This process represents a synthesis process and leads to an improved case understanding. From a functional perspective, the ratings of the case scenarios can be used to identify the most desirable scenarios for the final synthesis and the subsequent strategy development.

In AR, six ADNs were conducted in total, one for each facet group. To bring the scenarios into a form that can be easily compared and rated by non-experts, the facet groups prepared visualisations of the scenarios for the exploration parcours, in the form of posters and audiovisual presentations. Each participant of the ADN (about 120 in total) spent about 120 minutes on the parcours. First, he was welcomed by one of the members of the facet groups, who presented him with the visualised scenarios in random order. After a break, the participants rated the scenarios intuitively on a scale of 1–100. In a second step, the participants rated the scenarios for each sustainability indicator, also on a scale of 1–100. Finally, the participants weighed the sustainability indicators. In this way, an extensive quantitative data set is obtained, which allows comparing different stakeholder groups as well as differences between intuitive and indicator-based rankings. Additionally, the results can be compared to a scientific assessment, which is based on an expert rating of the sustainability indicator scores. The results of the exploration parcours served as a basis for discussion between the stakeholders, mapping out their differences in values and interests in a very clear and objective way.

The facet group concerned with textile industry came to the results shown in Fig. 14.4 and Fig. 14.5. In Fig. 14.4 it can be seen that the intuitive, indicator based and scientific assessments of the scenario 'Minimal cooperation' are very different. In this way, misperceptions of the stakeholders can be disclosed. Also, the scenario 'Full integration' was rated very low in the intuitive process, while it scored high in the indicator based assessment. This is a typical result of an exploration parcours. Feeding this result back to the stakeholders forces them to deal with their own cognitive dissonance and misperception, initiating very strong learning processes. In Fig. 14.5 it can be seen that the different groups of stakeholders rate the scenario 'Full integration' very differently, while the scenario 'Sharing of resources' ranks very high with all groups. This data provides important information, about which possibilities of consensus exist, and aids in the following negotiation and implementation process.

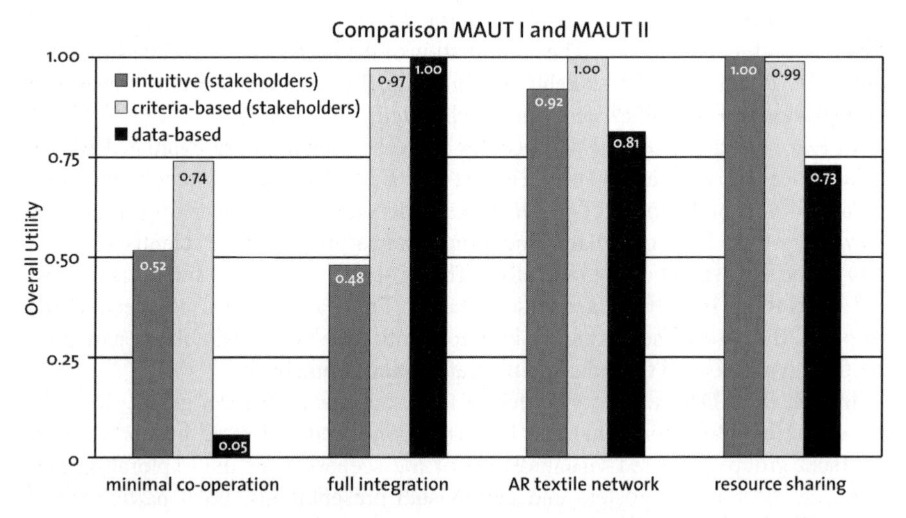


**Fig. 14.4** Results of the preference evaluation during the area development negotiations in AR (Scholz et al., 2003, p. 99, translated by the authors)

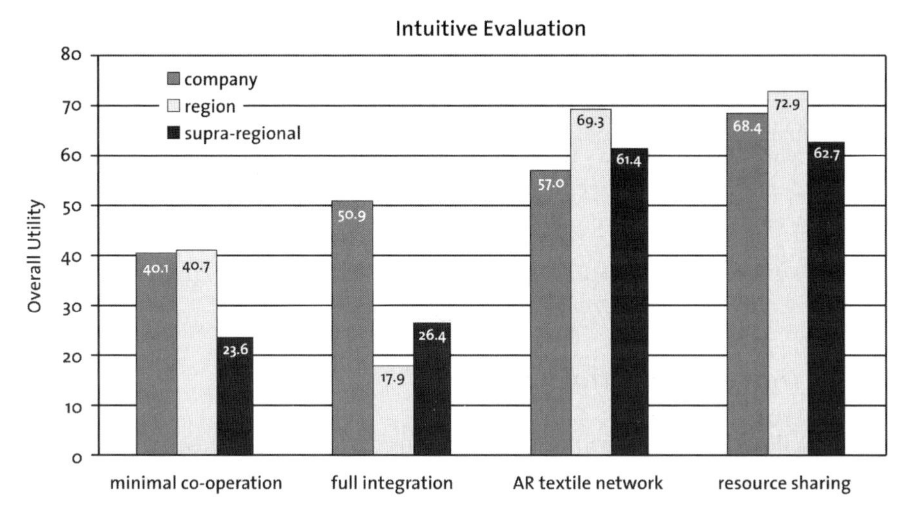


**Fig. 14.5** Results of the preference evaluation during the area development negotiations in AR (Scholz et al., 2003, p. 96, translated by the authors)

## *14.4.2 Facet Integration*

The regional development case study in AR was the first TIPS case study with a full integration of the results (Wiek and Walter, submitted). The final step of the methodology consisted of the derivation of strategies for cantonal development, as agreed upon in the guiding question. In order to reach that goal, the results of the three different facet groups had to be integrated into one model of the entire case area. Furthermore, these results had to be extrapolated to the whole canton from the smaller case area. This final synthesis was taken out by the cross-sectoral integration group, relying on the results produced by the three facet groups. A great deal of coordination and standardisation was necessary to achieve the data quality that is needed for a final integration, a task that could only be accomplished through an accompaniment of the process and not as an ex-post synthesis. Therefore, the cross-sectoral integration group was constituted at the very beginning of the study and had the task of ensuring a working synthesis process of the results.

The standardisation took place at a formal level as well as at the content level. Parameters of formal standardisation were the level of detail of the results, the terminology used, the consistent use of the methodology and the form of the results of the different steps. The form of the results was clearly defined to avoid the typical situation of an ex-post synthesis where the acquired results are on the same subject, but different in their formal nature, so that an extensive translation process is needed to make them comparable. Regarding the content, the results had to be standardised if the facet groups worked simultaneously on the same subject, e.g. the assessment criteria: all facet groups used the same set of assessment criteria and therefore developed this set jointly. Other areas of coordination included the design and controlling of the timetables of the individual facet groups as well as the research design of specific sub-projects, which could benefit from the coordinated actions of the facet groups.

To increase the reliability of the results, the final synthesis was done in two parallel work flows, relying on intuitive and formative methods, respectively. The results of the sub-groups were combined in several sub-steps to form bundles of case scenarios, which were consistent, extrapolatable to the whole canton and desirable.

### *14.4.3 Development Strategies*

After the described steps the integration groups disposed of 24 case scenario bundles, each of which had an external robustness value, an extrapolation value and a desirability value. As expected in a process of this complexity, no specific bundle turned out to be an 'overall winner'. After clustering the 24 bundles according to similarity criteria, representative bundles for each cluster were selected. The goal of this selection process was to select a variety of diverse bundles, in order to include all possible development paths. The selection process was not simply criteria driven, but included considerations about the applicability of the different selection criteria: a bundle that had a very low extrapolation value could be a part of a more spatially diverse development strategy for the whole canton. Additionally, special synergies between case scenarios, which were not included in the internal consistency analysis, could increase this value (Wiek and Walter, submitted).

The selected bundles were then worked out in detail, using the mathematical description as a basis for a 'thick' description of the possible future state. This step again relied on the input of the stakeholders. Following a general framework, strategies were described to reach these scenarios. The framework consisted of different important aspects of strategy derivation, based on a functional view of development. It comprised the main function of the landscape; possible actuators of the development; specific roles that had to be taken on by key actors in order to push forward the development; necessary decisions; the identity of the city region in this strategy; as well as a general description illustrating the strategy for laypersons. These strategies were finally presented and discussed with a wider audience. So far, they have led to several scientific follow-up projects with accompanying implementation of parts of these strategies (Scholz and Stauffacher, 2007; Binder et al., 2004) as well as to important political decisions about the future of tourism in the case area: a positive decision for a familiy holiday resort was made with clear reference to the case study results. An evaluation of the social impact of the case studies on AR further clarifies how strong the impacts of the TIPS methodology were and which mechanisms produced this impact (Walter et al., 2007). In earlier evaluation studies it could be demonstrated that a high diversity of the participants together with an intensive process of knowledge integration is an important success factor for multi-actor, complex decision- making processes (Walter and Scholz, 2007).

## 14.5 Recommendations

The TIPS approach is best suited for problems that could otherwise not be treated by scientific research, because they are too complex and inherit too many uncertainties and diverging interests (Lawrence and Desprès, 2004; Pohl, 2005). This applies to transdisciplinary research in general as well as to the TIPS approach.

Due to its synthesis module, the TIPS approach is especially suited for situations in which the problem is not easily addressed by traditional organisational or economic structures, such as economic sectors, branches of industry or policy areas. It is also suitable for situations with multiple stakeholders with different interests and perspectives on the problem.

The TIPS approach is a tool for producing long-term strategic orientations in development questions. It is not suited for tightly defined problem situations, e.g. the choice between two different measures of road pricing. Also, in problem areas of high political activity and therefore highly controversial discussion it is difficult to bring together all the involved stakeholders. This is a general problem of transdisciplinary research and should be subject of further investigation.

We strongly recommend that transdisciplinary projects be based on a clearly structured project architecture, to ensure that the complex real world problem stays in the focus of the project. The strength of the TIPS approach lies in (i) treating the case from an integrated perspective, (ii) through clearly defined project stages, and (iii) while integrating the knowledge and values of stakeholders throughout the process.

When applying the TIPS approach, the overarching project architecture and the embedding into the societal context are the two crucial success factors regarding the transdisciplinary interaction between scientists and stakeholders.

The process of embedding has to be carefully planned and flexible enough to handle unforeseen developments, such as a fundamental change of objectives by the stakeholders. Stakeholders should be included in the process as early as possible. Scientists have to be aware that their opinion is just one among many. Negotiation between scientists and stakeholders requires a great deal of mediation skill, which probably could also be made available through an external mediator.

An important role for the transdisciplinary interface is also that of the Pivot. This person is an expert of the case, living or working in the area, but with as few official institutional memberships as possible. On the other hand, he or she should be an excellent networker and connect the key players of the case to the scientific team and vice versa. Recently retired key players of the case are ideal for this position.

When negotiating, it is wise to be demanding of the stakeholders, keeping in mind what they get from the research project. Transdisciplinary research is cooperative; it is not a consultancy process. The interests of the scientists have to be respected as well, e.g. the acquisition of data, the write-up of research papers, and full access to important information.

It is important that the organisational structure of the study is fixed before starting with the subject matter. Likewise, the project architecture and the guiding question have to be agreed upon in writing (through letters of agreement) before the beginning of the actual research process. It is very difficult and time-consuming to change these fundamental structures during the case study process, and can lead to an abortion of the project. After the project has gone through the basic steps and reached specific strategies, the discussion on organisation and methodology can once again be opened up. Independent from that, the subject matter can be discussed controversially through the whole process.

When dealing with the design of the guiding question, it is important to abstract from the daily routine and decision structures of the involved stakeholders. By expanding the planning horizon towards long-term, strategic planning, fundamental constraints that are in effect today can be overcome more easily. Otherwise, the stakeholders can become lost in detail and are unlikely to reach a consensus on the guiding question.

During the process of problem solving, it is important for the scientist to always consciously differentiate between descriptive and normative knowledge. One of the main tasks of the scientific team is to keep a minimum scientific standard for the process, data and products throughout the whole process. This has to be defended against the interests of the stakeholders, who are often more interested in a quick solution than an exact one. Therefore, minimum quality standards should be agreed upon from the beginning.

## References

Bammer, G.: 2005, Integration and implementation sciences: Building a new specialization, *Ecol Soc* 10(2), Art. 6.

Bazerman, M. and Neale, M.: 1993, *Negotiating rationally*, The Free Press, New York.

Berger, P. and Luckmann, T.: 1966, *The social construction of reality: A treatise in the sociology of knowledge*, Anchor, Garden City.

Binder, C., Hofer, C., Wiek, A., and Scholz, R.W.: 2004, Transition towards improved regional wood flows by integrating material flux analysis and agent analysis: The case of Appenzell Ausserrhoden, Switzerland, *Ecol Econ* 49(1), 1–17.

Checkland, P.: 1999, *Soft systems methodology a 30-year retrospective*, Wiley, Chichester.

Clark, W.C. and Dickson, N.M.: 2003, Sustainability science: The emerging research program. In: *Proceedings of the National Academy of Sciences* 100, 8059–8061.

Hirsch Hadorn, G., Bradley, D., Pohl, C., Rist, S., and Wiesmann, U.: 2006, Implications of transdisciplinarity for sustainable research, *Ecol Econ* 60(1), 119–128.

Holmberg, J. and Robèrt, K.: 2000, Backcasting – A framework for strategic planning, *Int J Sustain Dev World Ecol* 7(4), 291–308.

Lawrence, R.J. and Desprès, C.: 2004, Futures of transdisciplinarity, *Futures* 36, 397–405.

Loukopoulos, P. and Scholz, R.W.: 2004, Sustainable future urban mobility: Using "area development negotiations" for scenario assessment and participatory strategic planning, *Environ Plann A* 36, 2203–2226.

Thompson-Klein, J., Grossenbacher-Mansuy, W., Häberli, R., Bill, A., Scholz, R., and Welti, M. (eds): 2001, *Transdisciplinarity: Joint problem solving among science, technology, and society. An effective way for managing complexity*, Birkhäuser, Basel.

Meppem, T. and Gill, R.: 1998, Planning for sustainability as a learning concept, *Ecol Econ* 26(2), 121–137.

Mingers, J. and Rosenhead, J.: 2004, Problem structuring methods in action, *Eur J Oper Res* 152(3), 530–554.

Pohl, C.: 2005, Transdisciplinary collaboration in environmental research, *Futures*, 37, 1159–1178.
Pruitt, D. and Carnevale, P.: 1993, *Negotiation in social conflict*, Brooks/Cole, Belmont.
Ravetz, J.: 2000, Integrated assessment for sustainability appraisal in cities and regions, *Environ Impact Assess Rev* 20, 31–64.
Resnick, L.: 1987, Learning in school and out, *Educ Res* 16(9), 13–20.
Scholz, R.W., Bösch, S., Koller, T., Mieg, H.A., and Stünzi, J. (eds): 1996, *Industrieareal Sulzer-Escher Wyss: Umwelt und Bauen: Wertschöpfung durch Umnutzung [Industrial area Sulzer-Escher Wyss: Environment and construction: Value added through re-use]*, Vdf, Zurich.
Scholz, R.W., Lang, D.J., Wiek, A., Walter, A.I., and Stauffacher, M.: 2006, Transdisciplinary case studies as a means of sustainability learning: Historical framework and theory, *Int J Sustain High Educ* 7(3), 226–251.
Scholz, R.W. and Stauffacher, M.: 2007, Managing transitions in clusters: Area development negotiations as a tool for sustaining traditional industries in a Swiss pre-alpine region, *Environ Plann A* 39, 2518–2539.
Scholz, R.W., Stauffacher, M., Bösch, S., and Krütli, P. (eds): 2003, *Appenzell Ausserrhoden – Umwelt Wirtschaft Region [Environment Economy Region]*, Ruegger, Zurich.
Scholz, R.W., Stauffacher, M., Bösch, S., and Wiek, A. (eds): 2002, *Landschaftsnutzung für die Zukunft: der Fall Appenzell Ausserrhoden. ETH-NSSI Fallstudie 2001 [Utilisation of landscape for the future: The case of Appenzell Ausserrhoden]*, Ruegger. Zurich.
Scholz, R.W. and Tietje, O.: 2002, *Embedded Case Study Methods: Integrating quantitative and qualitative Knowledge*, Sage, Thousand Oaks.
Stauffacher, M., Walter, A.I., Lang, D.J., Wiek, A., and Scholz, R.W.: 2006, Learning to research environmental problems from a functional sociocultural constructivism perspective: The transdisciplinary case study approach, *Int J Sustain High Educ* 7(3), 252–275.
Steiner, G. and Posch, A.: 2006, Higher education for sustainability by means of transdisciplinary case studies: An innovative approach for solving complex, real-world problems, *J Cleaner Product* 14, 877–890
Susskind, L.: 1999, *The consensus building handbook a comprehensive guide to reaching agreement*, Sage Publications, Thousand Oaks.
Tietje, O.: 2005, Identification of a small reliable and efficient set of consistent scenarios, *Eur J Oper Res* 162(2), 418–432.
Tress, B., Tress, G., and Fry, G.: 2005. Integrative studies on rural landscapes: policy expectations and research practice. *Landscape Urban Plan* 70(1-2), 177–191.
Walter, A.I., Helgenberger, S., Wiek, A., and Scholz, R.W.: 2007, Social impact evaluation of transdisciplinary research, *Eval Program Plan* 30, 325–333.
Walter, A.I. and Scholz, R.W.: 2007, Critical success conditions of multi-actor transport planning: A comparative evaluation using rough set analysis, *Transportation* 34(2), 195–212.
Wiek, A., Binder, C., and Scholz, R.W.: 2006, Functions of scenarios in transition processes, *Futures* 38(7), 740–766.
Wiek, A. and Walter, A.I.: submitted, A formalized transdisciplinary approach to integrate sectoral planning and decision-making in complex systems, *Eur J Oper Res*.
Wilson, B. and Myers, K.: 2000, Situated cognition in theoretical and practical context. In: D. Jonassen and S. Land (eds), *Theoretical foundations of learning environments*, Lawrence Erlbaum, Mahwah, pp. 57–88.
Yin, R.: 1984, *Case study research: Design and methods*, Sage, Thousand Oaks.

# Chapter 15
# Evaluating Landscape Governance: A Tool for Legal-Ecological Assessments

**Marianne Penker and Hans Karl Wytrzens**

**Abstract** If, and to what extent, landscapes can be governed by legal and social norms was the guiding question of a transdisciplinary research project funded by the Austrian Cultural Landscape Programme. The project involved three research organisations and several partners in public administration, politics and business. In a common effort, they developed a new methodological tool for the evaluation of legal effects on landscape development. The legal-ecological assessment draws on the basic assumption that legal regulations cannot impact on the ecological sphere directly, but can only intend to influence human behaviour.

Deliberately or not, human activities shape landscapes and impact on their aesthetic, recreational and ecological functions. The sociology of law and anthropology of law are scientific disciplines that are dedicated to the analysis of the interdependencies between law and society. Interestingly, the relationship between law and landscape has not yet attracted major scientific attention. Nevertheless, numerous legal regulations are not only supposed to control the social system, but in the case of environmental law, they are also intended to influence ecosystems, biodiversity, or landscape development. The extent to which law can help to govern landscape development is examined in the remainder of this chapter, using the agricultural landscape and the associated socio-economic and ecological processes as an example.

The tool for systematic impact assessments is based on a framework of crucial determinants that describe legal effects on human behaviour and landscape development and on procedures to assess the extent of legally induced landscape effects. The assessment tool has been applied for different types of regulations in three Austrian case studies. The empirical research indicates varying gaps between the intended and actual landscape effects, which might be explained by different determinants, such as the actual knowledge and acceptance of norms by the land users or frequency and average severity of controls and actual penalties.

✉ M. Penker
Department of Economics and Social Sciences, University of Natural Resources and Applied Life Sciences, Vienna, Austria
e-mail: marianne.penker@boku.ac.at

G. Hirsch Hadorn et al. (eds.), *Handbook of Transdisciplinary Research*, 

Based on the theoretical and empirical insights into the mechanisms of how law impacts on landscape, an 'outlook' section reflects on opportunities for more effective landscape governance in the future. In the context of this project, the transdisciplinary approach proved a successful procedure, which was based on the personal dedication of all people involved inside and outside academia and their willingness to contribute to integrative intellectual dialogs.

**Keywords:** Law · Implementation · Landscape · Evaluation · Effective governance

## 15.1 Background

The transdisciplinary research project 'Law and its Impact on Cultural Landscapes' analysed the interrelationship of law on landscape development in order to understand more about the mechanisms of landscape governance by law (Wytrzens et al., 2001). It involved experts in sociology, political sciences, environmental and international law, agricultural economics and landscape planning (affiliated with the 'University of Vienna', 'The Interdisciplinary Centre for Comparative Research in the Social Sciences' or the 'University of Natural Resources and Applied Life Sciences Vienna') and numerous representatives from administration, legislation and agriculture. The project was carried out between 1997 and 1999 and financed by the Austrian Federal Ministry of Education, Science and Culture within the Austrian research programme 'Cultural Landscapes'.

For centuries, cultural landscapes have been shaped by agricultural land use, and thus by the management decisions of farmers in local rural communities. Today's society, however, tries to govern landscape development centrally. Countless environmental regulations, contracts with landholders, agri-environmental schemes, and nature reserves are only some examples of how society tries to prevent unintended landscape change and thereby intentionally intervenes in landscape

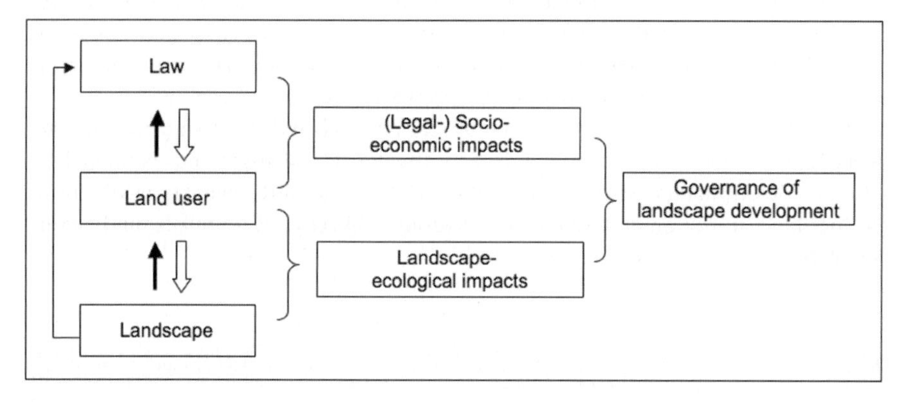


**Fig. 15.1** Landscape governance (adapted from Wytrzens et al., 2001)

development. Because of this, there is a need for general knowledge about the interrelation between law and landscape development, the mechanisms and outcomes of landscape governance.

To meet this need, the team followed a deductive (top–down) approach (Fig. 15.1). The (legal) socio-economic analysis examines the impact of law on the individual or collective activities of land users and distinguishes between the normative level (as reflected in the legal text) and the actual impact of law on social life (compliance with law and administrative execution). How the land users' actions influence the landscape is part of the second step of analysis: the landscape-ecological investigation. Only the integration of legal-sociological and landscape-ecological results promises information about the legal impact on landscape and can provide insight into the governance of landscape development.

## 15.2 The Concept of Legal-Ecological Assessments

'Legal-ecological assessments' draw on scientific concepts and theories from the sociology of law (Opp, 1973; Rehbinder, 1995; Rottleuthner, 1987), legal anthropology (Benda-Beckmann, 1995; Benda-Beckmann, 2001), and landscape ecology (Bastian and Schreiber, 1999; Naveh and Lieberman, 1994). In the empirical investigation, this interdisciplinary background is combined with the lay knowledge and expertise of those who are actually affected by legal institutions, in our case, farmers and those in charge of implementing agri-environmental and landscape law.

A literature review revealed factors that might determine the impact of law on society and the landscape. The complex of determinants illustrated in Table 15.1 differentiates between (legal) socio-economic and landscape ecological impacts. The (legal) socio-economic determinants define the application of (e.g. probability of sanctions), and compliance with the law (e.g. understanding of regulations, acceptance of the goals driving laws), while landscape ecological determinants define the environmental impacts (e.g. on the type and size of material and energy flows) of behaviour that has been induced through legislation. These determinants form a set of hypotheses, which on the one hand, provide a reference point for the empirical surveys. On the other hand, the empirically tested set of determinants might serve as an evaluation structure for assessments of other laws.

The success of a particular regulation might well be fundamentally determined through the way in which its text and meaning are presented, for example through clear and understandable language or through unambiguous and consistent objectives. However, its practical impacts are only brought about through the intermediary activities of those individuals who are responsible for implementing legislation (as representatives of the public administration), or those who are the targets of this legislation and can transform laws into concrete activities with relevance for the landscape. The motives and activities of these individuals were surveyed empirically in case studies. The results might shed some light on those factors determining why some laws are successfully implemented and others not.

**Table 15.1** Determinants representing the impacts of legislation on farmers and the landscape (adapted from Wytrzens et al., 2001)

| Det. class | Determinant group | Determinant | Operationalisation (indicators) |
|---|---|---|---|
| Determinants at the normative level | Impact dimension | Functional | Number and relevance of management measures subject to legislation |
| | | Personal | Number and relevance of affected groups of farmers |
| | | Temporal | Duration of obligations |
| | | Spatial | Size and agricultural relevance of the land area subject to legal obligations |
| | Willingness to implement | Intended administrative input | Resources allocated for enforcing the regulation |
| | | Control and monitoring | Control and monitoring measures defined in the legislative text itself |
| | | Definition of clear objectives | Set targets (clearness, limits, indicators) |
| Determinants of actual socio-economic impacts | Compliance | Subjective motivations | Legal knowledge |
| | | | Legal awareness (farmer's acceptance of the objectives and instruments used to achieve them) |
| | | | Legal ethos (identification with the principles of the legal system) |
| | | Economic motivations | Subjective expectations regarding the pros and cons of policy-driven behaviour (increased expenditure, lower yields, reduced likelihood of penalties, incentives, lower tax levels) |

**Table 15.1** (Continued)

| Det. class | Determinant group | Determinant | Operationalisation (indicators) |
|---|---|---|---|
| | | Technical motivations | Ease of integration within existing agricultural activities, technical equipment, required knowledge and skills |
| | Enforcement | Organisational structure and resources | Average duration of administrative files |
| | | | Financial and human resources |
| | | Administrative customs | Number of administrative files |
| | | | Penalties: frequency and average severity |
| | | | Frequency of control |
| Determinants of actual landscape-ecological impacts | Type and extent of influence on flows of materials and energy | Material and energy throughput | Quantity of enterprise inputs used |
| | | | Environmental harmfulness of enterprise inputs used |
| | | Reversibility | The chances of re-establishing original conditions (duration and costs of the required process) |
| | Type and extent of influence on spatial and structural aspects | Intervention affecting the distribution of different land uses | Changes in the relative areas occupied by different land uses as a result of behaviour induced by legislation |

**Table 15.1** (Continued)

| Det. class | Determinant group | Determinant | Operationalisation (indicators) |
|---|---|---|---|
| | | Intervention affecting habitat diversity | Changes in diversity (number of different habitats and relative areas occupied by each) as a result of behaviour induced by legislation |
| | | Intervention affecting biotope networks | Changes in the ratio between edge lengths and total area as a result of behaviour induced by legislation (number of landscape elements) |
| | | Reversibility | The chances of re-establishing original conditions (duration and costs of the required process) |
| | Robustness of the affected part of the landscape | | Robustness specific to landscape type (dependent on soil properties, climate, hydrology, geomorphology) |
| | Endangered status of the affected part of the landscape | | Endangered status of the landscape elements affected by the management measure (e.g. proportion of red list species) |

If we look at those targeted and affected by the legislation, the farmers, then we see that there are various key factors determining whether standards are actually implemented in practice. If the farmer is to be able to behave in the way intended by legislators, then he or she must first be aware of the regulations in question. According to Opp (1973), the 'level of knowledge' about a regulation can be measured as a person's ability to classify behaviour into that which is allowed by law, and that which is forbidden by law. This ability should also be based on an understanding, which is itself based on the original law (as a direct or indirect source of information). The survey captured the legal knowledge of farmers by recording their basic awareness of individual regulations and their understanding of the particular contents of regulations, the likely sanctions for illegal activities and procedural issues.

An understanding of the law alone is insufficient guarantee for compliant behaviour. The perception of the law is also of key importance, i.e. whether the law is perceived as useful and effective or not. If the farmer identifies with the objectives of the regulation, and with the suitability and appropriateness of the measures used to bring about the intended effect, then it can be assumed that this farmer will be more likely to be willing to act in a way that complies with this regulation. In this context, Rottleuthner (1987) talks of 'internalised objectives', which means that the land user shares the legislators' objectives. Conformist behaviour can therefore mean the simultaneous achievement of personal goals, which has obvious positive implications for compliance. This perception of the law is a measure of the degree to which the land user identifies with the relevant regulations, and was recorded in the research as the extent of the subjectively perceived usefulness and appropriateness of selected regulations.

If a legal system is effective, then it is necessary to ensure that regulations are also followed when individuals do not consider them worthwhile. The willingness of those people to observe a particular law or to identify with the legal framework and guiding principles of the political system, despite whatever criticisms they may have, can be described as their legal ethics (Rehbinder, 1995). The legal ethics of the individual farmer can only be indirectly operationalised as the degree of inconvenience (the 'ethical sacrifice') the farmer is willing to suffer by obeying a regulation which he or she opposes in principle.

We can also take an economic approach that focuses on the costs and benefits (or disadvantages and advantages) incurred by the individual. Becker (1968) was one of the first scholars to regard economic motivators as critical in determining compliance with a regulation. Compliant behaviour generally means some kind of economic benefit or cost, since it is rare to find any changes to production behaviour that have been induced by legislation and have no impact on economic success. Compliance with a regulation can be associated with some benefit, such as a subsidy, premium or tax break. Avoiding the risk of those sanctions that might follow non-compliant behaviour can also be considered an indirect benefit. However, compliance with some regulations may also involve extra expenses in terms of farm inputs, labour or additional taxes. From an economic point of view, it seems reasonable to assume that legal regulations will be followed when the associated benefits of compliance outweigh the associated costs. If the extra expense is higher than the expected benefits, then we can expect farmers to try and get round the regulations; they may well act in a manner quite opposite to that which the regulations were intended to induce. Executive bodies are just as important as the targets of legislation in determining the successful implementation of some regulations. An appropriate and efficient administrative organisation with sufficient financial and human resources is a prerequisite for the successful implementation of a regulation. The type, extent, reversibility or persistence of landscape changes are examples of landscape ecological determinants that describe the landscape impacts of those activities induced through legislation.

## 15.3 Problem Structuring

The above outline indicates the complexity of the research task. The governance of landscape development is so multifaceted that the knowledge of one single discipline is insufficient and cannot provide an appropriate explanation. Therefore, the project team integrated scientists from various disciplinary backgrounds: three experts in environmental and international law, two in politics and sociology, one in agricultural economy and one in landscape planning. The challenge for scientific cooperation was to find a common understanding and language and the flexible handling of disciplinary boundaries. We decided on an eclectic handling of concepts and methods from law, sociology of law, legal anthropology, law and economics, agricultural economics and landscape ecology. The project rested on the broad involvement of numerous experts in public administration (Federal Ministry of Agriculture, Forestry, and Water Management, Federal Ministry of Environment, Federal Ministry of Finance, experts in agriculture and nature conservation from the nine provincial governments, street-level bureaucrats on the district level in Upper Austria) and practical land management (farmers in Upper Austria, representatives of farmers' unions in the nine provinces, NGOs engaged in conservation issues). Their expert knowledge was indispensable for understanding the complex interrelations between law, land use activities and the resulting landscape impacts.

Figure 15.2 gives the basic structure of the empirical surveys. The figure clarifies the key issues and the design of the case studies. The case studies address the actual impacts of legislation on management decisions and thus on the managed landscape thereby focusing on 'how' legislation has an influence. At the same time, cross-case conclusions provide information on 'why' aspects, i.e. they provide insights into the determinants underlying the impacts of legislation on farmers and the landscape and help to test the literature-based set of determinants.

In order to answer the two pronged question at the core of this research, the case studies draw on the theoretical complex of determinants (Table 15.1) and on two conceptual impact models describing socio-economic and landscape ecological processes (Wytrzens et al., 2001). The complex of determinants forms a set of hypotheses, and therefore provides a reference point for the empirical surveys in the context of the 'why' issue and the factors determining the impacts of legislation. Research into the 'how', dealing with the influences of legislation, is based on the

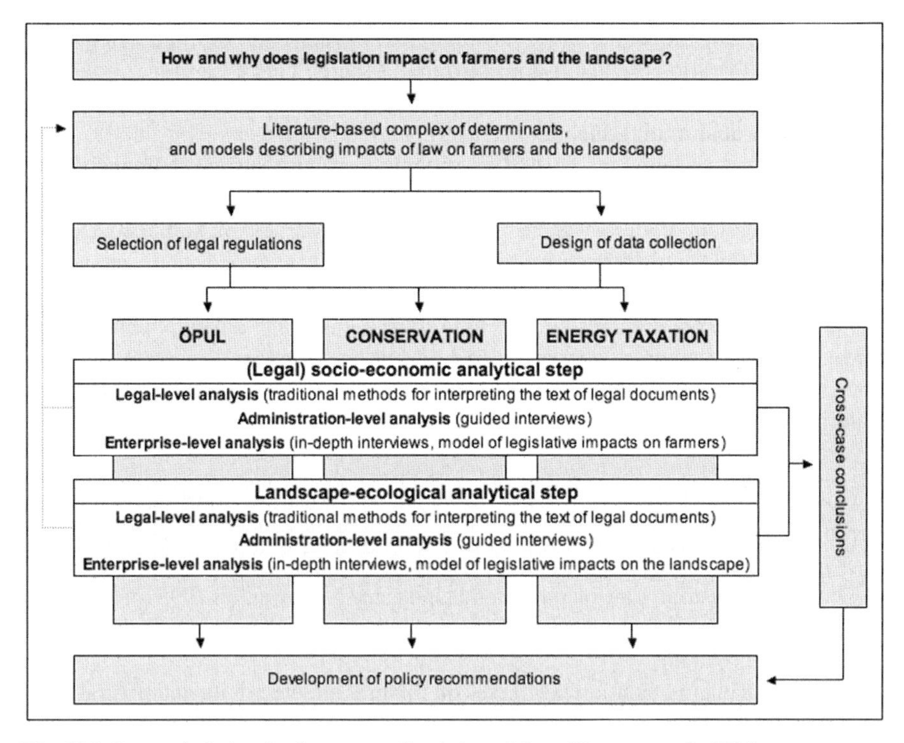


**Fig. 15.2** Research design for the case studies (adapted from Wytrzens et al., 2001)

models that address the legal impacts on farmers and the impacts on the landscape of changes in agricultural activity as brought about by legislation.

The theoretical framework of determinants provided a directional framework for the empirical surveys, clarified the theory-based intentions underlying the research approach and yet still gave the researchers enough flexibility to examine other aspects of legal impacts not yet formally accounted for.

The case study was designed so that surveys were carried out at three different levels: the legal level, the administration level and the farm level. A 'legal analysis' using the full range of methods applied in traditional legal text interpretation provided insights into the outcomes intended by legislators. This analysis was based solely on the evaluation and interpretation of legal texts and their accompanying documentation. The 'administrative' and 'farm' analysis then looked at the actual socio-economic impacts brought about by the regulations. Both analyses were primarily based on the results of guided interviews.

## 15.4 Investigation

The empirical analysis focused on agri-environmental regulations and their impact on agricultural landscapes. The rather confusing, disparate and extensive range of

Austrian agri-environmental laws and regulations were surveyed and categorised with the help of various legal and administrative experts. The result was a systematic collection of documents covering nearly 900 relevant laws and regulations applying at EU, federal and state levels (Penker, 1997).

It would not of course be possible to examine all the regulations in any depth, so a few standards were selected for use as exemplary, qualitative case studies. The selection process was based on a written survey of some 40 experts involved in the implementation of legislation, and recorded their subjective evaluations.

Particular relevance to sustainable agricultural landscape development was accorded to laws of nature conservation and to agri-environmental schemes. Accordingly, the study used the following regulatory and legal documents as the research material for the detailed case study analyses:

- Guidelines covering the EU co-financed Austrian agri-environmental programme – ÖPUL 95 (particularly the measures referring to 'stabilisation of crop rotations' and 'care of ecologically valuable areas');
- The Upper Austrian Law on nature and landscape conservation 1995 (particularly the regulations dealing with the licensing of clearing landscape elements and drainage activities);
- The federal energy taxation law (taxes on mineral oil, electricity and natural gas). This law acts as a kind of control, since it was expected to have no impact on the landscape.

The empirical surveys looked at both the subjective opinions held by farmers with regard to the objectives of legal regulation as well as the costs and benefits of compliant behaviour and the risk of sanctions, and at the number and extent of the actual control and punitive measures applied by public administrators in this regard. The interviews with the farmers also involved detailed discussion of qualitative and quantitative enterprise changes between 1994 and 1997, which were brought about through the legislation in question. Interviews with experts in public administration were used to identify the level of resources available for administrative tasks and services, for providing face-to-face advice, land assessments or control inspections.

The scope of the project was such that it was not possible to carry out independent landscape ecological surveys. In order to make empirically meaningful statements, surveys were used as a substitute for direct landscape ecological analyses. This means the study used expert surveys of key players in public administration and representative bodies, as well as farmers, to gain an understanding of landscape ecological scenarios. The survey results were then subjected to expert appraisal and interpretation using agri-ecological and landscape-ecological literature. The landscape-ecological conclusions drawn from this process only describe potential, or probable, landscape effects, because they do not take account of the natural characteristics of the landscapes concerned.

## 15.5 Results

The empirical surveys were carried out between June and November 1998, and identified various differences in the environmental and agricultural impacts of the regulations analysed. In general, however, the results broadly confirmed the theoretical assumptions concerning the determinants of these impacts. As the case study research was concluded several years ago, the following discussion of results will focus on more general issues and give only a summary of the outcomes regarding the individual regulations. A short outline of recent developments should provide the reader with some extra insight into what has happened in the meantime.

The ÖPUL measure, 'rotation stabilisation', led to rapid take-up (the programme only started in 1995). The case study results indicated positive impacts on the landscape due to extended rotations and green cover over winter and consequently reduced soil erosion. The ÖPUL measure, 'care of ecologically valuable areas', has been seeking to preserve those semi-natural habitats (e.g. traditional orchards 'Streuobstwiesen' and wet meadows) that are dependent on ongoing agricultural activity, and therefore threatened by the abandonment of agricultural land use, by intensification, and by fragmentation. The empirical results revealed that the subsidies could at best contribute to the broad preservation of these important landscape elements, but rarely led to the establishment of new habitats and landscape features (e.g. re-flooding of former wetland habitats, planting of traditional orchards). After our research project in 2000 with the Austrian Rural Development Programme 2000–2006, a new ÖPUL-measure was introduced: the conservation plan. This instrument compensates farmers for their efforts in developing individual conservation plans. The close cooperation of the farmers and conservation experts might provide the extra motivation needed to create new landscape features and habitats on farms.

Despite high compliance, none of the energy taxes analysed (mineral oil tax, electricity and energy tax) induced any of the farmers surveyed to save energy or produce (or use) more environmentally friendly forms of energy. It seems that the tax levels involved were too low and failed to properly differentiate between forms of energy or energy production in terms of their environmental consequences. Since 2005, the federal government has granted farmers the legal opportunity to have taxes on diesel oil refunded (by means of diverse flat ha-rates for grassland, tillage, wine-growing, forestry and other forms of land use). So, farmers have lost any tax induced incentive to reduce diesel oil consumption. Regarding general ecological differentiation, the legal situation improved because comparatively lower tax levels for non-sulphur fuels and fuels with a minimum of 5% bio-fuels were introduced in 2005.

It is particularly interesting to note that contrary to expert predictions, the conservation legislation examined had very little impact on the landscape. The analysis indicated several reasons for this (Wytrzens et al., 2001). One was the farmers' lack of knowledge concerning the law. But the farmers interviewed also indicated a lack of willingness to obey: they did not accept the underlying objectives of nature conservation or weighed individual economic benefit higher than the risk of being caught. These results reflect the general experience of massive implementation deficits regarding environmental policies all over the world (Carlman, 2005). It seems rather unlikely that the new Upper Austrian conservation law 2001 has improved the actual protection of valuable landscape features and habitats.

In general, it can be assumed that the documented excess of 900 agri-environmental regulations (Penker, 1997) is largely responsible for the legal insecurity that characterises social, economic and political conditions in rural areas, and that is part of the everyday context in which farmers and bureaucrats live and work (Benda-Beckmann, 2001). Furthermore, the density and abundance of legal regulations inevitably result in a partitioned approach when dealing with legislation. Interviews with relevant legal experts confirmed the impression that both practitioners and theorists tend to specialise in their own, usually very narrow, area of law. There is little understanding or awareness of the complex of sometimes conflicting regulations as a whole and of the interactions between the individual regulations within this complex. In this context, harmonisation between different policy fields and instruments becomes a quixotic endeavour. We know very little about the actual interplay between legislation, actual land use activities of landholders and landscape development. Individuals responsible for designing or implementing laws as well as researchers analysing the institutional dimension of landscape change, however, have to cope with this degree of complexity. Thus, there is definitely a demand for theoretical concepts, practical tools and a procedure which can facilitate a structured and integrative approach to the analysis of impacts of existing and proposed regulations, on agriculture and the landscape.

## 15.6 Recommendations

The 'legal-ecological assessment' presented in this paper is a multistage analytical tool based on a framework of relevant determinants that describe the interrelations between law, human behaviour and landscape development. The tool was applied in three case studies for different regulations. It would seem reasonable to suggest that the proposed methodology could also be applied to the assessment of other landscape-relevant regulations, since the hypotheses concerning the factors that determine the impacts of legislation were broadly confirmed by the case studies.

The integrative approach, combining theoretical concepts of social and natural sciences with the expert knowledge of those affected by legislation in administration and agriculture, also gives new and insightful perspectives on landscape governance. The framework of determinants (Table 15.1) could provide a suitable structure for the development of agent-based scenarios that should help to evaluate the effects of new regulations. Such participatory scenario workshops with farmers and experts from administration might help to anticipate likely problems in implementing the new laws. An institutionalisation of agent-based strategic assessments would have another welcome side effect: it would slow down the process of policy making and thus, reduce the excessive and inflationary output of legislation. Thus ideally, such transdisciplinary assessments might result in fewer, but better regulations, i.e.

regulations which are better integrated with other policy fields, better adjusted to farm operations and administrative necessities and therefore also more effective in terms of landscape governance.

Regarding the transdisciplinary process, the lesson to be learned from this project is that the first phase of 'project definition' is very crucial for the success of a research project involving scientists of different disciplines as well as stakeholders outside academia. Due to a delay of the actual start of the project after its general approval and to related adaptations of the personnel structure, the newly composed group used the extra time to redefine the project aims, to find a common language and to refine the methodological framework (Fig. 15.2 and Table 15.1). Retrospectively, this extra effort put into the project definition was of great benefit to the project. Anyone designing transdisciplinary research programmes should be aware that the initial phase of transdisciplinary projects takes significantly longer compared to disciplinary focused studies. They need extra time, personnel and financial resources dedicated to the first phase of project definition.

The final synthesis was driven by the broadly shared ambition to write a common publication (Wytrzens et al., 2001), which forced the authors to streamline the results and to refine the language and the crucial arguments. Individual writing, reading and revising tasks with several common writing workshops to discuss and (re-)phrase the common structure and logic of arguments as well as single sentences proved a useful structure for this final synthesis. Vis-à-vis the external stakeholders, periodic workshops, requests for comments, appraisals or information, formal interviews and informal meetings allowed for several checks and balances and a fruitful cooperation. Apart from their willingness to listen, criticise and contribute, dedicated stakeholders outside academia also require a suitable organisational framework, in order to be actually able to act as guarantors for the practical relevance of transdisciplinary projects.

**Acknowledgments** We gratefully acknowledge Ronald Pohoryles as project leader and thank Roman Tronner, Gerhard Loibl, Markus Reiterer, and Stephan Wittich for their inspiring and fruitful collaboration in the transdisciplinary research project. Our very special thanks go to the numerous experts outside academia: in legislation, public administration, private business, politics and non-governmental organisations. We are also grateful for the helpful comments provided by the anonymous reviewers.

## References

Bastian, O. and Schreiber, K.F. (eds): 1999, *Analyse und ökologische Bewertung der Landschaft*, Spektrum, Heidelberg., 564pp.

Becker, G.S.: 1968, Crime and punishment: An economic approach, *J Polit Economy* 76, 169–217.

Benda-Beckmann, F. von: 1995, Anthropological approaches to property law and economics, *Eur J Law Econ* 2, 309–336.

Benda-Beckmann, F. von: 2001, Between free riders and free raiders: Property rights and soil degradation in context. In: N. Heerink, H. van Keulen, and M. Kuiper (eds), *Economic Policy Analysis and Sustainable Land Use: Recent Advances in Quantitative Analysis for Developing Countries*, Physica Verlag, Heidelberg, pp. 293–316.

Carlman, I.: 2005, The rule of sustainability and planning adaptivity, *Ambio* 34(2), 163–168.

Naveh, Z. and Lieberman, A.S.: 1994, *Landscape Ecology: Theory and Application*, Springer, New York, 360pp.

Opp, K.-D.: 1973, *Soziologie im Recht*, Rowohlt, Reinbeck, 264pp.

Penker, M.: 1997, *Zusammenstellung des österreichischen Agrarlandschaftsrechts: Bestandsaufnahme landeskultureller Normen auf Europa-, Bundes- und Landesebene*, Österreichische Gesellschaft für Agrar- und Umweltrecht (ÖGAU), Vienna, 61pp.

Rehbinder, M.: 1995, *Abhandlungen zur Rechtssoziologie*, Duncker & Humblot, Berlin, 268pp.

Rottleuthner, H.: 1987, *Einführung in die Rechtssoziologie*, Wissenschaftliche Buchgesellschaft, Darmstadt, 200pp.

Wytrzens, H.K., Penker, M., Reiterer, M., Tronner, R., and Wittich, S.: 2001, *Der Rechtsökologische Befund – Ein Instrument zur Erfassung von Landschaftswirkungen des Rechts*, Facultas Universitätsverlag, Vienna, 152pp.

## Legal Sources

EU regulations on agri-environmental schemes (2078/92, 2772/95, 1962/96; 746/96, 435/97).

EU Rural Development Regulation (1257/99).

Guidelines covering the EU co-financed Austrian agri-environmental programme – ÖPUL (Sonderrichtlinie des Bundesministers für Land- und Forstwirtschaft für das Österreichische Programm zur Förderung einer umweltgerechten, extensiven und den natürlichen Lebensraum schützenden Landwirtschaft, Zl. 25.022/39-IIB8/95, as amended by Zl. 25.014/220-IIB8/96).

Austrian Rural Development Programme 2000–2006 (approved by the European Commission 14/07/2000, amended in 2001 and in 2002).

Upper Austrian Nature and Landscape Conservation Act 1995 (Oberösterreichisches Natur- und Landschaftsschutzgesetz 1995, LGBl. Nr. 1995/37 (wv), as amended by LGBl. LGBl. Nr. 1997/147).

Upper Austrian Nature and Landscape Conservation Act 2001 (Oberösterreichisches Natur- und Landschaftsschutzgesetz 2001, LGBl. Nr. 2001/129, as last amended by LGBl. Nr. 2005/61).

Federal law on the taxation of mineral oil (Mineralölsteuergesetz 1995, BGBl. Nr. 1994/630 as last amended by BGBL I Nr. 2004/180).

Federal law on electricity taxation (Elektrizitätsabgabegesetz, as last amended by BGBl. I Nr. 71/2003).

Federal law on the taxation on natural gas (Erdgasabgabegesetz, BGBl. Nr. 1996/201; as last amended by BGBl. I Nr.71/2003).

Federal law on the refund of energy tax (Energieabgabenvergütungsgesetz, BGBl. Nr. 1996/201, as last amended by BGBl. I Nr. 92/2004).

# Chapter 16
# Children and Divorce: Investigating Current Legal Practices and their Impact on Family Transitions

**Heidi Simoni, Pasqualina Perrig-Chiello and Andrea Büchler**

**Abstract** The research project 'Children and Divorce – Current Legal Practices and their Impact on Family Transitions' is part of the Swiss National Research Programme 'Childhood, Youth and Intergenerational Relationships in a Changing Society' (NRP 52). Its objective is the scientific analyses of the amended Swiss divorce law, in order to get, on the one hand, an estimation of its outcomes on children's well-being and, on the other, to find out if it takes into account and encourages modern forms of familial allocation of duties and responsibilities in an appropriate manner. The research project is embedded in a societal evaluation process of law amendment, for which the dialogue between science and practice is fundamental.

The project, 'Children and Divorce', aims at following a transdisciplinary approach, integrating various disciplinary perspectives and methods. It is managed by a jurist and a psychologist. The research team includes jurists, psychologists and sociologists and is supported by a group of external experts working in the field.

To elaborate the research questions, various mutually complementary qualitative and quantitative methods of empirical social research have been applied. Data was collected synchronically at three levels: (a) analysis of court files and interviews with judges, (b) written interviews with divorced mothers and fathers, (c) in-depth interviews with children, mothers and fathers. The analyses of the data focus on the following main questions:

- The experiences with the revised Swiss divorce law focusing on the well-being of the affected children.
- The decisions taken in divorce proceedings such as the award of custody (sole and joint custody), the settlement of visitation rights and maintenance, the right of the child to be heard, and the representation of the child.
- The everyday life of divorced parents and their children.

✉ H. Simoni
Institute Marie Meierhofer, Zurich, Switzerland
e-mail: simoni@mmizuerich.ch

G. Hirsch Hadorn et al. (eds.), *Handbook of Transdisciplinary Research*, 

- Interrelations between legal context, resources of divorced families and the best interest of the child.
- The participation of children of divorced parents in the reorganisation process of the family.

As to the inter- and transdisciplinary process, the elaboration of a common theoretical framework and the integration of the results, provide the most interesting and challenging parts. The aim of the following chapter is to focus on these aspects, without going into the results.

**Keywords:** Welfare of the child · Divorce · Family transition · Legal practice · Social change

## 16.1 Background

### *16.1.1 Social Change and Participative Rights of Children – a Challenge for Social Science Research*

The demand, over the last decades, for cross-discipline cooperation in research, teaching and practice of social science comes as a reaction to the unbridled increase in knowledge within the special disciplines on the one hand, and to the growing complexity of social problems on the other. One example of these problems is the current demographic change with its far-reaching implications (associated with a fundamental change in values) particularly for the way people live together in a society (Perrig-Chiello and Darbellay, 2002; Perrig-Chiello and Arber, 2002). This comes out most clearly when we take the example of familial inter-generation relationships. Like most European countries, Switzerland over the past decades has been witnessing a process of disintegration of the bourgeois marriage and family model with profound repercussions for family life. This fact is reflected in the following indicators (Eidg. Departement des Innern, 2004; Perrig-Chiello, 2004):

- The marriage rate (the number of individuals out of 100 individuals who marry at least once during their lifetime) of women dropped from 87–63 from 1970–2005 and of men from 83–59 in the same period.
- The average number of children a woman gives birth to in Switzerland in the course of her lifetime totalled 1.39 in 2003, the birth rate having tended to drop continually over the preceding years. In order to secure the continued existence of the population, a birth rate amounting to 2.1 would be needed. The birth rate has remained below that figure since 1970.
- The percentage of children born out of wedlock, although increasing, is still relatively low. In 2003, 12.4% of children born alive were born to unmarried women. From the beginning of the 90s (1992: 6.2%), this figure has been rising steadily.

- Switzerland displays a high divorce rate by European standards (percentage figure of divorced marriages over time). It has seen a more or less continual increase over the last decades (1960: 13%, 2004: 44%). The figure reached a climax in 1999 with the divorce rate reaching 51% as the result of an above average number of divorce cases. When the new law came into force, the incidence dropped to one half. Within one year (2000: 26% and 2001: 39%) the divorce rate had again increased by 50%. At the same time we are witnessing an increase of one-parent families, patch-work families, and one-person households.

These changes have a direct impact on the life and potential development of children and adolescents who, quite obviously, still grow up in families. A growing number of children and adolescents experience change in their family relationships in the course of their development (divorce or new partnership of one parent); moreover, in most families both parents work full or part time. In this context, the questions arise of how children and adolescents perceive and cope with these circumstances; how these affect their development in the long term; and in what way they have a right to participate in deciding and shaping events in these matters. How does the changed situation affect the development of children and adolescents, in particular their attachments, well-being and health? As far as Switzerland is concerned, there are hardly any valid answers that can be generalised. This complex of themes has not been sufficiently studied although there exists considerable research in the field. More generally, we can observe that children and adolescents constitute a grossly neglected section of the population in relation to social issues and social analysis, having no lobby and insufficient means and possibilities to make their needs and interests heard and taken notice of in public. Moreover there is a deplorable lack of scientifically ascertained knowledge of the concrete structure of everyday life in which they live and grow up. This is true to a large extent for Switzerland, too.

### *16.1.2 National Research Programme 52 'Childhood, Youth and Intergenerational Relationships in a Changing Society'*

It is against the backdrop of the social changes referred to above, on the one hand, and the insufficient data base, on the other, that the National Research Programme 'Childhood, youth and intergenerational relationships in a changing society' (NRP 52; www.nfp52.ch) of the Swiss National Science Foundation took shape. This NRP, endowed with 12 million CHF, was embarked on in 2003. It aims at gaining scientifically established insights into the present and prospective living conditions as well as the needs of children and adolescents in Switzerland. Special attention is supposed to be given to both intergenerational and legal aspects, given that considerable gaps in research regarding this subject exist in Switzerland. On this basis, approaches shall be made to clarify the question of whether any action, and if so what kind, needs to be taken within politics, the administrative bodies of the federation, cantons and municipalities, corporate economy and non-governmental welfare organisations and what kind of reaction is appropriate with regard to the issue. Given the complexity of the issues involved, research projects using interdisciplinary and transdisciplinary approaches were suggested. In the light of these objectives and intentions, NRP 52 shows all the typical characteristics of a National Research Programme, in particular:

- transdisciplinary approaches to questions and interdisciplinary research methods
- problem-related research in between pure basic and applied research
- validation of methodology and translation of the research results into practice
- preset financial limits and time table (5 years; deadline of NRP research: end of 2007).

The standard of the National Research Programmes predetermines these to be a suitable interface between science and society. What matters, on the one hand, is to investigate socially relevant problems using a transdisciplinary approach and to develop effective strategies for politics, economy and society and, on the other hand, to give the public access to these research results.

The 29 research projects in the framework of NRP 52 had three research foci: (1) living conditions of children and adolescents in the intergenerational context, (2) family and systems complementary to the family, (3) learning, leisure, media and consumption. It was suggested that researchers not only consider diversity and variety as well as intergenerational interdependence, but also take into account the children's and adolescents' competence to act (within the context of the research topic). As a matter of fact, children from an early age are capable of actively participating in shaping their own development. When it comes to social relationships and way of life, children and adolescents have creative potentials that can be encouraged or inhibited depending on each individual's everyday world. The well-adapted encouragement of the competences to act and their translation into forms of shared decision-making or participation are relevant to research work on children and adolescents. The children's and adolescents' reliance on parents and other adults requires from the latter a sense of responsibility and a continuing adaptation to mutual needs. In this context, particular emphasis has to be placed on the role of the family and the question: what can society and the law do to strengthen the parents and thus to help children and adolescents? This complex of themes is tackled mostly in research focus 2 'Family and systems complementary to the family' of NRP 52. One of the key questions raised is: how far are the rights of children and adolescents to participate, translated into practice? According to the UN convention on the rights of the child, these rights have to be honoured in all legal actions involving children and adolescents. What deserves particular attention in this context is the question of the juridical and psychosocial determinants and consequences of this in real life practice. The project 'Children and Divorce' is meant to serve as an example in this respect.

## 16.2 Children and Divorce – an Inter- and Transdisciplinary Challenge

### *16.2.1 Starting Point and Development of the Project*

When the National Research Programme 52 'Childhood, adolescence and intergenerational relationships in social change' requested research proposals both managers of the project discussed in this chapter, independently from one another, were motivated to submit research proposals focusing on 'the amended Swiss divorce law' as well as 'rights of the child and welfare of the child'. Rather remarkably, one of them is a jurist, holding a chair for private law at Zurich University, while the other has graduated in psychology and works as the head of the practical research department of the Marie Meierhofer Institute in Zurich, which deals with conditions of development of children and families in our society. This initial situation is enough to make it clear that quite similar questions ought to be empirically investigated starting from each individual discipline. Obviously, we are dealing with a complex of themes that suggests both an interdisciplinary (legal science & psychology) and a transdisciplinary (theory & practice; university & institute of applied science; science & society) approach. Taking a look at the initial situation of the scheduled projects with regard to their research topic is intended to make this point come out more clearly.

The amended divorce law, which came into force in 2000, introduced some substantial legal amendments equally affecting the children and adolescents involved and the divorcee parents. The most substantial amendments deal with the competence of the court:

- to either directly hear the children in the course of the divorce proceedings of their parents and/or to have them represented in order to safeguard their interests
- to leave parental responsibility with the mother and the father jointly upon application by the parents.

This means that the amended divorce law has given children the status of legal subjects, thereby honoring a principle of the UN Convention on the Rights of the Child (UN CRC, 2007). In addition to this, an attempt was made to take into consideration the fact that children benefit from both parents feeling responsible for their children's wellbeing despite them being divorced.

The legal principles regarding this issue explicitly provide for the matter to be based on other disciplines. To illustrate this point, the related articles of the Swiss Civil Code (ZGB) as well as an article of the UN CRC will be quoted. Passages such as 'welfare of the child', 'form/express an opinion', 'appropriate', 'commensurate to age' refer to topics the clarification of which, when dealing with general and particular cases, requires specialist knowledge and the expert knowledge of specialists trained in developmental psychology and educational theory.

*Art. 133 Swiss Civil Code [ZGB]*

1. The court attributes parental responsibility to one parent, regulating the claim to personal access and the contribution to the child's maintenance by the other parent according to the provisions on the effects of the parent and child relationship. The amount of the maintenance payment may be fixed beyond the attainment of legal age.
2. All circumstances relevant for the welfare of the child have to be considered when attributing parental responsibility and regulating personal access; joint application by the parents and, as far as it is feasible, the opinion of the child have to be taken into account.
3. If the parents in an approvable settlement have agreed on their shares in the maintenance of the child and the sharing of the maintenance costs, the court, upon joint application, leaves parental responsibility with both parents, insofar as this can be reconciled with the welfare of the child.

*Art. 144 Swiss Civil Code [ZGB]*

1. If orders concerning children have to be issued, the court hears the parents in person.
2. The children are heard in person in an appropriate manner by the court or by a mandated third person, unless their age or other substantial reasons make it seem inadvisable to do so.

*Article 12 UN CRC*

1. States Parties shall assure to the child who is capable of forming his or her own views the right to express those views freely in all matters affecting the child, the views of the child being given due weight in accordance with the age and maturity of the child.
2. For this purpose, the child shall in particular be provided the opportunity to be heard in any judicial and administrative proceedings affecting the child, either directly, or through a representative or an appropriate body, in a manner consistent with the procedural rules of national law.

Article 12 of the UN CRC makes it clear that the participative rights of children in court proceedings are taken to constitute a mere specification of a more fundamental right to have their voices heard. This means that in the case of a divorce, besides the hearing of children in the course of the divorce proceedings, what matters is whether they can make their own contribution to the reorganisation process of the family and if so, how they can contribute. On the international level, the altered legal position of the child is indeed undisputed in principle. Honouring the right of the child to participate, however, still falls short of constituting an established practice. The regulation of parental responsibility is currently the subject of a controversial and highly emotional debate. This is true for Switzerland as well as for other Western countries where differing legal models are being applied. The fairly heated debates show that children's welfare and parental divorce is still a highly explosive issue for society. All in all, it is not only a matter regarding questions of legal practice, but involves more fundamental issues: what kind of legal model regarding parental responsibility really makes everyday life of divorcee families and the shaping of relationships of children with their divorced mothers and fathers easier? Or to put it the other way round: are there legal guidelines regularly providing additional fuel for conflict, thereby making the reorganization process of the families following the separation of the parents even more difficult? Furthermore, it is still not clear what relation the rights of children have with the rights of adult women and men, for questions of equality and the sharing of duties between the sexes are a further issue regularly mixed up with questions of parental responsibility and the welfare of the child.

The legal system reacts to social change by amending the law, while at the same time acting upon it. This suggests that an analysis of interpreted and applied legal practice is necessary to answer questions of how smoothly and effectively it can contribute to the solution or defuse actual problematic situation. Besides this, the question arises of whether legal science adapts to the diversity of real situations and how consistent it is perceived to be by the specialists and individuals involved. This is particularly interesting where it touches upon fields of tension in which various discursive levels intermingle, as is the case of parental responsibility. From a psychological perspective, the questions that come to the fore are whether legal processes correspond to the perception and behaviour of the adults and children involved and are felt to make sense; whether they are helpful in potentially conflictive or critical situations in life; and finally, whether regulations can be found that are well adapted to the realities of everyday life and that have a relieving effect on the psycho-social situation.

Pursuing a transdisciplinary approach was intended in both research programmes as an exchange between science and practice. Science that claims to provide potentially relevant results has to rely on questions and suggestions from practice. On the other hand, science has an advantage over the latter in so far as the scientific perspective makes it possible to study the research object systematically, objectively and in detail at the same time. Empirical research steps back to observe both social and individual events from a distance. This makes it possible to adopt various perspectives and to distinguish – expected or unexpected – patterns from a bird's eye view, as it were. The systematic view allows for in-depth study of selected aspects. Both taken together helps the researcher stay detached from preconceived opinions and convictions and to start searching for contexts on the assumption that things may well be completely different from what they were supposed to be.

The management team of NRP 52 has appreciated the outlined transdisciplinary dimension of the subject and the need for an interdisciplinary empirical approach. They have also recognised the potential for synergy provided by both project proposals submitted. This is why both researchers were asked jointly to work out a research application. Relying on international experts' reports, the management team of NRP 52 granted the application 'Children and divorce; the impact of legal practice on familial transitions'. The crucial factor in the decision, besides the scientific quality, was the interdisciplinary and transdisciplinary approach intended in the project. The research project aims at understanding the psycho-social situation of children during divorce proceedings as well as following the divorce of the parents and to relate this reality to the 'child's welfare'. Science can take up past or urgent questions of practice, creatively reformulating them, preparing them for further investigation, and in this way contribute substantially to clarifying them in the best of cases. This is where the potential of the proposed project lies with regard to the subsequent usability of its results.

### *16.2.2 Elaboration of a Common Theoretical Framework*

The project 'Children and Divorce' is part of a social process involving the evaluation of an amendment of law, while going far beyond it. In order to back up and ground the project broadly, exchange with further experts from research and practice had already been sought at a very early stage of its realisation and is maintained in the form of a supporting team of specialists and of individual experts. This team is made up of specialists from all over the world, including experts pursuing scientific and practical activities in the fields of family law as well as child and family psychology.

It was a vital condition for the realisation of the project to develop a common basis or a common theoretical framework notwithstanding the differing perspectives of the disciplines involved. The particular challenge in this process was to connect the potentials of the special disciplines with each other and make them interact. The specific perspectives and analytical levels of the disciplines involved were not intended to be dissolved, but to be explicitly made use of. The following aspects constitute the basis of the research project:

- juridical, sociological and psychological knowledge of familial transitions
- legal comparisons regarding the regulation of the interests of children in the course of divorce proceedings of the parents
- psychological knowledge of risk or defensive factors, processes of coping, development of the child, development during the course of life
- data on demographic development in Switzerland.

The overall concepts of 'divorce law, child law and welfare of the child' were fixed as a common denominator. It soon became evident that 'process orientation' represented an additional common interest with three differing aspects connected to it: one, from a perspective of legal science and another from the perspective of psychology. The third one is related to the former two, concerning the temporal structure of the period in question during which changes are to be investigated. The first concerns the aspect of 'translation into practice', and the second that of 'familial transition'.

From a primarily juridical perspective, the core issue of an evaluation of an amendment of law concerns its translation into practice: how are the guidelines being taken into consideration in judicial custom and the options provided for fully made use of? Psychology, on the basis of the idea of translation into practice, may ask whether, when and how judicial custom is reflected in the everyday lives of the families.

The term 'familial transition' denotes that the nuclear family may see changes involving fundamental role reversals (birth of a child, separation, divorce) over time. These typically act both 'from inside out' and 'from outside in'. Familial transitions are coped with in different ways according to the resources available. Although the reorganisation processes related to this can provide an opportunity for the persons affected (from an individual and social aspect), they also have considerable potential for crisis in the short and middle term (Cowan, 1991; Perrig-Chiello and Perren, 2005). What matters in the long term, however, is whether actual adverse consequences could be overcome and whether the changes could be coped with. In connection with separation and divorce, legal practice and the wellbeing of the individuals affected are dynamically interrelated. Environment, judicial custom, societal value judgments and social resources, besides personal psychological resources, affect coping with divorce and thus the wellbeing of the individuals involved. Their perceptions, feelings and attitudes in their turn influence the everyday life and behaviour of the environment, i.e. external conditions. The concept of transition in the familial life cycle perfectly serves the purpose of empirically utilising the variety of perspectives adopted by jurists, sociologists and psychologists to view the 'inside' and 'outside', understand structural and psycho-dynamic characteristics of the transition 'parental divorce' and thereby overcome the additional juxtaposition of the differing approaches.

The temporal structure given to the period under investigation is based both on juridical and psychological considerations. What matters from the juridical perspective is the division into the separation, divorce and post-divorce phases. On the one hand, this is because the phases referred to differ with the competence of the court and of the guardianship authority to regulate the child's interests as well as in other respects. What matters on the other hand, is the question of continuity of regulations at different points in time such as the relevance of regulations applying to the period of separation and everyday life as experienced by the individuals concerned with the divorce decree. An important issue from a psychological perspective is determining where, when and how often the investigation is conducted. Taken together these determine what can be understood in retrospect only and what can be understood actually or prospectively. It has to be taken into account that retrospective interviews are valid only with qualification and with regard to selected questions because objective human memories do not exist. The coming to terms with events and the course things have taken always affect memory. Furthermore, assuming that divorce is a transition period, existing knowledge of the working of such reorganisation processes has to be included. As a case in point, it had to be taken into consideration, at about what point following a divorce decree can coping with the altered family situation be expected in most families.

Methodologically speaking, the common basis is provided by the scientifically backed up agreement that data collection using multiple informants and perspectives, promises to serve the complex of topics best (Kraemer et al., 2003).

### 16.2.3 Drafting of Common Research Questions

Finally, the following research questions were drafted and made serviceable for empirical treatment:

- What are the experiences Switzerland has had with the amended divorce law with special regard to the welfare of the children involved?
- What is the legal practice for dealing with divorce with regard to: parental responsibility (sole and joint parental responsibility), care and custody, visiting rights, hearing of the child, representation of the child, child maintenance?
- What is the situation in life of divorced parents and their children?
- How are divorce decrees translated into the everyday life of children and parents?
- What changes and constants become visible in the course of time, as for example in the contact between the child and both parents?
- Are there any connections between overall legal conditions, resources of the family and welfare of the child?
- How can children affected by divorce appropriately participate in the reorganisation process of the family?
- What are the experiences and the condition of children and adolescents during and following the divorce of their parents?
- Do gender-specific differences emerge in the context of the issues concerning children and adolescents?

This is to say that the research project 'Children and Divorce', on the one hand, investigates the translation of the divorce law in force in Switzerland from 2000 into court custom, and on the other, investigates the translation of the divorce decrees into the everyday life of the families involved. The catchphrase 'welfare of the child' focuses on the regulation of the child's interest in the course of divorce proceedings and in everyday life, the participation of the children and adolescents affected in the reorganisation process of the family as well as the satisfaction and condition of children and parents in the post-divorce period. Special attention is given to answering the question of whether including the perspective of the child as laid down by law can in fact be found to be part of the divorce situation.

### 16.2.4 Interdisciplinary Empirical Methods and Perspectives Concerning Data Collection

Using an interdisciplinary approach to the research topic requires the sustained communication of explicit and implicit assumptions related to specific subject areas. This leads to the specifying and clarifying of concepts, research questions and interpretations within and between disciplines at every stage of the project. The empirical phase has particular significance in this process: what matters is to make sure that questions and perspectives in the fields of legal science, psychology and sociology are given due consideration in all parts of the investigation, and in the related instruments of data collection. They have to be interconnected in a way that will allow for the interconnection of the maximum amount of collected and evaluated data and the integration and communication of the results. Making optimum use of interdisciplinary resources and establishing the connections between the collections of data, suggested the following procedure: overall control of two of the three parts of the investigations was to be with the jurists (analysis of records) or the psychologists (interviews). The instruments used for data collection were complemented, whenever necessary, by the other part of the team and all of them coordinated with each other. The third level of data collection, which is the large scale written interviews, was designed to serve as a hinge for method and subject matter: the data collected helps make it possible to inter-relate the quantitatively evaluated juridical-social data of the record analysis and the qualitatively evaluated psycho-social data of the interviews.

What follows is a description of the methods and categories used to collect data. The resources available did not allow for a comprehensive investigation covering the whole of Switzerland. Since both rural and urban contexts had to be taken into consideration, all data collection was conducted in the three cantons of Basel-Land, Basel-Stadt and Zurich.

*Record analysis and interviews with female and male judges*: Analysis of a random sample from one out of five court records dealing with divorce decrees from 2002 to 2003 is designed to yield information on agreed regulations regarding children at the moment of divorce. In the process, divorce records are investigated systematically for substantial and procedural measures involving children in the course of divorce proceedings (attribution of parental responsibility to one parent or leaving joint parental responsibility with both parents, regulation of care and custody and of personal access, payment of maintenance for the child, hearing of the child, appointment of a representative for the child, etc.). Record analysis was mainly developed and conducted by the legally trained members of the research team. The trained psychologists and the team of experts supported the process in an advisory capacity. In order to compare the regulations found in the decrees with the perception of court custom by the court officials, the project managers conduct interviews with divorce judges parallel to record analysis. These interviews are expected to give us some clues as to the applicability and the smooth translation of the law into practice. A semi-structured interview guideline was jointly worked up by the team as a whole.

*Written interviews with divorced mothers and fathers:* A structured questionnaire was used to interview all parents of under-age children who got divorced in the three cantons in 2002 and 2003 concerning the regulation of the child's interests in the divorce decree and its translation into practice. The question of how families organise their everyday life after divorce and how children, mothers and fathers feel in this situation is of particular interest. The net feedback of 40% makes it safe to say that, with regard to subject matter and form, we have most happily succeeded in establishing contact with the parents affected. The drafting of the questionnaire was done by the whole team, in cooperation with two scientifically experienced sociologists who carried out the preliminary evaluation of the data. The data was returned to the research team for in-depth analysis. The scientific team of experts supported the process in an advisory function. Drafting and conducting the written interviews jointly has proved highly helpful for interdisciplinary communication and continuous exchange.

*In-depth oral interviews with families:* In order to connect our understanding of familial reorganisation processes as closely as possible with the experience and behaviour of divorced mothers, fathers, children and adolescents affected, interviews focusing on specific topics were conducted. We aimed at having at least one parent and one child per family tell us about their personal experiences and wellbeing in connection with separation and divorce. This method makes available information based on self-perception as well as on the perceptions of others. The interviews are the only category of data collection allowing children to speak up for themselves directly. This will make it particularly interesting to compare their view of the situation and their experiences of participation in the family reorganisation process, with information from court records, parents and judges. Special importance was given to including children of various ages and parents involved in various regulations concerning legal parental responsibility. Special attention is given to the question of coping with changes and coming to terms with the situation. All the elements referred to provide clues to adverse and protective processes affecting the development of the children and adolescents interviewed. The oral interviews with children and parents are mainly being developed and conducted by the psychologists on the research team. The legally trained members of the team and the fellow team have been assisting the process in an advisory function.

### 16.2.5 Integration of the Results: The Major Challenge with Transdisciplinary Consequences

As the research process has shown, the phases of data evaluation and especially the integration of the results call strongly for interdisciplinary cooperation and communication. Of course, the research questions drafted and the method used in data collection set the direction to be followed. This still did not specify sufficiently how the scientific answers to the societal discourse dealing with the topic of divorce and welfare of the child could be brought to fruition. The questions drafted at the beginning of the research process moreover still draw upon the interdisciplinary potential in a supplementary rather than an integrative manner while both the specification of the questions and utilisation of the interdisciplinary potential are vital for the project to succeed. What mattered at that was to formulate and balance the crucial questions' so that the scarce resources of time and personnel could be maximised.

The difference in focus between 'events and decisions' and 'condition and behaviour of individuals' presented a challenging obstacle at this stage of research: legal science is primarily concerned with the legal institute of the right to be heard, the regulation of parental responsibility, maintenance and their effects, in a manner geared towards translation into practice, in order to deduce recommendations regarding court custom. On the other hand, it is more in line with the psychological concept to start from the persons involved, trying to understand the multiple experiences of children, parents and judges in the light of certain aspects related to the topic. This means that fusing together analyses and results once more required an interdisciplinary effort and exchange, highlighting the need to keep working from our common base. This, however, proved yet another obstacle which turned out to be an opportunity once the team succeeded in integrating both preferences referred to above. The higher degree of orientation toward practical application shown by the jurists proved to be a strong point at this stage of the project because it helped use the abundant data base in an object-directed way. Eventually, the overall guideline for the focusing of the evaluation and integration of the results was phrased to read: '*Separation and divorce as a (re)organization process of family life from the perspective of children, parents and judges*'. In a further step, we jointly defined with which priority to analyse the data of the various collection categories with regard to which research questions. The priorities have been ranked as follows:

1. 'parental responsibility and care and custody as experienced in real life' from the perspective of children, parents and judges and in the conflicting relationship of arrangements agreed on in court (Büchler and Simoni, 2006)
2. participation of the children in separation and divorce from the perspective of children, parents and judges
3. state intervention in the interests of the child from the aspect of the welfare of the child.

The method agreed on is designed to make sure that evaluation can be carried out with all necessary focus and care while nevertheless making the results available for intra-, inter- and transdisciplinary use as soon as possible. Once the available data referring to the first complex of themes have been analysed, the results are due to be communicated in the form of papers in scientific journals, articles in the daily print media and in lectures, etc. Concentrating data analysis on these three foci regarding the research is in line with the intention of the project to pursue a transdisciplinary approach. Besides this, it will provide jurists and psychologists with data specific to their disciplines and create interdisciplinary 'heightened value'.

On the whole, the experiences made so far with the realisation of the project 'Children and Divorce' have confirmed the researchers in their conviction that by using the potential of an interdisciplinary approach they can in fact contribute to clarifying questions. The answers are eagerly awaited by the specialists of each discipline, the experts in applied science and the wider public. At the same time, the inspirational effect of inter- and transdisciplinary dialogue on any individual specialist field ought not to be underestimated.

## 16.3 Recommendations

The project 'Children and Divorce' claims to clarify socially relevant questions that were formulated by the scientists as the result of a dialogue with experts in the field of social practice. After the empirical evaluation, the research results will be communicated to the scientific community and also fed back into practical application in the form of recommendations and implementations.

The research level itself requires the utilisation of various scientific perspectives and methods in order to make the topics empirically manageable without sacrificing its complexity. An important precondition for this is that it remains possible for the different research collaborators to stay rooted in their own disciplines, adopting a methodical and substantially precise working 'state of the art' process. In this respect, the project 'Children and Divorce' represents an interdisciplinary endeavour attempting explicitly to utilise the professional idiosyncrasies of each of the disciplines involved. The boundaries of each discipline are not removed but overlap in the process: for the disciplinary identity and the curiosity to go beyond it are equally indispensable. From the very outset, the working out of a common theoretical framework for the subject of parental divorce and welfare of the child proved both challenging and rewarding. Interdisciplinary communication has been upheld during the entire period of the project so far, constantly resulting in adding to its accuracy. The fascinating point is that the specific assets of the disciplines involved make themselves felt in accordance with the requirements of the individual stages of the project. Examples are the contribution of the powerful concept of 'biographical transition' and of methodological know-how of psychology, and the effectiveness of the high degree of systematic and practical orientation of jurisprudence.

The present chapter allows insights into a process of research still under way, which therefore cannot yet be conclusively evaluated. However there is much evidence suggesting that the inter- and transdisciplinary approach adopted for the project 'Children and Divorce' has an advantage over a monodisciplinary method in content as well as structure. Although the project is still in its final stage, the experiences so far in connection with the most important milestones are sufficient and convincing enough to draw some conclusions concerning the heightened value of the scientific approach adopted.

- The interdisciplinary orientation allows for the treatment of complex and multifaceted research topics on the one hand, thereby contributing to a comprehensive perception of the facts and to clarifying complex research questions. On the other hand, generating interdisciplinary knowledge does not prevent the scientific discourse (related to a special field) from making headway. This project involves – besides the interdisciplinary points of intersection – very specific clusters of disciplinary topics that could be followed up.
- The transdisciplinary orientation contributes to gathering valid and socially relevant results. Furthermore, it encourages the formation of networks, especially the improvement of cooperation within and among research institutions, inside and outside academia, as well as between science and practice.

To develop and conduct projects with a complex design, the researchers are in need – apart from their commitment – of overall conditions and infrastructure allowing them, firstly, to pursue a dialogue among themselves and with the experts in the practical fields, and secondly, to bring the project to fruition with some form of cooperation. This necessarily includes subsidies that encourage research earmarked for that very goal, duly valuing, honouring and promoting the transdisciplinary intentions and efforts in evaluating research results. In order to allow inter- and transdisciplinary projects to shape and maximise the research process, the timetable fixed for the project must not be set too tight. The integration of the project 'Children and Divorce' into the National Research Programme 52 shows how this complementary cooperation, between research and subsidies encouraging research, can work out in a way that adds value at the project and programme level. This manifests itself in results like:

- the stimulation of national and international research cooperation
- the formation of junior scientists (PhD students)
- an issue-related and more comprehensive dialogue between science and society.

**Acknowledgments** The project 52-103348 of Prof Andrea Büchler and Dr Heidi Simoni was supported by the National Research Programme 'Childhood, Youth and Intergenerational Relationships in a Changing Society' (NRP 52) of the Swiss National Science Foundation.

## References

Büchler, A. and Simoni, H.: 2006, July 15, Scheidungskinder zwischen Wohnmüttern und Besuchsvätern. Elterliche Sorge und Kindeswohl – Modell und Alltag (Children of divorced parents between residental mothers and non-residental fathers. Parental responsibility and welfare of the child – models and daily life), *Neue Zürcher Zeitung*, Nr. 162, S. 65.

Cowan, P.: 1991, Individual and family life transitions: a proposal for a new definition. In: P. Cowan and M. Hetheringthon (eds), *Family transitions: a proposal for a new definition*, Lawrence Erlbaum Associates, Hillsdale, pp. 3–30.

Eidg. Departement des Innern (ed.): 2004, *Familienbericht 2004: Strukturelle Anforderungen an eine bedürfnisgerechte Familienpolitik*, Bern.

Kraemer, H., Measelle, J., Ablow, J., Eessex, M., Boyce, W, and Kupfer, D.: 2003, A new approach to integrating data from multiple informants in psychiatric assessment and research: mixing and matching contexts and perspectives, *Am J Psychiatr* 160, 1566–1577.

National Research Programme 52: n.d., Retrieved June 11, 2007, from http://www.nfp52.ch.

Perrig-Chiello, P. and Perren, S.: 2005, Biographical transitions from a midlife perspective, *J Adult Dev* 12, 169–181.

Perrig-Chiello, P.: 2004, Soziale Integration im Spiegelbild lebenszyklischer Übergänge (Social integration as reflected in life-cycle transitions). In: Ch. Suter (ed.), *Sozialbericht 2004* (Social Report), Seismo, Zurich, pp. 129–154.

Perrig-Chiello, P. and Arber, W.: 2002, *Inter- und Transdisziplinarität – zwischen akademischem Anspruch und gesellschaftlichem Bedürfnis* (Inter- and transdisciplinarity – between academic intention and social need), Réalités Sociale, Lausanne.

Perrig-Chiello, P. and Darbellay, F. (eds): 2002, Qu'est-ce que l'interdisciplinarité?, *Réalités Sociale*, Lausanne.

Schweizerisches Zivilgesetzbuch (Swiss Civil Code, ZGB) of May 1, 2007, Retrieved June 11, 2007, from http://www.admin.ch/ch/d/sr/c210.html.
UN Convention on the Rights of the Child, Office of the United Nations High Commissioner for Human Rights, Retrieved June 11, 2007, from http://www.unhchr.ch/html/menu3/b/k2crc.htm.

# Part IV
# Bringing Results to Fruition

# Chapter 17
# Towards Integrated and Adapted Health Services for Nomadic Pastoralists and their Animals: A North–South Partnership

**Esther Schelling, Kaspar Wyss, Colette Diguimbaye, Mahamat Béchir, Moustapha Ould Taleb, Bassirou Bonfoh, Marcel Tanner and Jakob Zinsstag**

**Abstract** Mobility of pastoralists in arid and semi-arid zones renders access to primary social services difficult. The experiences, local concepts and propositions of nomadic communities of Chad were essential to fill the information gaps on how to provide adapted health services to mobile communities. In Chad, we have taken an iterative, corkscrew-like research and action strategy: a better understanding of the determinants of health and communities' health priorities – obtained by interdisciplinary collaborations between medicine, anthropology, epidemiology, social geography and microbiology – were integrated in the participatory identification of intervention options out of a range of possible responses by the health and veterinary services. Recommendations from national stakeholder workshops paved the way for implementing and testing new interventions. All stakeholders reviewed outcomes of interventions periodically. The programme provided opportunities for participatory processes and actions that were defined in an open way at the beginning. An appropriate North–South research partnership framework and the long-term commitment of all partners have been crucial in the process of building stakeholders' ownership. University curricula rarely enable scientists to communicate with other disciplines, and researchers first needed to acquire skills in crossing the boundaries between human and natural sciences and between sectors. We describe here in a chronologic way the elements that led to innovative health and veterinary services for nomadic pastoralists of Chad; such as joint vaccination services of the public health and the livestock sectors and subsequent initiatives that were initiated by the nomadic communities once they began to trust the programme.

**Keywords:** Inter-sectoral public health and veterinary services · Stakeholder assessment · Access to health care · Remote rural zones · Mobile pastoralists

✉ E. Schelling
Swiss Tropical Institute (STI), Basel, Switzerland; International Livestock Research Institute, Nairobi, Kenya
e-mail: esther.schelling@unibas.ch

G. Hirsch Hadorn et al. (eds.), *Handbook of Transdisciplinary Research*,

## 17.1 Background

Nomadic and transhumant pastoralists, in arid and semi-arid zones, use vast dry grasslands and crop residues productively. Their mobility is a strategy to manage efficiently the uncertainty in a fragile environment where rain-fed agriculture is risky (Niamir-Fuller, 1999). However, mobility renders access to preventive and curative treatments, and information difficult. Poor health is often the ultimate consequence of poverty and the limited access to education, sanitation and health services, combined with societal change and political unrest. Health is a sensitive indicator of development and health outcomes but can only be interpreted if the social, cultural, political and economic backgrounds are understood. Studies to determine health indicators among nomadic pastoralists are rare as there is little experience of delivering health services to them (Swift et al., 1990).

Mobile pastoralists are the subject of an unusually large number of myths and misunderstandings so policy makers give them a low priority and this leads to inadequate policies. It is thought that pastoralists need to settle to benefit from social services. Indeed, rural health dispensaries are typically located in agricultural villages to which access of pastoralists with their animals is often hindered. There are also socio-linguistic barriers at rural dispensaries between nomadic groups and health workers (Loutan, 1989). Nomads may be treated with disrespect or have to pay under-the-table sums. A mobile way of life jeopardises treatments requiring follow-up, e.g. against tuberculosis and HIV infection. Lack of maternal health services including attended skilled birth delivery is associated with a high pregnancy related morbidity and mortality. The access of sick women to health services depends on the network they can mobilise in order to receive treatment (the necessary resources and a male chaperone) (Hampshire, 2002). This may be especially problematic in the nomadic context because families are every so often separated.

Mobile health services that reached nomadic camps were avidly used by nomadic pastoralists; however, due to their high costs they were not sustained. The training of community health workers (CHWs) was promising in Niger (Loutan, 1989). Women need no special permission to consult a nomadic CHW as they do for outside practitioners.

Traditional healers in many pastoral societies treat people and animals. Besides providing pastoralists with their main source of subsistence, livestock is the basis of economic wealth and social respect. The good health of their animals is pivotal and therefore, animal health may provide a key entry point for the provision of both human and animal health understanding and services. Pastoralists, for example in Chad, seek efficacious western-type drugs for particular human and animal health problems, e.g. chloroquine for malaria, antibiotics for respiratory diseases and treatments against trypanosomosis of cattle and camels.

The American epidemiologist Calvin Schwabe – inspired by Sudanese Dinka pastoralist healers – comprehensively outlined the synergies of human and veterinary medicine and the added values to public health of a 'one medicine' (Schwabe, 1984). Majok and Schwabe (1996) observed that veterinarians are the most extensively distributed individuals with a higher educational degree in many African rural areas and advocated intersectoral collaboration between the health and veterinary services.

Although approximately ten percent of the Chadian population are nomads, like most sub-Saharan African countries, there are no available definitive demographic enumerations. In Chad, livestock is the second most important export after petrol. Three quarters of the cattle are kept in the Sahelian zone, about 80% of these in mobile systems.

In the rural Chari-Baguirmi of Chad, in the early 1990s, the staff at one health centre observed that nomadic pastoralists passed by without visiting their centre. They asked the provincial primary health care programme to develop more appropriate services for the mobile populations. The health programme was executed by the Centre de Support en Santé International (CSSI), the local branch of the Swiss Tropical Institute (STI) in Chad. The STI is a public health institute associated with the University of Basel. Its mandate is to contribute to the improvement of the health of populations internationally and nationally through excellence in research, services, teaching and training. With its broad network it can cover and link health topics from the laboratory bench to the patient. The majority of collaborators in the research and service departments have a professional background in natural and social sciences rather than in the health sciences.

In 1996, the STI commissioned an exploratory pre-study of access to health services by nomadic pastoralists in Chad. Based on its recommendations and those of a first national workshop, the research and action programme 'Improved access to health services for nomadic pastoralists of Chad' was launched in 1998 (Wiese and Tanner, 2000). The programme wanted to overcome barriers to access to health care by nomadic people by generating a scientific basis for the development and validation of adapted, efficient and innovative strategies. Table 17.1 shows the chronology and elements of the programme as is described in the text below.

## 17.2 Element 1: Launching the Programme

A new institutional research partnership was setup with the Chadian Veterinary Laboratory (Laboratoire de Recherches Vétérinaires et Zootechniques de Farcha, LRVZ) and a research collaboration was established with the Ministry of Health. The Centre de Support en Santé Internationale in N'Djaména served as a base for all researchers. The principles of the Swiss Commission for Research Partnerships with Developing Countries (KFPE, 1998) were accounted for in the interinstitutional and interpersonal North–South collaborations for capacity building in Chad and in Switzerland. Shared responsibility, transparency and formalisation were a prerequisite for the excellent institutional framework. A better understanding between partners from different cultural backgrounds was possible once the differing motivations for the research work were made explicit (Zinsstag, 2001). The programme in its first year brought together the following disciplines: anthropology

**Table 17.1** Chronology and elements of the programme 'Improved access to health services for nomadic pastoralists of Chad'

| Element | Time span | Activities | Outcome | Function in the research process |
|---|---|---|---|---|
| Launching of research collaborations | 1996–1998 | - Health staff called for adapted health services to nomadic pastoralists<br>- Exploratory study with site visit and extensive literary research<br>- Setup of North-South research collaborations | - Initiation of interdisciplinary research programme | - First available information and propositions for future research, e.g. integration of animal health<br>- Establishment of institutional framework |
| Stakeholder involvement | 1998 | - 1. national stakeholder workshop: setting the stage | - Setting of priorities and goals of research and action programme | - Recommendations were integrated into research protocols |
| Interdisciplinary research | 1998–1999 | - Execution of research studies | - Identification of actors<br>- Evaluation of morbidity patterns, risk factors and communities' health service priorities<br>- Setup of mycobacteriology laboratory | - Baseline studies<br>- Establishment of contacts with stakeholders |
| Stakeholder involvement | 1999 | - 2. national workshop: formulation of priority actions | - Review of intermediary findings<br>- Participatory identification of pilot-interventions<br>- Publications | - Consent on interventions and attribution of intervention zones by ministries |

**Table 17.1** (Continued)

| Element | Time span | Activities | Outcome | Function in the research process |
|---|---|---|---|---|
| Testing of interventions | 2000–2001 | - Training of barefoot veterinarians<br>- 13 vaccination rounds in two zones | - Feasibility of joint human and animal vaccination campaigns | - Evidence of appreciation of health and veterinary interventions |
| Interdisciplinary and transdis-ciplinary research | 2000–2004 | - Operational research | - Assessments and costing study of interventions<br>- Academic degrees (4 PhDs, 1 MPH, 1 MSc) | - Recommendations for improved and new interventions |
| Stakeholder involvement | 2002 | - 3. national workshop: identification of new research and development issues | - Initiation of new research projects<br>- Publications | - Review of the programme's first phase with all stakeholders<br>- Setting the stage for the second phase |
| Launching of research col-laborations | 2002 | - Integration of the programme in NCCR North-South | - South-South collaborations<br>- Testing of new interventions | - Comparative and complementary studies in other settings |

**Table 17.1** (Continued)

| Element | Time span | Activities | Outcome | Function in the research process |
|---|---|---|---|---|
| Interdisciplinary and transdisciplinary research | 2002–2006 | - Study on essential drugs for nomadic communities<br>- Comparative study between Mauritania and Chad on perception of tuberculosis within NCCR North-South<br>- Molecular epidemiology of tuberculosis | - Complementary social and molecular studies to microbiological study on tuberculosis | - Refined focus of research activities |
| Testing of interventions | 2002–2004 | - Training of traditional midwives<br>- Extension of vaccination campaigns to two new zones<br>- 2 NCCR North-South Partnership Actions for Mitigating Syndromes on 'Information, Education and Communication' and treatment of tuberculosis | - Feasibility of interventions in new zones<br>- Assessment of vaccination coverage<br>- Adapted information campaigns | - Strengthened involvement of authorities and communities in interventions |

**Table 17.1** (Continued)

| Element | Time span | Activities | Outcome | Function in the research process |
|---|---|---|---|---|
| Subsequent activities | 2003–2006 | - Rehabilitation of livestock vaccine production<br>- Collaboration with UNICEF to setup nomadic schools and initial funding for new nomadic school | - Better acceptance of livestock vaccines<br>- Access to education for nomadic children | - Integration of development goals |
| Stakeholder involvement | 2005–2007 | - 4. national workshop: ownership building and policy and 5. national workshop including 8 ministries<br>- Extension of vaccination campaigns with new health and veterinary interventions<br>- Ownership building | - Recommendations on the process of ownership building and scaling-up of interventions | - Handing over of interventions to communities and authorities |
| Launching of research col-laborations | 2005 | - NCCR North-South transversal topic on extensive semi-arid land use | - New PhD study on assessment of health indicators and ownership building<br>- Joint protocol with environmental and sustainability sciences | |

(STI), social geography (Geography Department of the University of Freiburg im Breisgau), biology (CSSI), medicine (Ministry of Health), veterinary epidemiology (STI), and microbiology (LRVZ). Physicians and veterinarians established protocols at the interface between human and animal health using the concept of 'one medicine'. The research hypothesis and objectives of all studies were guided by the recommendations of the first national workshop and were carried out in a common study zone (among Fulani, Arab and Dazagada nomadic communities of the Chari-Baguirmi and the Kanem). The projects were complementary, although each project had its own funding, disciplinary methodological basis and developed its own specific research components. Researchers coordinated and integrated the disciplines after defining the complementary relationships and synergies between the research studies; however, research protocols were also driven by the demand for solid disciplinary work and the validation of hypotheses.

All research studies have contributed to the implementation and testing of priority actions that were negotiated among stakeholders. In the course of the programme, stakeholders were continuously contacted and information was shared. The programme's aim was to show the relevance of the results, to provide authorities with information needed for decision making relating to (limited) resource management and to bridge the 'applicability gap' (Lawrence, 2004). This could only be done by involving stakeholders from the community and the policy levels (as described below).

## 17.3 Element 2: Inter- and Transdisciplinary Research

Studying the epidemiological pathways of disease and disease patterns was essential for developing desirable direct causative interventions. Particularly in the international public health field, epidemiologists need a comprehensive understanding of disease determinants rather than an out of context focus on individual risk factors. Transmission of brucellosis – a severe disease in animals and people – from livestock to nomadic pastoralists could be partly prevented by boiling milk to kill the causative bacteria (Schelling et al., 2003). Yet, the wood that is needed for cooking is limited in the Sahel and cutting of wood may put families in conflict with sedentary communities and authorities from the Ministry of Forest. The anthropological study integrated the local understanding, attitudes, beliefs, knowledge, illness experience and priorities of nomadic communities with regard to health related problems. Pastoralist groups had no concept of diseases that are transmitted from animals to people (zoonoses) with the exception of anthrax (Krönke, 2004). The fact that livestock's milk should be boiled to prevent disease transmission was thus not evident to the communities until adequate information was provided. The first mycobacteriology laboratory in Chad was established for the study of another important zoonosis – bovine tuberculosis caused by *Mycobacterium bovis*. The Chadian public health authorities gained an important tool in the fight against tuberculosis (Diguimbaye, 2004). Population based laboratory studies can lead to the identification of appropriate strategies of control. However, for effective implementation of control measures in the field, the socio-cultural context of a disease has to be known. For example, it is reported that nomadic pastoralists in Mauritania consider tuberculosis as inherited and without cure (M. Ould Taleb personal communication). By way of culturally adapted pictograms, informing about tuberculosis treatment, members of these communities started to make better use of available treatment facilities.

During clinical surveys in 1999 and 2000, almost all participants complained about health problems (Schelling et al., 2005). Treatment seeking behaviour is strongly influenced by cultural norms. For example, Fulani pastoralists' system of norms and values – *pulaaku* – includes the fulfilment of duties and expectations between Fulani and encompasses a high degree of self-control to not express discomfort in public. This may result in attendance of health services only at an advanced stage of disease (Krönke, 2004). The aim of the socio-geographical surveys was to provide a basis for understanding the ecological, socio-economic and political dimensions of nomadic livelihood in the complex crisis situation of Chad. This study complemented the research studies on perceived ill health (emic view) and health risks (etic view) at the intersection between medical, environmental and social sciences. One major outcome was an understanding that nomads' priorities are every day concerns related to securing their livelihood. Access to key pastoral resources and related conflicts with sedentary communities also strongly influenced care seeking behaviour (Wiese, 2004). Severe vitamin A deficiency among nomadic women may also show the range of topics that required consideration for mitigating actions. Vitamin A levels in women's sera were directly correlated to the vitamin A levels in the livestock's milk (Zinsstag et al., 2002). Indeed, the highest milk levels were found in herds grazing on green pastures around Lake Chad, but access to these pastures is difficult due to increased agricultural activities (Wiese, 2004).

We used the community based research results to initiate broader collaborations with authorities and scientific experts (see below). Results were also reviewed in the communities during focus group discussions and regional workshops to obtain a broader perspective of nomadic men and women. At this stage, the overall goal and expectations of the programme were clarified with all stakeholders and a basis of trust and mutual respect was established. These operational research parts provided a most fruitful ground for transdisciplinary processes as shown in the following example of a livestock vaccine against anthrax.

Anthrax is a fatal bacterial disease of people and livestock, and thus, a public health threat and an economically important disease. Pastoralists knew that people can be infected by contact with contaminated animal carcasses. Control of anthrax in an endemic setting such as Chad includes annual compulsory vaccination of livestock, meat inspection and education about the risks of handling and consuming contaminated livestock products. Still, anthrax outbreaks continue to occur. Insufficient vaccination coverage due to a lack of veterinary infrastructure and refusal of livestock owners to have their animals vaccinated are commonly thought responsible for these outbreaks. In view of future vaccination campaigns, we have evaluated the satisfaction of livestock owners with veterinary vaccination using over 100 interviews. Cattle and camel owners acknowledged decade long good experiences with vaccination of livestock. However, the majority reported that vaccines were no longer efficacious and they provided many explanations for this phenomenon. Their arguments gave an impressive insight into the detailed observations and validation criteria they used – many were plausible to scientists with regard to biomedical concepts. A quality control of the anthrax vaccine, which had not been previously considered by researchers, showed that the vaccine was indeed contaminated. The laboratory examination confirmed the concerns of the livestock owners. Documentation of the pastoralists' observations led to laboratory results that in turn attracted the interest of a donor agency and led to the rehabilitation of the livestock vaccine production – to the advantage of the whole country.

## 17.4 Element 3: Stakeholder Involvement

Stakeholders were the academic institutes within the research partnerships, the Ministries of Health (MoH) and Livestock Production (MLP) (with the veterinary services); NGOs working with pastoralists; international bi- and multilateral organisations such as WHO, UNICEF and the Swiss Agency of Development and Cooperation; and donors; on the community level nomadic families (men, women and children), their representatives and associations; as well as the staff of local health and veterinary posts.

Subsequent national stakeholder workshops (in 1998, 1999, 2002 and 2005) were organised to inform interest groups about research results, to formulate health service priorities from a range of options, and to readjust ongoing interventions. Nomads could express their concerns and needs directly to the authorities and also voice non-health related demands such as requests for new legislations on land use. The national workshops in the period 1998–2005 bridged research and practical application and overcame conceptual, institutional and social barriers which Lawrence (2004) identified as crucial elements in transdisciplinary practice. Given the diversity of interests among the stakeholders, the priority setting process started from health system driven approaches – pragmatic in the sense that interventions could be carried out by the health and veterinary services and could be validated by involved scientists – and moved incrementally towards the inclusion of other communities' priorities.

The first national workshop set the overall aims and approaches of the programme. The second workshop recommended testing of joint human and animal vaccination campaigns, better access to essential drugs, and strengthening the training of barefoot veterinarians and traditional midwives. The proposition to join public health and veterinary services for simultaneous vaccination was based on the research results showing that most animals were vaccinated but no nomadic child was fully immunised – although the communities wanted access to vaccination for children. During a third workshop, further research results from the first phase and evaluations of ongoing interventions were presented. The participants identified new research and intervention objectives: in-depth studies on social organisation; poverty reduction strategies (using a conceptual framework of vulnerability and resilience);

a multidisciplinary population based study on human and bovine tuberculosis, for treatment of mobile livestock holders; and quality control of locally produced veterinary vaccines. A recommended development goal – based on community demand – was the strengthening of local and national legislative capacities to design a pastoral code that would secure their transhumance routes. The forth national workshop addressed policy issues almost exclusively. It came up with the next steps for building ownership – by communities and governmental organisations – of the pilot-interventions, and their translation into national public health and veterinary strategies. In conjunction with the last stakeholder workshop in 2005 a process of interministerial planning of a national action plan to support nomadic communities in Chad started under the leadership of the Ministry of Planning in collaboration with eight other ministries. In autumn 2006, the national action plan will be validated during a workshop with the concerned population. The proposed funding should be covered by Chad and international donors.

## 17.5 Element 4: Testing Interventions

As recommended at the 1999 workshop, nomadic barefoot veterinarians and traditional midwives were subsequently trained by professionals. Members with a strong commitment were selected by the communities for training. One midwife reported during the most recent workshop that women now ask about the use of contraceptives. The midwives could in future peer-educate mothers in early management of sick children at the household level. In Chad with its serious shortage of qualified health personnel, capacity building needs to be considered continuously (Wyss et al., 2003).

Joint vaccination campaigns for people and livestock were implemented and evaluated for their feasibility and costs. Between 2000 and 2004, 4,022 nomadic children were fully immunised (three repetitive contacts with the vaccination team) and 6,284 women received at least 2 doses against tetanus. A total of 103,500 livestock were vaccinated. The capacity of existing mobile veterinary infrastructures was extended for simultaneous vaccination during every third vaccination round. Besides support and follow-up of activities, the project played a facilitating role to harmonise the schedules of the public health and veterinary services. It made use of all existing infrastructure (cold chain, transportation) to avoid the establishment of parallel structures. The National Expanded Programme on Immunization (EPI) provided the vaccines and consumables and was involved in the evaluations which further strengthened the good collaboration between the national programme and the project. They appreciated the regular detailed intervention and evaluation reports. The costing analysis showed that the public health sector saved 15% of its operational costs (Bechir et al., 2004).

The campaigns were set up with the local health and veterinary personnel. The vaccination team was highly motivated as their arrival in a nomadic camp was a social event. Half of the nomads had never visited a health centre before. Therefore, a first contact with the health personnel was established during the vaccination activities. Nomads appreciated the quality and the potential of health services for the first time and started to trust the providers. The public health services were able to build on this important gateway opened up by the project. After years of compulsory vaccination of livestock with vaccines that were perceived ineffective by the communities, veterinarians regained trust during the joint vaccination campaigns. They were most willing to share their infrastructure for a more sustainable use.

Simultaneous offering of human and animal health services is particularly suited to pastoralist populations who care equally for the health of family members and for their livestock. Some families travelled from far away to arrive at the vaccination site. Due to communities' participation in the planning, they became aware that vaccination is not a governmental matter alone. Indeed, they asked for sufficient information on vaccination and other health topics. The project developed pictograms and short movies for convening health messages to beneficiary groups. An artist with nomadic background elaborated images adapted to the cultural and educational background. One movie was shot with nomadic actors, a second with professional comedians. In a third movie, highly respected people from the three ethnic groups explain in their mother tongues the importance of vaccination for people and animals. Images and movies were shown before und during vaccination campaigns with ample time for questions and clarifications. Because communities were informed about the nature of vaccines, a rumour that seems to be spreading in Sahelian countries, on a so-called 'sterilising virus' disseminated by vaccination, did not affect our campaigns. The information campaigns can be further capitalised by adding information on livestock management (e.g. diseases, sustainable grazing management) and by encouraging people to seek professional care when they and their children require it.

## 17.6 Element 5: Subsequent Activities

Research and action activities triggered community initiatives on education. Once the Fulani had acknowledged that education facilitates knowledge transfer and negotiations of their own interests, they were encouraged to build a school and appoint a teacher. The programme was asked for support to strengthen the education. An ongoing cooperation with the UNICEF Chad project on education for nomads and funding to help start up, provided by STIs support group, allowed the expansion of this project.

Fund raising for research and intervention, and report writing was time consuming for all researchers involved. The driving force was the stimulating motivation of all stakeholders, enhanced by the mutual learning feedback from stakeholders. A holistic approach is increasingly considered essential for successful medical practice (Flinterman et al., 2001) and appreciated by funding agencies. The Swiss National Center for Mycobacteria in Zurich and the Institute for Veterinary Bacteriology in Berne openhandedly provided needed external expert knowledge and capacity building.

The National Centre of Competence in Research (NCCR) North–South is a joint venture of eight Swiss research institutes and their international partners. In 2001, the extended programme on health of nomads was integrated within the individual project 4, 'Health and Well-being', of this new research programme. It promotes explicitly transdisciplinary research and multilevel stakeholder approaches to increase the acceptability of solutions in concrete contexts. The Swiss National Science Foundation (SNF) and the Swiss Agency of Development and Cooperation (SDC) each fund half of the budget. SDC specifically asks for societal consent and options of action are tested within research projects. The research programme focuses on similar combinations of core problems of non-sustainable development (a total of 18) that can be found in different contexts and thus facilitates the extrapolation of results obtained from one context to others (Hurni et al., 2004). In semi-arid areas, increasing pressure on land and competition for fertile spaces and water resources between nomadic and sedentary populations leads to conflicts that can no longer be resolved by the traditional legal systems alone. By networking within the NCCR North–South with its broad development oriented research alliances, the study by Fokou et al. (2004) showed the required institutional determinants for securing access to health care for nomadic pastoralists. Since only pastoralists with enough means can maintain a mobile lifestyle, the study by Ould Taleb et al. (2006) examines social networks and access to social services of impoverished and settled pastoralists in Mauritanian shanty towns. Chadian, Cameroonian, Malian and Mauritanian researchers perceived South–South collaborations as very motivating and fruitful. Core problems raised by the communities will be central to identifying avenues for sustainable development.

## 17.7 Recommendations

The problem oriented programme aiming at improving access to health care for the nomadic pastoralists of Chad started with very little information on important health issues or daily use of health services. The health and veterinary authorities welcomed and assisted the search of innovative ways to reach more efficiently the 'hard to reach' communities that were virtually excluded from primary social services. Nomadic communities wanted better access to particular western-type medical care for their families and their animals. The simultaneous inclusion of the health of livestock – the most important element in the livelihood of nomadic pastoralists – was certainly an advantage of the new initiative in comparison over other projects reaching pastoralists.

The objectives in the research protocols were formulated flexibly so that the first baseline studies would provide an overview of the involved health related topics and elucidate the relevant issues from the perspectives of public health and the communities. Important public health diseases may not be perceived as such by individuals and communities, while less important diseases may be seen as important. Facilitating elements for the interdisciplinary research group were the formalised institutional arrangements with emphasis on capacity building in Chad and in Switzerland as well as on transparency between partners. Researchers were challenged to cross the boundaries between human and natural sciences to generate research results that could not have been attained using a disciplinary or sectoral approach alone. University curricula rarely enable scientists to communicate with other disciplines and researchers first needed to acquire their own experiences and skills.

Once first actions of the programme became visible, a true basis of trust was established between communities and researchers. The programme became the interlocutor between them and the authorities and the communities were empowered to take their own initiatives. In parallel, the openness of researchers and authorities to the problem stating and organisational skills of communities increased. An external review with emphasis on roles and benefits of all stakeholders helped to identify gaps in the ownership building process.

The long-term commitment of all partners, the transdisciplinary framework of the NCCR North–South, an interdisciplinary North–South research group with formalised institutional agreements allowed: (1) the generation of generic new knowledge on ill health and daily use of health and veterinary services by hard to reach groups; (2) the sharing and disseminating of research results among all stakeholders; (3) the establishment of bridges between research results and their application; (4) the improvement of primary service delivery to nomadic people; and (5) the raising of awareness among policy makers and donors of livelihoods and problems of the poor and vulnerable groups. The last point led to an upcoming interministerial workshop, led by the Ministry of Planning, inviting representatives of eight ministries, to elaborate the technical, institutional, financial, and political factors needed for sustainable implementation of intersectoral services.

**Acknowledgments** The programme received funding from the Swiss National Science Foundation (SNSF 3233.52202.97) and the National Centre of Competence in Research 'Research Partnerships for Mitigating Syndromes of Global Change' (NCCR North–South), the Swiss Agency for Development and Cooperation (NCCR North–South and 7F-03676.01.01); Lotteriefonds beider Basel; OPTIMUS Foundation; and UNICEF. We would like to thank Prof. J. Frey and PD Dr. H. Hächler of the Institute for Veterinary Bacteriology in Berne, the team of the CSSI, the Chadian public health and veterinary authorities, and td-net (network for transdiscplinarity of the Swiss Academy of Sciences) for their continuous support.

## References

Bechir, M., Schelling, E., Wyss, K., Daugla, D.M., Daoud, S., Tanner, M. and Zinsstag, J.: 2004, An innovative approach combining human and animal vaccination campaigns in nomadic settings of Chad: experiences and costs, *Med Trop (Mars)* 64, 497–502.

Diguimbaye, C.: 2004, *La tuberculose humaine et animale au Tchad: Contribution à la mise en évidence et caractérisation des agents causaux et leur implication en santé publique*, Université de Bâle.

Flinterman, J.F., Teclemariam-Mesbah, R., Broerse, J.E.W. and Bunders, J.F.G.: 2001, Transdisciplinarity: The new challenge for biomedical research, *Bull Sci Tech Soc* 21, 253–266.

Fokou, G., Haller, T. and Zinsstag, J.: 2004, A la recherche des déterminants institutionnels du bien-être des populations sédentaires et nomades dans la plaine du Waza-Logone de la frontière comerounaise et tchadienne, *Med Trop* 64, 464–468.

Hampshire, K.: 2002, Networks of nomads: negotiating access to health resources among pastoralist women in Chad, *Soc Sci Med* 54, 1025–1037.

Hurni, H., Wiesmann, U. and Schertenleib, R. (eds): 2004, *Research for Mitigating Syndromes of Global Change. A Transdisciplinary Appraisal of Selected Regions of the World to Prepare Development-Oriented Research Partnerships.* Perspectives of the Swiss National Centre of Competence in Research (NCCR) North–South, University of Berne, Vol. 1, Geographica Bernensia, Berne, pp. 468.

KFPE: 1998, *Guidelines for Research Partnerships with Developing Countries,* Council of the Swiss Scientific Academies (CASS), Berne.

Krönke, F.: 2004, Zoonosen bei pastoralnomadischen FulBe im Tschad, *Zeitschrift für Ethnologie*, 129.

Lawrence, R.J.: 2004, Housing and health: from interdisciplinary principles to transdisciplinary research and practice, *Futures* 36, 487–502.

Loutan, L.: 1989, Les problèmes de santé dans les zones nomades. In: A. Rougemont and J. Brunet-Jailly (eds), *La santé en pays tropicaux*, pp. 219–253.

Majok, A.A. and Schwabe, C.W.: 1996, *Development among Africa's migratory pastoralists*, Greenwood Publishing Group, Westport.

Niamir-Fuller, M.: 1999, *Managning mobility in African rangelands: the legitimization of transhumance,* Impressum London: Food and Agriculture Organization of the United Nations, London.

Ould Taleb, M., Schelling, E., Essane, S., Cissé, G., Lô, B., Obrist, B., Wyss, K. and Zinsstag, J.: 2006, Le desert existe aussi dans la ville: regard sur la lutte contre la maladie chez des populations défavorisées en milieu périurbain de Nouakchott (Mauritanie), *La revue électronique en sciences de l'environnement VertigO*, Hors Série 3, December 2006.

Schelling, E., Daoud, S., Daugla, D.M., Diallo, P., Tanner, M. and Zinsstag, J.: 2005, Morbidity and nutrition patterns of three nomadic pastoralist communities of Chad, *Acta Trop* 95, 16–25.

Schelling, E., Diguimbaye, C., Daoud, S., Nicolet, J., Boerlin, P., Tanner, M. and Zinsstag, J.: 2003, Brucellosis and Q-fever seroprevalences of nomadic pastoralists and their livestock in Chad, *Prev Vet Med* 61, 279–293.

Schwabe, C.: 1984, *Veterinary medicine and human health,* Williams and Wilkins, Baltimore/London.

Swift, J., Toulmin, C. and Chatting, S.: 1990, Providing services to nomadic people: A review of the literature and annotated bibliography. *UNICEF Staff Working Papers No. 8.* UNICEF New York, USA.

Wiese, M.: 2004, *Health-vulnerability in a complex crisis situation – Implications for providing health care to nomadic people in Chad*, Verlag für Entwicklungspolitik Saarbrücken GmbH.

Wiese, M. and Tanner, M.: 2000, A preliminary study on the health-problem in nomadic communities: A case-study from the prefecture of Chari-Baguirmi, Chad, *Aktuelle Beiträge zur angewandten physischen Geographie der Tropen, Subtropen und der Regio TriRhena* 60, 79–102.

Wyss, K., Moto, D.M. and Callewaert, B.: 2003, Constraints to scaling-up health related interventions: the case of Chad, Central Africa, *J Int Dev* 15, 87–100.

Zinsstag, J.: 2001, *Nord-Süd-Forschungspartnerschaft zur Gesundheit von Pastoralnomaden im Tschad – eine Herausforderung für die Zusammenarbeit von Natur- und Geisteswissenschaften*, SAGW and KFPE, Bern.

Zinsstag, J., Schelling, E., Daoud, S., Schierle, J., Hofmann, P., Diguimbaye, C., Daugla, D.M., Ndoutamia, G., Knopf, L., Vounatsou, P. and Tanner, M.: 2002, Serum retinol of Chadian nomadic pastoralist women in relation to their livestocks' milk retinol and beta-carotene content, *Int J Vitam Nutr Res* 72, 221–228.

# Chapter 18
# Sustainable Prevention of Water Associated Infection Risks: An Awareness Campaign Using Visual Media

**Anne Luginbühl**

**Abstract** Water-skin contact associated infections constitute a considerable portion of the most important parasitic infectious diseases in tropical and subtropical environments, with significant public health and economic consequences. Though these diseases have been recognised as 'diseases of behaviour' for many decades, an essential approach to reduce their incidence, namely changing attitudes to infection risk situations, has been largely neglected.

It was thus the aim of this project to develop an awareness campaign to minimise infection risk situations at the waterside. The awareness campaign is focused in a local setting endemic for water-skin contact associated infections, mainly schistosomiasis and soil-transmitted helminthiasis (STH). It is based on the assessment of environmental features, socio-cultural and behavioural factors of local population groups, which affect the transmission of schistosomiasis and STH. Local communication patterns and information sources are evaluated, to help the awareness campaign reach its target. Accordingly, a transdisciplinary approach, taking into account all these diverse aspects, has been adopted. This required the collaboration of researchers from the natural, socio-cultural and communication sciences, local health and media workers, and members of the community at risk.

The studies revealed that schistosomiasis and STH are not known to the targeted population groups, though they are endemic to the region. Therefore, defecating at the waterside where people swim, fish or wash clothes and dishes, is common. Using sanitary structures is not perceived as an efficient health care strategy. The visual medium of TV turned out to be a suitable way of bringing the awareness campaign to the mostly illiterate members of the target population. Animated cartoons in the form of 'TV-spots', alerting people to the infection risk situations caused by particular activities at the waterside and the possibilities for avoiding these activities have been developed.

The project was run in cooperation with a comprehensive programme for integrated control of human parasitoses in the region of Man in western Côte d'Ivoire.

✉ A. Luginbühl
Institute of Infectious Diseases and Institute of Geography, University of Berne, Berne, Switzerland; Federal Veterinary Office (FVO), Berne, Switzerland
e-mail: annelugin@bluewin.ch

G. Hirsch Hadorn et al. (eds.), *Handbook of Transdisciplinary Research*,

**Keywords:** Awareness campaign · Visual media · Prevention · Communication · Water associated Infection risks

## 18.1 Background

Worldwide, one third of all deaths are due to infectious diseases. After malaria, schistosomiasis and STH are the second most important parasitic infectious diseases in tropical and subtropical regions with significant public health and economic consequences (WHO, 2002). Schistosomes and soil-transmitted helminths (STHs) are passed with human faeces or urine, develop in the water and in the humid ground along the shore and penetrate the human body, on contact. Environmental changes and ecological transformations, due to the construction of water resource schemes to meet the rising energy demand and agricultural requirements for development in tropical and subtropical countries, have led to the increase in transmission of schistosomes and STHs and have expanded their range into new areas (Chitsulo et al., 2000).

Control of disease transmission used to focus on environmental management and the control of morbidity was through patient treatment. Integrated control strategies, including the control of biological agents, treatment of patients, as well as environmental improvements, basic sanitation and health education have been recommended for a long time. In 1995, Kloos pointed out: 'The prevalence and intensity of schistosomiasis and the potential for control are largely functions of the interaction of social, cultural, behavioural, geographical and economic factors in a given area with local and regional ecological and environmental factors'. Preventive and control measures that are restricted to eliminating the biological agents, treating patients and implementing sanitary structures are often unsuccessful because they require investment in education and socio-economic improvements. As human parasitoses are considered 'diseases of behaviour', changing attitudes to infection risk situations is essential. Therefore health education, increasing the awareness of infection risk situations and the possibilities for avoiding them, are essential for effective and sustained control programmes. Nevertheless, chemotherapy is still seen as the backbone of morbidity control for schistosomiasis and STH, today and for the future, as Utzinger and Keiser show in their review paper (Utzinger and Keiser, 2004).

This project contributes to the longer-term priority of a more effective control strategy: the avoidance of infections, only achievable through health education – communication, information and awareness – in order to promote behaviour change and to develop preventive attitudes. This must happen because the dependence on a few drugs is an alarming situation – particularly in view of growing resistance – and because the never ending struggle with drugs against symptoms cannot be the aim of sustainable disease control. Chemotherapy will remain an important possibility for controlling acute mortality cases, but it should not remain the backbone of morbidity control for some of the most important 'behavioural diseases' that exist today. Chemotherapy as the backbone of disease control should be replaced by health education (communication, information and awareness), aiming at a prevention of infections and – complemented with socio-economic improvements and sanitary structures – making the treatment with drugs unnecessary.

Few previous projects have succeeded in developing integrated socio-cultural and educational frameworks to promote awareness, prevention and control of water-skin contact associated infections. The beliefs, attitudes, behavioural patterns, knowledge and misconceptions of local populations and the ecological features inherent to each environment were not taken into consideration (Kloos, 1995). Moreover, already existing health education material hasn't been successful in limiting of infection risk situations, as the targeted population groups often have no access to education and are unfamiliar with the interpretation of health education materials (based on an explanation of the transmission cycle or in the measures people should adopt to avoid contact with schistosomes and STHs). In these situations the information becomes useless (Schall and Pinto Diniz, 2001).

It was thus the aim of this project to develop an awareness campaign, aimed at changing water use behaviour in order to minimise infection risk situations at a specific local setting. At the same time, the project offers a complementary possibility of preventing human parasitoses (especially schistosoma and STH infections) in the framework of an integrated control.

The awareness campaign, educating people about infection risk situations and their avoidance, is thus the synthesis of the natural and socio-cultural background of the complex of human parasitoses. The natural background – the ecological and medical context – leading to an expansion of schistosomes and STHs, is well known in the literate world. However, epidemiological studies show that millions of people in the high risk areas become infected again and again, with significant health and economic consequences. Therefore, the population's socio-cultural, behavioural and economic background has to be considered in the development and implementation of health education material. Only the essential interaction of all these diverse disciplines can lead to a reduction of the infection risk situations that confront people.

The project, described in detail by Luginbühl (2005), is part of a comprehensive programme concerning an integrated control of human parasitoses in the region of Man in western Côte d'Ivoire. Besides the development of an awareness campaign, it includes epidemiological surveys and studies regarding an enhancement of the understanding of demographic, ecological, environmental and socio-economic factors that influence disease distribution in the whole region. This knowledge base will facilitate the creation of risk maps and predictions of parasitic infections. A comprehensive geographical information system for the region of Man, including environmental data obtained from satellite imagery and digitised maps, has been established. The findings contribute to the planning of integrated control strategies of several human parasitoses, in particular schistosomiasis and STH, by providing risk maps that can guide decision makers in the region of Man, western Côte d'Ivoire (Raso, 2004). The programme involves the collaboration of diverse targeted population groups, health and public media workers in the region of Man and in Abidjan; the University of Cocody (with the Institute of Ethno-Sociology and the Department of Biosciences) and the Centre Suisse de Recherches Scientifiques in Abidjan

(Côte d'Ivoire); the Swiss Tropical Institute in Basel (Switzerland) and the University of Bern, Switzerland (with the Institutes of Geography and Infectious Diseases). The project work lasted from 2002 to 2005.

## 18.2 Problem Identification and Problem Structuring

Due to the complexity of mutual relationships surrounding parasitical infectious diseases, the concept of the awareness campaign is based on a transdisciplinary approach, which considers the diverse environmental and socio-cultural aspects. This takes into account the local population's knowledge, explanations and perception of morbidity concerning parasitic infectious diseases related to water bodies. Other conditions to be considered are environmental factors: existing water points, water quality, and existing sanitary structures – as well as knowledge about infection risk situations – leading to a specific behaviour at the waterside. The following questions have to be assessed accordingly: How does the population concerned behave at the waterside – with special focus on infection risk situations – and what are the reasons for a specific behaviour? The local communication patterns, the chance of local media reaching the population and the population's perception of the health education material and information strategies in general, have to be assessed in relation to whether the awareness campaign will reach the focus group. Suggestions from the population groups concerned, local health authorities, teachers and media workers, regarding an improvement of the water situation and the quality of information concerning health care strategies for the community at risk, have to be considered and integrated.

Considering all these diverse aspects is essential for an effective control programme, and presupposes the collaboration of all the social players involved in the network of transmission. The community at risk is one of these social players involved in disease transmission and in the dispersal of knowledge, as well as the focus for the awareness campaign. The community participates by contributing its expertise, to the development of the programme: mothers of school and pre-school children, as well as older women, traditionally responsible for health, are a key target group for health education. Mothers' role in the care and education of children, gives women a central position as health promoters at a grassroots level. Moreover, women play a major role in domestic water management in areas where safe water and drainage are not available in the house. In these settings, women are typically responsible for collecting, storing, and using water and for disposing of wastewater (Watts, 2004). Another important social group with regard to health education are the children who are the most heavily infected group in most schistosomiasis endemic areas: children have not yet acquired good hygiene habits, and rivers, lakes etc. are intensively used by this age group, especially in economically deprived areas, where few leisure options are available. Children are the most effective group to target to bring about a sustainable change in human attitudes and behaviour in order to minimise infection risk situations at the waterside. Moreover, children act as links to their families and the community by passing on their knowledge and so helping to generate change. In communities with high illiteracy rates, school children can act as agents for the diffusion of health education messages. Accordingly, the rural school and above all the rural teacher is an important route for the dissemination of information (Schall and Pinto Diniz, 2001). The school and the community are crucial to the success of health promotion and contribute to the prevention and control of diseases. Due to its representativeness, the school atmosphere can offer favourable conditions for the changing of attitudes and the understanding of new alternatives by individuals and communities (Massara and Schall, 2004).

Other social players involved in the network of disease transmission are the policy makers: initially, it was agreed with the national health ministry of Côte d'Ivoire that the intention of our project would be coordinated with ongoing national health care projects. Contacts were established with the School of Arts, the Institute of Communication Sciences and Techniques (ISTC) and the national radio and TV station (RTI) in Abidjan, Côte d'Ivoire, regarding collaboration during the translation and implementation phase of the planned awareness campaign.

From the beginning of the project – besides the exchange with scientific African and European collaborators on the trail of the environmental, ecological and epidemiological aspects of the schistosomiasis and STH complex – a parallel and synchronised process involved the integration of local actors (mainly health providers, public media workers and representatives of the general population) as partners in critical discussions.

## 18.3 Problem Investigation

The study area was situated in the rural and hilly region of Man in western Côte d'Ivoire, West Africa. The whole region is characterised by a dense hydrological network of permanent and temporary rivers, shallows and damp zones. The population is composed of seven main ethnic groups and many others. Each of them uses its own local language. About 70% of the population lives from agriculture and the majority is illiterate. The town of Man and the surrounding villages and settlements are endemic for *Schistosoma mansoni* and STH infections, particularly hookworms (Utzinger et al., 2000, 2003). Consequently, since 2000, district health authorities have implemented a programme to improve access to treatment. The programme primarily targeted the school-age population, with the aim of controlling morbidity among this high-risk group. Due to socio-political unrest commencing in September 2002, treatment campaigns had to be interrupted.

In collaboration with regional health and community authorities, two villages (Mélapleu with the main ethnic group Yacouba, and Zouatta II with the main ethnic group Wobé) were chosen for in-depth study. The selection of the villages was based on their demographic, socio-economic, sanitary and water resource structures (which were representative for the region), on the community interest in the project and on the intention of the village authorities (the school directors, teachers and the villagers in general) to cooperate.

In an epidemiological survey, conducted in Zouatta II in May 2002, high frequencies of *S. mansoni*, hookworms, *P. falciparum*, and several intestinal protozoa were found, and polyparasitism was very common. The overall real prevalence of *S. mansoni* was 39.8% and the overall real prevalence of hookworms was 45% (Raso et al., 2004a).

To do justice to the knowledge gaps to be filled, diverse qualitative and quantitative methodologies involving the different social players and developed through different disciplines were used and integrated.

These were Focus Group Discussions (FGDs) and semi-structured interviews with different groups of women, school directors, teachers and traditional health workers about the local population's knowledge, explanations and morbidity perceptions concerning parasitic infectious diseases related to water bodies. The discussions also covered water use behaviour, communication patterns and local media as sources of knowledge. Discussions and interviews with different social groups allowed for the exploration of the same research topics from different perspectives.

Another method of assessing the local people's behaviour at the waterside was used in the 'Drawing Project': primary schoolchildren were asked to draw themselves and their family members at their different daily activities with particular consideration to water contact. Several children volunteered to explain their drawings in front of the camera. They were asked to explain the meaning of their drawings and the health problems they are confronted with. The drawings were analysed according the 'Content Analysis' by Rose (2001). The aim was to evaluate how the different family members and the children artists themselves were in contact with water, if and how they were protected against water-related infectious diseases and if different activities were taking place at the same water point. A further aim was to evaluate the effectiveness of using children's drawings as a method, in this specific context (Egger and Thoma, 2006).

Quantitative questionnaires to assess demographic, socio-economic (e.g. presence of sanitary structures and water and information sources), morbidity risk factors and perceived morbidity indicators were administered in Zouatta II. The questionnaires were developed and adapted to the current setting after discussions with local field assistants who were designated by the village chief and his authorities. During a one-day workshop, the assistants were trained to interview household members and to fill in the questionnaires. All individuals involved in the survey were screened for *S. mansoni*, STHs and intestinal protozoa as well as for malaria parasites by a collaborating epidemiology-research team. In cases where infection was present the individuals were treated according to current national public health guidelines of Côte d'Ivoire. The relationship between self-reported morbidity indicators and the effective infections of an individual have been assessed (Raso et al., 2004b).

Notes, photo and video recording documented all fieldwork. After having obtained the permission of the communities concerned, photos and video records of the village life with special focus on activities at the waterside and the environment were taken. The people welcomed being photographed and filmed and did not seem to be distracted from their natural behaviour by the photo and video camera. Existing health education material and advertising in the public space has been documented and analysed. Observations made in the field, information obtained by key informants and during spontaneous contact with the population, discussions with researchers from different disciplines familiar with the research context, artists and social assistants, familiar with the local culture, have been registered. These included regular meetings and coordination with local health workers of the regional health centre 'Grandes Endemies', the project 'Water – source of life and development, Man', and the regional radio station 'Radio Man', that produced broadcasts about health care in cooperation with the village communities. The field notes, photo and video documentation have been triangulated and integrated with all the other methodologies and data.

The different methods, approaching the same scientific objectives and directed towards the enrichment and completion of the findings, were administered simultaneously (Flick, 2000). However, the main part of the data acquisition is based on the qualitative methods rather than the highly structured methods. The qualitative methods permit the researcher to be guided by the logic of the target population groups (Groupe de recherche Chari, 2001).

## 18.4 Baseline Results

The following main conclusions can be drawn with respect to the questions addressed by the project: schistosomiasis and STH – with their specific transmission through skin contact with contaminated water bodies or soil – are not familiar to the population groups concerned, though they are highly endemic in the region. They are often diagnosed as diarrhoea, dysentery and 'general worm infections', diseases which are mostly linked to dirty drinking water, poor nutrition or witchcraft. People have already been made aware of the fact that clean drinking water is important for their health because it protects them from various diseases. If there is no pump providing clean water in the neighbourhood, and if time and personal conditions allow, people get their drinking water at so called 'water sources'. These provide relatively clean water, but are often far from the dwellings. As alternative, people filter or boil the water from the rivers and pools before drinking. Inertia, tiredness or lack of available time can prevent people undertaking the long trip to get clean drinking water or to process their drinking water, even when they understand the importance of these procedures to their health. Therefore, getting clean drinking water is a simple analysis of costs against benefits, depending on how inevitable, immediate and significant the health problems, resulting from drinking dirty water, are perceived.

However, people do not realise that clean drinking water alone does not protect them against schistosoma and STH infections. This means that there is no knowledge or awareness of the necessity of not defecating at the waterside or into the water where people swim, fish or wash clothes, even if they do not drink it. Accordingly, an analysis of costs against benefits can be made for the choice of place for defecating and urinating. If there is no advantage perceived in not defecating near or into a water body where people swim, fish or wash clothes – because people are not alert to the risk of infection resulting from this behaviour – no efforts are made to build more latrines or to maintain existing ones and it is considered unacceptable to travel a bit further to reach a latrine. Therefore, there should be information available to alert people about schistosoma and STH infections and how they are transmitted!

Existing health education (e.g. about Aids) reaching the rural villages through radio and TV programmes, is mostly communicated in the French language. Therefore, it doesn't reach the illiterate members of the target group. The people concerned wish to view health education programmes presented by their peers (for identification purposes) and in their own language. Moreover, images are perceived as more attractive than solely spoken or written information. A suitable method of carrying out an awareness campaign is through moving pictures, using visual elements that the local population can identify with. An appropriate medium for conveying these is the TV, which reaches a large proportion of the targeted population. The TV is not as widespread as the radio: 27% of the households in Zouatta II own a TV, 63% a radio (Luginbühl, 2005). According to a regional survey, one in five schoolchildren reported to have a TV and more than half of the schoolchildren reported to have a radio in their household (Raso et al., 2005). In fact more people have access to TV programmes: TVs are running in the evening when everybody comes back home from the daily work outside the villages. Parents and neighbours meet in front of a TV, which means that people have access to a TV even if they live in a household without one. TV watching is a social event.

## 18.5 Concept of an Awareness Campaign

In accordance with the baseline results and to an analysis of literature about existing health education material, the concept of an awareness campaign to influence the behaviour at the waterside of the population groups concerned, and to limit infection risk situations, has been developed. The main medium is animated cartoons in the form of TV-spots. Two main points fundamentally differ from previous health education material: (1) The animated cartoons are without vocal or written text. The information is conveyed by pictures only. Therefore, comprehension and identification work for illiterate people and for more than one ethnic and language group at the same time; (2) As the people at risk of infections with schistosomes and STHs do not see the parasite or its penetration of the intact skin, neither the parasite nor its biological development cycle is shown in the TV-spots. On the other hand, the cause and the consequences of an infection are shown through the different daily activities at the waterside that constitute an infection risk – performed by different actors and actresses – abstract cartoon figures that the local population can identify with.

As an awareness campaign is more effective when the messages are specific and unambiguous, the TV-spots are as simple as possible – graphically and in content: Each sequence shows the same simple and easily identifiable scene: a house, a latrine and a water point. The message is reduced to the two statements: defecating at the waterside means that people who come in contact with this water feel sick, and using the latrine means that people who come in contact with the water feel well (Fig. 18.1).

**Fig. 18.1** Sequence of one possible story out of the TV-Spot series: The left side shows a boy in his local school uniform walking to a water point to defecate. Later on his mother washes the laundry at the same water point. Returning home she feels sick. The right side shows the possible alternative to this hygiene behaviour. The boy running to the water point for defecating becomes aware that it would be better to use the latrine. He turns around and runs towards the latrine to defecate. Later on his mother washes the laundry at the water point, now not contaminated with faeces. Returning home she feels well

The actors and actresses enter the scene in independent sequences. Each sequence tells its own story (concerning the activity of the second figure entering the scene: swimming, fishing, washing clothes etc.), conveying one of the two messages mentioned above. A sequence lasts less than 60 seconds. The animated cartoons are broadcast individually and alternately on TV, before and after the 'news', in the course of an advertising block or programme break.

The TV-spots are designed to be broadcast over a period of time in order to achieve an understanding of the need to change behaviour – the importance of using latrines instead of defecating at the waterside – for the benefit of everyone's health. The long lasting TV-spot awareness campaign influences viewers subconsciousnessly, to change their behaviour. Once, people have accepted the need for latrines, a programme establishing sanitation facilities could be successful. In a further step, complementary print media – with additional information about the 'biological cause' of the specific 'health problems' and the course of the diseases – adapted to different local platforms such as schools and health centres, are to be added. They will address the conscious perception of the population and the cognitive thinking of the public. For the purpose of recognition the complementary media uses the same visual elements.

## 18.6 Recommendations

The project focused on the realisation of an awareness campaign, aiming to limit the infection risk situations at the waterside. It started with very little information about environmental, socio-cultural and behavioural factors, all of which affect water-skin contact associated infectious diseases. Therefore, community participation was essential to the development of an effective awareness campaign. The target population, acted as experts concerning the environmental, socio-cultural and behavioural factors that lead to infection risk situations at the waterside. Social workers and researchers, on the other hand, knew the ecological and epidemiological context that favours water-skin contact associated infections. Therefore, during the investigation phase, the collaboration between the population groups, social workers and researchers allowed the exchange of information and mutual learning, as well as an increase in awareness of the problem to be solved. Moreover, the community's suggestions and assistance in the design of health education material ensured that the information is understood to the target group. The contribution of women turned out to be very significant as they showed great interest in participating and helping to implement health education strategies. They were very receptive concerning health problems, the infection risks that their families are confronted with at the waterside, and to the respective prevention strategies. Children were another important social group in the implementation of health education programmes. They live in the heart of a community, perceiving all the different daily activities from the inside. As they are accompanying their mothers, fathers, brothers and sisters, grandparents and neighbours during their daily work in and around the villages, they become witnesses to activities that researchers only know through long term observations, or not at all. Therefore, the drawing project turned out to be an effective and fast method of assessing the waterside activities of different community members. Within the community, trust was established between the community members and the researchers allowing successful collaboration. Community authorities were convinced of the importance of the project's aim to reduce the parasitic disease burden and the subsequent improvement in living standards so the project workers were welcome in the community.

In order to implement the awareness campaign in different high risk cultural regions, the different visual elements are to be adapted to each cultural setting. The consideration of local visual perceptions is a prerequisite for a visual awareness campaign and the engagement of local designers and artists can contribute to the target public identifying better with the content.

An awareness campaign has to repeat its message over a long time period for it to have a subconsciousness influence on the target group, to achieve a sustainable behaviour change and thus to limit infection risk situations. Periodic evaluation of the effects of the awareness campaign can lead to required adaptations. Within the framework of a research programme lasting three to four years the effect of an awareness campaign to change behaviour at the waterside and a subsequent reduction of infections – through the successful implementation of sanitation facilities and socio-economic improvements – is difficult to evaluate. A long-term commitment by all stakeholders, including local policy makers, is needed to achieve a reduction of infection risk situations. This precondition for a successful implementation and outcome of an awareness campaign is not easy to achieve in a politically unstable region such as Côte d'Ivoire.

**Acknowledgments** I would like to thank Prof. Kurt Schopfer from the Institute of Infectious Diseases (University of Bern) for ideological and financial support, Prof. Doris Wastl-Walter from the Institute of Geography (University of Bern) for scientific support and Cinthia Acka Adjoua from the Institute of Ethno-Sociology (University of Cocody, Abidjan, Côte d'Ivoire) for excellent collaboration in the field. Thanks are due to the communities involved, health and media workers in the region of Man, western Côte d'Ivoire, the researchers from the University of Cocody (Abidjan, Côte d'Ivoire) and the Swiss Tropical Institute (Basel, Switzerland) for their scientific collaboration, as well as to Holger Hoffmann-Riem from the td-net (network for transdisciplinarity of the Swiss Academy of Sciences) for helpful comments to the manuscript.

## References

Chitsulo, L., Engels, D., Montresor, A. and Savioli, L.: 2000, The global status of schistosomiasis and its control, *Acta Trop* 77, 41–51

Egger, A. and Thoma, P.: 2006, *Evaluation of Human Behaviour at the Waterside in rural Côte d'Ivoire by Analysis of Children's Drawings*, MD Thesis, Institute of Infectious Diseases, University of Bern. Switzerland, 104pp.

Flick, U.: 2000, *Qualitative Forschung. Ein Handbuch* (E. von Kardoff and I. Steinke, eds), Rowohlt, Reinbek bei Hamburg, 768pp.

Groupe de recherche Chari: 2001, *Méthodes qualitative en recherche sociale: Eau et hygiène du milieu.* Rapport de recherche, 39 pp. TDR/RCS/MQRS/02.2

Kloos, H.: 1995, Human behaviour, health education and schistosomiasis control: A review, *Soc Sci Med* 40(11), 1497–1511

Luginbühl, A.: 2005, *Social Perception of Water Associated Infection Risks – Concept of an Awareness Campaign using Visual Media*, Ph.D. Thesis, Institute of Geography, University of Bern, Switzerland, 340pp., Retrieved June 11, 2007, from http://www.stub.unibe.ch/ download/eldiss/05luginbuehl_ak.pdf.

Massara, C.L. and Schall, V.T.: 2004, A pedagogical approach of schistosomiasis: An experience in health education in Minas Gerais, Brazil, *Mem Inst Oswaldo Cruz* 99(suppl. 1), 113–119

Rose, G.: 2001, *Visual Methodologies – An Introduction to the Interpretation of Visual Materials.* Sage Publications, London.

Raso, G.: 2004, *Assessment, Mapping and Prediction of the spatial distribution of parasitic infections in western Côte d'Ivoire and implications for integrated control*, Ph.D. Thesis, Swiss Tropical Institute, University of Basel, Switzerland, 183 pp.

Raso, G., N'Goran, E.K., Toty, A., Luginbühl, A., Acka, C.A., Tian-Bi, N.T., Bogoch, I.I., Vounatsou, P., Tanner, M. and Utzinger, J.: 2004a, Efficacy and side effects of praziquantel against Schistosoma mansoni in a community of western Côte d'Ivoire, *T Roy Soc Trop Med H* 98, 18–27

Raso, G., Luginbühl, A., Acka, C.A., Tian-Bi, N.T., Silué, K.D., Matthys, B., Vounatsou, P., Wang, Y., Dumas, M.E., Holmes, E., Singer, B.H., N'Goran, E.K., Tanner, M. and Utzinger, J.: 2004b, Multiple parasite infections and their relationship to self-reported indicators in a community of rural Côte d'Ivoire, *Int J Epidemiol* 32, 1–11

Raso, G., Utzinger, J., Silué, K.D., Ouattara, M., Yapi, A., Toty, A., Matthys, B., Vounatsou, P., Tanner, M. and N'Goran, E.K.: 2005 Disparities in parasitic infections, perceived ill health and access to health care among poorer and less poor schoolchildren of rural Côte d'Ivoire, *Trop Med Int Health* 10(1), 1–16.

Schall, V.T. and Pinto Diniz, M.C.: 2001, Information and education in schistosomiasis control: An analysis of the situation in the state of Minas Gerais, Brazil, *Mem Inst Oswaldo Cruz* 96 (Suppl.), 35–43

Utzinger, J. and Keiser, J.: 2004, Schistosomiasis and soil-transmitted helminthiasis: Common drugs for treatment and control, *Expert Opin Pharmacother* 5(2), Ashley Publications, Monthly Focus: Anti-infectives

Utzinger, J., N'Goran, E.K., and Ossey, Y.A.: 2000, Rapid screening for Schistosoma mansoni in western Côte d'Ivoire using a simple school questionnaire, *Bull World Health Organ* 78, 389–398

Utzinger, J., Müller, I., Vounatsou, P., Singer, B.H., N'Goran, E.K. and Tanner, M.: 2003, Random spatial distribution of *Schistosoma mansoni* and hookworm infections among school children within a single village, *J Parasitol* 89, 686–92

Watts, S.: 2004, Women, water management, and health, *Imerg Infect Dis,* Retrieved June 11, 2007, from http://www.cdc.gov/ncidod/EID/vol10no11/04-0237.htm.

WHO – World Health Organization: 2002, Prevention and control of schistosomiasis and soil-transmitted helminthiasis, *Tech Rep Ser* 912, 1–57

# Chapter 19
# Behavioural Sciences in the Health Field: Integrating Natural and Social Sciences

**Bettina F. Piko and Maria S. Kopp**

**Abstract** The main goal of this project has been threefold: first, to create an integrated course system under the name 'Behavioural Science' in which students of medicine and health sciences are provided with a set of social and behavioural sciences applicable to medicine; second, to develop a health status monitoring system by means of two surveys (Hungarostudy which collects data on the health status of the Hungarian adult population, and South Plain Youth Study which gathers data on the health status of the adolescent population); and third, to apply theoretical knowledge and empirical research results in the field of practice. In this case, practical prevention programmes and skills development training are planned and implemented. The phases are built on each other and multiple feedback systems are used to integrate them into a joint transdisciplinary project. The methods contained six main steps: (1) to collect the most relevant statistical data about the population's health status, health needs and professional and public knowledge; (2) to apply a biopsychosocial approach to evaluate the statistical data and understand its background; (3) to introduce the biopsychosocial model in medical/health science courses; (4) to carry out surveys to collect information on the psychosocial aspects of the population's health status, based on the biopsychosocial model; (5) to evaluate data from the surveys (Hungarostudy, South Plain Region Youth Study) and to develop prevention programmes and skills training; (6) to get continuous feedback from prevention programmes, including the updated statistical data.

A system based theory was applied which provided a line of theories which contributed to the development of a systematic concept of transdisciplinary research.

**Keywords:** Behavioural science · Biopsychosocial model · Medical training · Disease spectrum · Health development

✉ B.F. Piko
Department of Psychiatry, Behavioral Science Group, University of Szeged, Szeged, Hungary
e-mail: pikobettina@yahoo.com

## 19.1 Background

After a century of successive global transformations in mortality and morbidity schedules, new challenges to health have arisen to erode the vision of progressive health security (Worthman and Kohrt, 2005). There is a requirement for a new paradigm to handle transdisciplinarity in the health field. Our project 'Applying Behavioural Science in Medical Schools' was designed to meet this requirement by involving two different paradigms (natural and social/behavioural science paradigms) in both theory and practice – from research through education to health promotion. The transdisciplinary approach is based on the application of the biopsychosocial model. Programme evaluations are made at different levels of the project. Future research should use a longitudinal design to evaluate the role of a long term transdisciplinary appoach in the health field.

The biopsychosocial model has brought about changes in both medical education and medical practice. Hopefully, these changes are so intensive that they will lead to a new medical paradigm. Because of its complexity, behavioural medicine should become an integrated scientific field at the crossroads of the natural and social/behavioural sciences. The cultures of medicine (objectivity versus subjectivity, quantitative versus qualitative) meet here. In modern medicine, the introduction of the natural scientific paradigm in the 20th century was a great development. At present, a similar acceptance of social/behavioural paradigm should be necessary. Diagnostic procedures are based on the natural scientific mechanism which enables us to make an exact diagnosis of the body and bodily processes. However, the patient's body is also heavily influenced by his/her psychosocial processes. We need different types of investigations to detect the patient's psychosocial state.

Applying behavioural science in the health field has always been a challenge (Piko and Stempsey, 2002). Not only modern medicine but also traditional medicine see the process of healing and dealing with health and illness in general as belonging to natural and social/behavioural science paradigms. Our project 'Applying Behavioural Science in Medical Schools' deals with the implementation of this double-bind paradigm, which is rooted in both theory and practice (Weidner et al., 2002; Fig. 19.1). The project contained a research phase, measuring the health status of the population; an educational phase, applying research results in the courses for medical and other health science students; and a practice phase, developing skills and prevention programmes based on the research results. The transdisciplinary approach provides a good rationale and framework for this application. While the natural science paradigm suggests that we should find objective facts and quantifiable data, the social/behavioural science paradigm suggests that we should find additional, narrative, and in many cases non quantifiable data, which helps develop an integrated approach to health and illness in society (Wulff, 1999).

A system knowledge approach (Stokols, 2000) is applied, called a 'biopsychosocial' model of human processes. This holistic view had been introduced during the ancient Greek times but scientific investigations and empirical justifications were not provided until the 1970s (Engel, 1977). The biopsychosocial model emphasises that external and internal, genetic and environmental, somatic and psychosocial

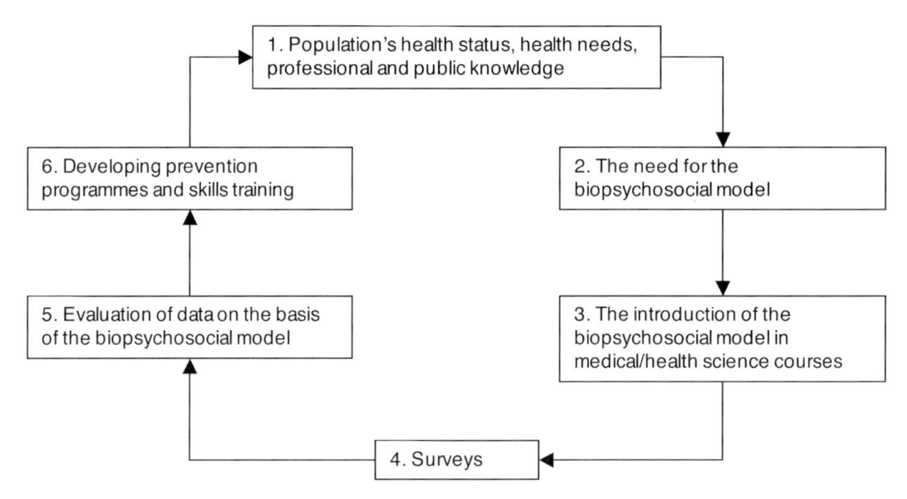


**Fig. 19.1** The six steps of the research process

factors are all important in determining health and inducing disease (Cohen et al., 2000). In addition, within this integrated system, each subsystem (biological, psychological and social) may affect, and may be affected by, every other subsystem. Applying the biopsychosocial model in medicine, and the health field in general, makes it possible to understand human beings from their molecular or cellular structure and psychological functioning to their social and socio-cultural background. Not surpisingly, the World Health Organisation (WHO, 1984) has defined human health in terms of combined biological, mental and social well-being with relation to environment.

Although this project is primarily based on a health promotion issue, it may also be of interest to others who carry out transdisciplinary projects in a broad context of the medical field or elsewhere. For example, the system based theory provides a line of theories which contributes to the development of a systematic concept of transdisciplinary research. The integrated system emphasises the need for a complex approach to human systems. In addition, the planning process involves several steps which would be useful in other areas. The multilevel evaluation process also provides a useful insight into the project, in terms of strengths and weaknesses, opportunities and threats.

Our project focuses on the health issues although with a different approach from the traditional biomedical model. In addition, our project is an experimental development in Hungary: it goes beyond the traditional views and represents an innovative development. Therefore, the development and introduction of the project is not without difficulties. In this case, the description of the project should also be of interest to those who carry out unusual, innovative work in a specific field, not only in the medical field.

Figure 19.1 shows the model of our project. Overall, the project bridges the gap between a current health problem (that is, poor health status of the population) and prevention measures, via training of medical students. It involves six steps:

1. Collecting the most relevant statistical data, about the population's health status, health needs, professional and public knowledge, which serves as a basis for prevention strategies.
2. Recognising the need for a biopsychosocial approach to evaluate the statistical data and understand the background.
3. The introduction of the biopsychosocial model in medical/health science courses for medical students.
4. Carrying out surveys to collect information about the psychosocial aspects of the population's health status, based on the biopsychosocial model.
5. Evaluating the data collected from these surveys (Hungarostudy, South Plain Region Youth Study) and basic statistical data about the population's health in the light of the previous model.
6. Based on the empirical data, prevention programmes and skills training are to be developed (partly for medical students and partly for the population).

In addition, there is continuous feedback from prevention programme and the updated statistical data (from steps 1 to 6): that is, a new cycle is to be started when the programmes are ready.

## 19.2 Health Status of the Hungarian Populations

In the late 1980s Hungarians had the lowest life expectancy in Europe and the highest mortality rate due to accidents, suicide, gastrointestinal disease, heart disease and neoplasm (Cockerham, 1999; Forster and Józan, 1990; Kopp et al., 2004a). Even when compared with other central or eastern European countries, the data from Hungary are a cause for concern. For example, the life expectancy of males in 1990, in selected European countries were: Hungary (65.1 years), Czech Republic (67.6 years), Bulgaria (68.0 years), Poland (66.5 years), Romania (66.6 years), Sweden (74.8 years), United Kingdom (72.9 years) (Piko, 2004). We assume that the lack of a behavioural science approach to both education and health promotion practices might have contributed to this situation in addition to the economic gap between 'east' and 'west'.

## 19.3 Need for the Biopsychosocial Model

The first Institute of Behavioural Sciences was founded in 1991 in Pécs. In 1993, the Institute of Behavioural Science at Semmelweis University of Medicine, the largest and most internationally recognised institution in Hungary, was founded in Budapest. The Institute has subdepartments for each area of behavioural science: psychosomatic medicine, health psychology, medical sociology, bioethics, medical anthropology, communication studies. These subdepartments teach the different branches of behavioural science separately but all apply the interdisciplinary, integrative, biopsychosocial paradigm, in a system approach. A Handbook of Behavioural Science was published in 2001, and provided information about the interdisciplinary approach to the public. This model is present at different levels of teaching: the undergraduate level (for medical and health science students), the post-graduate level (postgraduate training for specialists, and teachers) and the doctoral level (Ph.D. programmes in behavioural science for both medical and non-medical graduates).

These departments are integrative, not only in teaching, but also in their important transdisciplinary task of transferring survey results into practice. During our courses, research findings from the two presented surveys are usually used to demonstrate psychosocial aspects of the health status of the Hungarian population. There are specific requirements for students to develop small-scale projects, which include these data, as a method of practicing concrete prevention work. These cross-sectional surveys are repeated in every second or fourth year allowing monitoring system may also be set. This method of teaching project management means that transdisciplinarity is alive in everyday practice.

If we analyse the relationship between system parameters and norms across different problem contexts, we may conclude that the societal norms support such a transdisciplinary project. For example, there is rising public dissatisfaction with personal health and the health care services provided. Despite these values and norms, the acceptance of behavioural science by natural scientists working in medical schools is still low in some places (Brook et al., 2000; Margolis et al., 1987). All these problems highlight the limitations of the biomedical perspective.

The research process needs strong and continuous cooperation between behavioural scientists and non-scientists (teachers, public health experts, health policy experts), as well as between scientists from natural and social/behavioural sciences. This process involves multiple feedbacks and systematic evaluations (see Fig. 19.1).

## 19.4 Introduction of the Biopsychosocial Model

The biopsychosocial model has been a focal point for the development of a behavioural science approach to medicine (Engel, 1977). Engel, one of the prominent propagators of the biopsychosocial model, emphasised that external and internal, genetic and environmental, somatic and psychosocial factors are all important in determining health and inducing disease. His biopsychosocial model was a cornerstone in medical education and practice in a broader sense. Applying the biopsychosocial model in medicine makes it possible to understand human beings from molecular or cellular units to societal levels. Engel's model has led to an attitude change in medical practice; moreover this process should lead to a new paradigm in medical sciences (Sallis et al., 2000; Piko and Stempsey, 2002). Due to its complexity, medicine should become an integrated scientific field, at the crossroads of the natural and social sciences, needing a transdisciplinary approach.

In the project, both academic and programme experts are involved. Firstly, academics from faculties in the medical schools in Hungary, are active participants in the preparation of developing course curricula. Secondly, surveys are planned and carried out to get a data pool for the monitoring system. In this phase, academics play the leading role. Researchers in other fields, prevention programme managers and researchers, are also involved, e.g. public health experts working in the field of public health services at county levels, teachers and other health care workers. In the third phase, experts working in the health field play the most active and decisive role in implementing the theories and empirical data gathered in the surveys.

Various disciplines are involved in the project. All these are applied disciplines and all apply social and behavioural science theories, concepts and empirical findings integrated in the paradigm of medicine and health sciences (Piko and Stempsey, 2002).

- *Medical psychology*: applies psychological theoretical models and practical aspects in various fields of medicine and health care, e.g. doctor-patient encounters, health and illness behaviours, psychologically demanding situations such as death and bereavements, dealing with health risks, etc.
- *Medical sociology*: applies various sociological theories (e.g. social inequalities, deviance theories) in health and health care institutions.
- *Medical anthropology*: applies techniques of lay beliefs about health and illness in detecting different aspects of individuals and societies in modern and archaic cultures.
- *Medical communication*: communication studies applicable to doctor–patient and other healthcare worker–patient communication.
- *Medical ethics:* involving questions regarding bioethical problems in healthcare and everyday life. Many academics have an educational background in more than one field, e.g. a medical sociologist with an M.D. and a sociology degree; a medical psychologist who graduated from both medicine and psychology.

All these disciplines are integrated under the umbrella 'Behavioural Medicine', which includes various disciplines and approaches, from prevention to hospice care.

The project also involves the introduction of the biopsychosocial model to non-medical experts. This implementation needs an integration of different societal attitudes. Publications in the field of medical anthropology revealed that laypersons had particular beliefs about the cause, significance and treatment of their conditions: beliefs that sometimes lie outside the scientific paradigm of Western medicine (Helman, 1991). Likewise, in many cases, a considerable gap exists between medical and non-medical experts' perspectives on health and illness. Empirical findings show that laypersons and non-medical experts in the health field tend to apply the holistic (biopsychosocial) perspective while medical experts usually think of health and illness in a biomedical perspective.

## 19.5 Surveys

Two interdependent survey programmes were introduced to collect data about the health status of the Hungarian population. The main goal of these surveys is to detect the interrelationship between psychosocial factors and the health status of the population. The 'Hungarostudy' is a national, cross-sectional survey that is representative of the Hungarian adult population (over 18 years). It takes into account sex, age, and the 150 subregions in the countries (Kopp et al., 2004b). A clustered, stratified sampling procedure was developed by experts at the National Population Register. In 2002, 12,643 persons were interviewed in their homes, 0.16% of the population older than 18. The refusal rate was 17.7%. The interviewers were district nurses, and approximately one hour was spent in each home completing the questionnaires. Items varied from life meaning, psychosocial health measures (e.g. depression, WHO well-being scale, self-efficacy scale, ways of coping questions, social support, self-rated health and self-reported disability) to sociodemographics and socio-economic variables. In 2004 the 'South Plain Region Youth Study' collected data from students enrolled in the secondary schools of the South Plain Region (Bacs-Kiskun, Bekes and Csongrad counties) of Hungary. This representative sample of 1,200 students was based on randomly selected classes from three randomly selected high schools in each county. Of the 1,200 questionnaires sent out, 1,114 were returned and analysed, yielding a response rate of 92.8%. The age range of the respondents was 14–21 years of age (mean: 16.5 years of age, S.D.: 1.3); 444 (39.9%) were boys and 670 (60.1%) were girls. Parental permission was obtained so trained public health workers could distribute the questionnaires to students prior to the start of class. Students were given a brief explanation of the objectives of the study and instructions for filling out the questionnaire. Participation in the study was voluntary. Confidentiality of the responses was emphasised and that the data would be used for research purposes only. The response time ranged from 30 to 40 min. Completed questionnaires were placed in sealed envelopes and collected from each of the participating schools. The self-administered questionnaires were used to obtain information from students regarding family structure, psychosocial health, materialism and other value orientations, substance use, depression, satisfaction with life and sociodemographics.

## 19.6 Evaluation

This ongoing project should be investigated through an evaluation model of strengths, weaknesses, opportunities and threats (SWOT analyses):

*Strengths*: Despite the fact that training in behavioural sciences was not provided between the 1950s and 1989 in higher education in Hungary, the psychosomatic approach has a long tradition dating back to the 19th century when Ignaz Semmelweis discovered the cause of puerperal fever (poor hygiene). In addition, later in the 20th century, a number of Hungarian scientists, Sandor Ferenczi, Franz Alexander, Hans Selye (who described the stress model), and Mihaly Balint, significantly contributed to the development of psychosomatic and biopsychosocial medicine (Piko and Kopp, 2002). It is also a strength that by the time our departments were founded, the teaching of behavioural sciences in medical schools had been already accepted in both Western Europe and the United States. Another strength is that many of our behavioural scientists have degrees in both medicine and social/behavioural science, e.g. psychology, sociology or philosophy. Moreover, some of our scientists have a strong public health orientation and expertise in the field of health promotion. Our scientists usually publish their findings in both Hungarian and international journals and participate in international scientific societies. A number of research projects receive extramural grants. Finally, our departments have a number of international cooperations which may contribute to the acceptance of the field.

*Weaknesses*: First is the late introduction of behavioural science in medical schools. The lack of physicians', and other health professionals', knowledge, skills and attitudes in behavioural science may have contributed to the deterioration of the health status of the Hungarian population. Due to this long break, applied behavioural science still faces a number of problems and challenges. For example, students, particularly medical students, find this integrative approach unfamiliar since they face mainly biomedical subjects, which do not operate within this biopsychosocial paradigm. The lack of a common social science paradigm may contribute to the inadequate cooperation between behavioural science researchers with different disciplinary backgrounds.

*Opportunities*: More cooperation between behavioural science researchers and clinicians needs to develop. This may contribute to a greater acceptance of this approach. In addition, more integrative courses should be developed and introduced. For example, the first semester of medical education begins with a course 'Introduction to Preventive and Behavioural Medicine'. Later, students may choose the courses 'Behavioural Science Approach to Substance Use' and 'Behavioural Science Approach to Health Promotion' both of which provide an integrated system approach to health problems.

*Threats*: The biggest threat to our faculty is an official comparison between the scientific accomplishments of behavioural science researchers and those working in the biomedical field. In many cases, behavioural science researchers are underachievers in terms of their scientific evaluation (e.g. Cumulative Impact Factors or Science Citation Index) when comparing their contributions with those of biological researchers. This stems from the low number of behavioural science publications by biomedical researchers in medical schools (Diehl and Perkins, 2000).

## 19.7 Prevention Programme

Another example of transdisciplinarity is the implementation in the survey results and the educational goals in the public health field. This may happen at two levels. First, our scientists participate and integrate various prevention and (mental) health promotion programmes (e.g. life skills training, coping skills, stress management training, substance use prevention programmes). In these programmes both survey results and the information based on the biopsychosocial approach are transferred to the public. In this process NGOs and government agencies (such as regional public health offices) are active participants in both planning, developing and implementing. A good example of transdisciplinary cooperation is the so called Szécheny Plan: a foundation for research of transdisciplinary cooperation between academic, government and non-government research projects and practical implementation. The Szécheny Plan is directed to map the biopsychosocial influences of premature death among middle-aged men. It has a number of goals: it aims to develop health promotion protocols; it aims to develop indicators for measuring health-related quality of life; and it aims to build a stronger connection between the health care system and the relatively new private health insurance companies. All in all, it also aims to support a transdisciplinary research cooperation.

## 19.8 Recommendations

Our transdisciplinary project suggests that the integration of these scientific paradigms may lead to the implementation of the biopsychosocial paradigm in the medical practice of both prevention and clinical care. The former area in particular is a relatively underinvestigated field of research. Therefore, future research should apply the biopsychosocial approach to human functioning, e.g. how the ideas, beliefs and other psychosocial factors of individuals influence their health consciousness and attitudes towards health promotion and disease prevention.

The transdisciplinary approach, however, does not mean an interchangeability of the methodological guidelines of research in the natural and social sciences but rather a synthesis of research findings into one complex view of life. The methods used in natural and social sciences are sometimes quite similar but they also have important differences (Kopp and Rethelyi, 2004).

Future research should concentrate on the development of a transdisciplinary resource methodology. Cooperation among transdisciplinary researchers may help a lot with this development. Future projects should involve the introduction of the biopsychosocial model to non-medical experts. This implementation needs an integration of different societal attitudes. More applied disciplines, which apply social and behavioural science theories, concepts and empirical findings integrated in the paradigm in medicine and health sciences, should be involved in these projects.

## References

Brook, D.W., Gordon, C., Meadow, H. and Cohen, M.C.: 2000, Behavioral medicine in medical education: Report of a survey, *Soc Work Health Care* 31 (2), 15–29

Cockerham, W.C.: 1999, *Health and social change in Russia and Eastern Europe*. Routledge, New York.

Cohen, J., Krackov, S.K., Black, E.R. and Holyst, M.: 2000, Introduction to human health and illness: A series of patient-centered conferences based on the biopsychosocial model, *Acad Med* 75 (4), 390–396

Diehl, A.K. and Perkins, H.S.: 2000, Beyond the hematocrit and $pO_2$: A symposium on teaching humanities in academic medical centers, *Am J Med Sci* 319, 271–272

Engel, G.L.: 1977, The need for a new medical model: A challenge for biomedicine, *Science* 196, 129–136

Forster, D.P. and Józan, P.: 1990, Health in Eastern Europe, *Lancet* 335, 458–460

Helman, C.G.: 1991, Limits of biomedical explanation, *Lancet* 337, 1080–1083

Kopp, M.S. and Rethelyi, J.: 2004, Where psychology meets physiology: Chronic stress and premature mortality—the Central-Eastern European health paradox, *Brain Res Bull* 62, 351–367

Kopp, M. S., Csoboth, C.T. and Rethelyi, J.: 2004a, Psychosocial determinants of premature health deterioration in a changing society: The case of Hungary, *J Health Psychol* 9 (1), 99–109

Kopp, M.S., Skrabski, A., Rethelyi, J., Kawachi, I. and Adler, N.E.: 2004b, Self rated health, subjective social status, and middle-aged mortality in a changing society, *J Behav Med* 30 (2), 65–70

Margolis, R.B., Duckro, P.N., Sata, L.S. and Merkel, W.T.: 1987, Status of behavioral medicine in American and Canadian medical training, *Int J Psychiatr Med* 17 (3), 249–260

Piko, B.F.: 2004, Interplay between self and community: A role for health psychology in Eastern Europe's public health, *J Health Psychol* 9 (1), 111–120

Piko, B.F. and Kopp, M.S.: 2002, Behavioral medicine in Hungary: Past, present and future, *J Behav Med* 28 (2), 72–78

Piko, B.F. and Stempsey, W.E.: 2002, Physicians of the future: Renaissance of polymaths?, *J Roy Soc Promot Health* 122 (4), 233–237

Sallis, J.F., Owen, N. and Fotheringham, M.J.: 2000, Behavioral epidemiology: A systematic framework to classify phases of research on health promotion and disease prevention, *Ann Behav Med* 22 (4), 294–298

Stokols, D.: 2000, Social ecology and behavioral medicine: Implications for training, practice, and policy, *J Behav Med* 26 (3), 129–138

Weidner, G., Kopp, M.S. and Kristenson, M.: 2002, *Heart disease: Environment, stress and gender*. IOS Press, Amsterdam

World Health Organisation: 1984, *Health promotion: A WHO discussion on the concept and principles*, World Health Organisation, Geneva

Worthman, C.M. and Kohrt, B.: 2005, Receding horizons of health: Biocultural approaches to public health paradoxes, *Soc Sci Med* 61, 861–878

Wulff, H.: 1999, The two cultures of medicine: Objective factors versus subjectivity and values, *J Roy Soc Med* 92, 549–552

# Chapter 20
# Sustainable Coexistence of Ungulates and Trees: A Stakeholder Platform for Resource Use Negotiations

**Karin E. Hindenlang, Johannes Heeb and Michel Roux**

**Abstract** Browsing by ungulates is broadly seen as a major problem for tree regeneration in Alpine forests. At the regional or local level, a resource management problem arises because there is still a lack of scientific knowledge about the long-term importance of herbivore impact on forest dynamics and because conflicting interests between different stakeholders such as foresters, hunters, farmers and nature conservationists persist. A common understanding of the problem and an agreement on the management aims are needed before an effective and broadly accepted wildlife and forest management strategy can be established.

Within the framework of the Swiss National Research Programme 48 (NRP 48, Landscapes and Habitats of the Alps) we developed instruments and procedures for solving a regional forest–wildlife conflict in a mountainous environment by means of a 'platform for resource use negotiation' and collaborative learning. A management concept has been developed, in consultation with all the relevant stakeholders, defining the most appropriate measures for improving the situation and based on a common understanding and common objectives. Particular emphasis has been given to involving the scientists of two projects of the NRP 48 as stakeholders in the platform. The active involvement of scientists, and mutual learning between scientists and practitioners, facilitated the conflict-solving process and produced an added value as revealed by an external evaluation of the learning process. The platform project was carried out in four conceptual phases and contributes new transdisciplinary knowledge about how to structure and implement a process of problem solving in the field of resource use negotiations.

**Keywords:** Forest–wildlife–conflict · Learning process · Collaborative learning · Mental modelling · Integrated management strategy

✉ K.E. Hindenlang
Swiss Federal Institute for Forest, Snow and Landscape Research (WSL), Birmensdorf, Switzerland; Grün Stadt Zürich, Zurich, Switzerland
e-mail: Karin.Hindenlang@zuerich.ch

## 20.1 Background

### *20.1.1 The Mountain Forest–Ungulate Conflict*

Ecosystem management and in particular forest landscape management has emerged as a new challenge to resource managers (Baskent and Yolasigmaz, 1999). Landscape management entails a demanding choice between alternatives, which can lead to conflicts between resource user groups who claim a stake in the outcome (Staehelin-Witt et al., 2005). One such resource management conflict exists between forest managers and other interest groups. The browsing of young trees by wild ungulates, in particular red deer (*Cervus elaphus*) and chamois (*Rupicapra rupicapra*), is widely considered by forest managers to be the most important problem affecting the future of Swiss Alpine forests (Brändli, 1995). Although there are few precise ideas on how forests will develop under the influence of browsing ungulates over the next century (Senn and Suter, 2003; Wehrli et al., 2005; Weisberg et al., 2005), a general agreement exists among forest managers that tree regeneration is insufficient to ensure the continuous existence of mountain forests and that the protective function of the forests will, at least locally, be severely compromised (Kupferschmid, 2005). Hunters and conservationists disagree about the importance of ungulate browsing for tree regeneration. This leads to fruitless disputes on silvicultural practices, hunting schemes and yearly hunting quota, and a general feeling of helplessness among local stakeholders involved in forest management or biodiversity conservation. Such forest–wildlife conflicts have increased in parallel with the increase in populations of game species during the 20th century, in response to their protection, and there is still a lack of scientific knowledge about the long-term importance of herbivore impact on forest dynamics. Therefore, wildlife management in terms of forest–wildlife conflict management has become an emerging field for forest and wildlife conservation agencies, private organisations and professionals (Messmer, 2000).

### *20.1.2 Stakeholder Involvement in Conflict Resolution Processes*

Various approaches have been used in attempting to solve conflicts associated with the use of natural resources in landscape and wildlife management (Skutsch, 2000). They can be grouped in two general categories: ‘Conflict management’ focuses on controlling the conflict and recognises the positive effects that conflicts can provoke between different user groups; and ‘Conflict resolution’ attempts to terminate the conflict by facilitating a consensus decision. In the following we will focus on a conflict resolution process. Public participation and stakeholder involvement are tools used in both conflict management and conflict resolution. A wide range of techniques exist, differing in the degree of stakeholder participation from public information and public hearings, collaborative problem solving, assisted negotiation and mediation, to joint decision-making. In Switzerland, the federal forest legislation of 1991 demands public participation in the sustainable management aspect of forest development planning. The regional planners are encouraged to inform the public, accept suggestions, release draft development plans for public scrutiny and answer objections.

Solutions to conflicts associated with the use of natural resources such as the forest–ungulate conflict are often hampered by a lack of scientific understanding (e.g. how herbivory by ungulates interacts with other factors driving forest dynamics) and by the fact that stakeholders often fail to discriminate between scientific facts, value judgements and intention. Successful conflict resolution in forest and wildlife management requires a detailed assessment of the particular situation, a common understanding of the situation and an agreement on the management aims, before an effective and broadly accepted management strategy can be established. Thus arises the demand for a systemic approach, social learning, and the active engagement of stakeholders and authorised management agencies in the conflict resolution process. The establishment of 'platforms for resource use negotiation' is a way of dealing with complex natural resource management problems (Buck, 1999; Steins and Edwards, 1999). Therein applied collaborative methods put special emphasis on joint learning between the concerned stakeholders. Forest management agencies increasingly implement collaborative methods of public participation (LeMaster and Huebner, 1997; Selin et al., 2000) but the effectiveness of these collaborative processes is rarely evaluated.

### *20.1.3 Solving a Regional Mountain Forest–Ungulate Conflict by Means of a Communication Platform*

Within the framework of the Swiss National Research Programme 48 (NRP 48 Landscapes and Habitats of the Alps, www.nfp48.ch) we developed instruments and procedures for solving a regional forest–ungulate conflict in an mountainous environment by means of a 'platform for resource use negotiation' and collaborative learning. Engaging all the relevant stakeholders, an integrated management concept was built, which is based on a common understanding and common aims and defines the most appropriate measures to improve the situation in concern. In order to amplify the general understanding of the system and to promote better acceptance of scientific results in practice, two scientists, leaders of two NRP 48 research projects (Brang et al., 2001; Senn et al., 2001), have been involved as stakeholders in the platform. The following questions were addressed by the transdisciplinary project:

- Which instruments and procedures, developed by different stakeholders associated in a platform for resource use negotiation, are appropriate for solving a regional mountain forest–ungulate conflict and establish a successful forest and wildlife management strategy?
- Can mutual learning between scientists and practitioners facilitate the conflict-solving process in a platform for resource use negotiation and ensure the implementation of scientific results?

The case study is located in Gurtnellen, Canton Uri, Switzerland, where the 'Stotzigwald' forest protects the A2 motorway, a major regional road and the railway from natural hazards such as rockfall and avalanches. The forest is part of a federal game reserve and is used by chamois (*Rupicapra rupicapra*) as winter habitat. The protective function of the forest is negatively affected by a severe lack of tree regeneration, particularly of *Abies alba*, due to unfavourable local environmental conditions and ungulate browsing. Protective structures against falling rocks have been established along the motorway but can never replace the protection of the forest for economic reasons. Various attempts to mitigate the problem of lack of regeneration have been made (e.g. restricted hunting and silvicultural measures) but have not succeeded. Furthermore, the perceptions of various stakeholders (e.g. foresters, hunters, civil engineers) about how to deal with the situation differed widely. Therefore, the communication platform 'Stotzigwald' was established in 2002 involving 27 stakeholders from various user groups: land and forest owners, hunters and game wardens, foresters, nature conservationists, transport and tourism officers, responsible authorities and researchers from the Swiss Federal Institute for Forest, Snow and Landscape Research WSL. They agreed to develop common goals and measures in order to ensure the protective function of the Stotzigwald into the future. Within the project period of $2\frac{1}{2}$ years the learning process has been evaluated twice.

## 20.2 Collaborative Learning on Platforms for Resource Use Negotiation

The concept of *Collaborative Learning* was used in the platform 'Stotzigwald' to facilitate the conflict-solving process. Collaborative Learning actively involves people in discussion, learning and decision making about land management. It is a hybrid approach developed from soft systems analysis (Checkland and Scholes, 1990) and conflict management (Wilson and Morren, 1990). Collaborative learning does not stress consensus between the different user groups, but emphasises learning, understanding and developing improvements for a particular problem. It focuses on the concerns and interests of the stakeholders, encourages systemic thinking and takes into account that considerable learning about different values and scientific results must occur before management improvements can be implemented. Communication and negotiation-interaction are the means by which learning and progress in a particular situation occur. Collaborative learning processes require skill, commitment and perseverance by the participants. In complex natural resource controversies, facilitation of collaborative learning usually requires help from an outside party. This involves a mixture of mediation and process consultation strategies and tactics applied by skilled facilitators (Wilson and Morren, 1990). Collaborative learning design is based on Soft Systems Methodology (SSM) and aims to create a temporarily shared culture, in which conflicts can be accommodated in a way that allows action to be be taken (Flood and Jackson, 1991). SSM aims to improve situations of social concern by activating a hopefully long-lasting learning cycle in

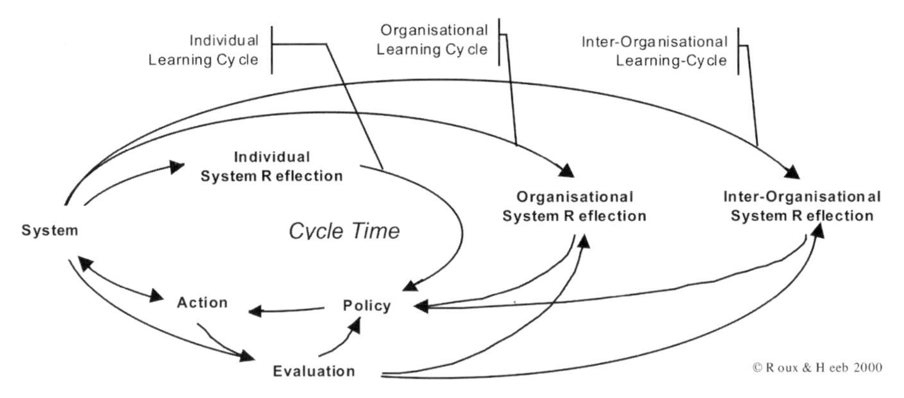


**Fig. 20.1** Concept for collaborative learning in platforms for resource use negotiation (Heeb and Roux, 2002; Roux and Heeb, 2002)

the people involved. The learning works through an iterative process using model concepts to reflect and debate the problem situation and its perception (Fig. 20.1, Heeb and Roux, 2002).

In this context, *platforms* are understood as loosely structured social networks in which representatives of a particular actions system are brought together for a particular purpose such as a landscape management concept. Platforms create a space for communication that is based on confidence. By applying collaborative learning methods, such as the elaboration and discussion of the mental models of the involved stakeholders, platforms allow:

- for achieving a joint understanding of the problem system
- for developing a set of goals as the basis for system development
- for designing concrete project ideas and measures
- for assessing the actions taken, by applying suitable evaluation tools.

## 20.3 Participatory Elaboration of Target and Transformation Knowledge on the Platform 'Stotzigwald'

The platform project 'Stotzigwald' was carried out in four conceptual and temporal phases relating to problem investigation, problem structuring and implementation:

1. Preparation phase (2002): formation of the platform including all relevant stakeholders; formulating basic principles and rules for collaboration and communication;
2. System Reflection phase (2002): elaborating a common understanding of the system and objectives – the first evaluation of the learning process;
3. System Development phase (2003, 2004): elaborating a common strategy and appropriate measures in order to achieve the objectives – second evaluation of the learning process;

4. Knowledge Transfer phase (from 2005 ongoing): application of the measures in the field and the establishment of a control system for measuring success.

From February 2002 to September 2005 ten daylong meetings have so far taken place. A local forestry consultant was in charge of the moderation of the meetings and an expert in change management and platform processes monitored the working programme.

### *20.3.1 Formation of the Platform*

In the *Preparation Phase* 27 stakeholders from all relevant interest groups (forest managers, hunters, farmers, nature conservationists, tourist officers, road planners, community representatives, cantonal and federal administrative bodies, researchers) participated at a kick off meeting in February 2002 where the authors presented the goals and methods of the platform process. The participants agreed on several principles of cooperation, which defined the general aim of the platform process, the duration of collaboration and the transparent and nondiscriminatory communication between the members of the platform.

### *20.3.2 Developing a Common Understanding*

At the beginning of the *System Reflection Phase* an inventory of perceptions of the problem situation in Stotzigwald was created. On this baseline 13 different *mental models* were built by interviewing individual actors (Fig. 20.2). The interviewed stakeholders put forward their individual problem perceptions, their goals and suggestions for approriate measures with respect to the forest–wildlife conflict in Stotzigwald following the procedure of Model Moderation (Heeb and Roux, 2002; Hindenlang et al., 2005):

1. 10–15 most important key terms regarding the system were quoted for each stakeholder.
2. The key terms were structured according to their causal interrelations. It is advisable to classify the terms into groups such as restricted framework conditions, steering factors, system inherent factors and control factors.
3. The stakeholders explained their model and gave it a short and concise title.

Each stakeholder presented his own mental model in the plenum of the platform, emphasising his perception of the key factors for steering the system in question and the most relevant causal interrelations and external factors that influence the system. The individual mental models were discussed in the platform, compared with each other and complemented. Similarities and deviations were marked in order to develop a common functional understanding. The individual models were then combined to form a synthesis model in September 2002 allowing each stakeholder to recognise his or her individual perception within the synthesis model. Such procedure facilitates the relief of existing fears and inhibitions. The synthesis

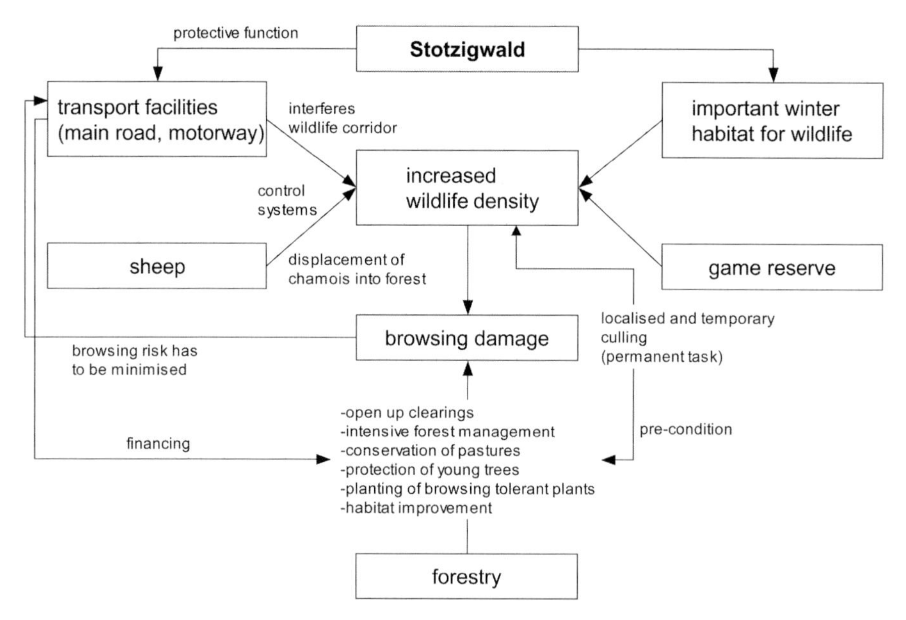


**Fig. 20.2** Visualised Mental Model of an individual actor participating in the platform 'Stotzigwald' (graphic by J. Heeb)

model represents the jointly developed common understanding of interrelations and interdependencies of physical factors and management measures on which all further working steps are based. It includes restricted framework conditions (such as historical and environmental conditions), system inherent factors (such as wildlife density and natural hazards) and steering parameters (such as current management measures and legal obligations). Objectives and appropriate control parameters have been defined in order to assess the effect of future measures.

### *20.3.3 Elaborating Common Objectives and Measures – An Integrated Management Strategy*

Within the different working groups (forestry, agriculture, wildlife and nature management, public relations) detailed strategies and measures for meeting the formulated objectives, aimed at sustainable habitat management, were compiled during the *System Development Phase* in 2003. The results were presented in the plenum of the platform and discussed. The participants prioritised the suggested measures and actions and defined impact related objectives and target values for each strategy. The thematic working groups were further responsible for scrutinising the technical and political feasibility of the selected measures. The suggestions were verified during a joint excursion to the Stotzigwald. Furthermore, in the phase of system development, selected members of the platform and researchers from the NRP 48 projects gave thematic input-presentations referring to specific questions, which had

**Table 20.1** Measures and activities on which the platform 'Stotzigwald' agreed in the integrated action plan

| Subject and objectives | Measures | In coordination with |
|---|---|---|
| ***Forestry***: Improvement of tree regeneration in order to ensure protection function of the forest 'Stotzigwald' | Management measures (thinning, fostering) to facilitate tree regeneration and ensure manifold forest structure, establishing monitoring plot | Agriculture, Wildlife and Nature Management, Public Relations |
| ***Agriculture***: Stabilisation of traditional cultivation in the area in order to improve forage availability and quality for wildlife | Promotion of traditional cultivation of meadows and pastures by integral concepts and financial incentives | Forestry, Wildlife and Nature Management, Public Relations |
| ***Wildlife and Nature Management***: Reduction of wildlife population until tree regeneration has been restored and contemporaneous improvement of wildlife habitat quality | Localised and temporary culling of chamois in tree regeneration plots, amelioration of wildlife habitat inside and at the edge of the forest by means of structural forestry measures | Forestry, Agriculture, Public Relations |
| ***Public Relations***: Increase of the awareness in the parties involved and in the public for the problem and for the proposed solutions | Informative meetings with all parties involved (land owners, beneficiaries, politicians, broader public), periodical reporting in print media | Forestry, Agriculture, Wildlife and Nature Management |

arisen during discussions thereby further amplifying the common knowledge in the platform and stimulating the exchange of information between practitioners and scientists. In June 2004 a first draft of the integrated action plan 'Stotzigwald' for sustainable forest and wildlife management, which included financial perspectives, was presented and discussed in the platform. By combining different measures belonging to different sectors the platform members succeeded in overcoming sectoral thinking and operation (Table 20.1). All members of the platform agreed on the final version of the integrated action plan in November 2004. They further confirmed their willingness to continue their work in the platform 'Stotzigwald' promoting the implementation of the suggested actions and initiating the development of an appropriate control system for success.

In order to assess the effect of the applied collaborative learning methods on the progress of the platform process, the learning process was evaluated twice during the first three conceptual phases by means of questionnaires; once at the end of the System Reflection Phase in summer 2002; and once at the end of the System Development Phase in autumn 2004. After the first evaluation the recommendations made by the external evaluators were implemented and improved the platform process in the System Development Phase.

### 20.3.4 Presentation and Implementation of the Action Plan

The fourth and final *Knowledge Transfer Phase* of the process was and still is dedicated to the implementation of the action plan and to its performance control. The goal was to establish a governing body which will press ahead with the execution of the actions proposed by the platform 'Stotzigwald' on its own initiative and without the support of our project (Fig. 20.3). The required knowledge transfer was initiated in February 2005 at a press conference when the platform achievements so far and the action plan were presented to the public (Hindenlang et al., 2005). Since summer 2005 a task force including local actors belonging to the platform, together with cantonal representatives, started to implement the action plan within their own sphere of authority. In parallel an evaluation system is being developed in the platform providing appropriate indicators and methods for monitoring the changes in Stotzigwald in response to the management measures taken.

### 20.3.5 Evaluation of the Management Strategy with Respect to New System Knowledge

The adopted management plan integrates objectives and measures relating to forestry, farming, wildlife management, habitat management, and public relations (Table 20.1). The objectives and measures were developed by means of a dialogue between practitioners and scientists. The practitioners mainly contributed traditional

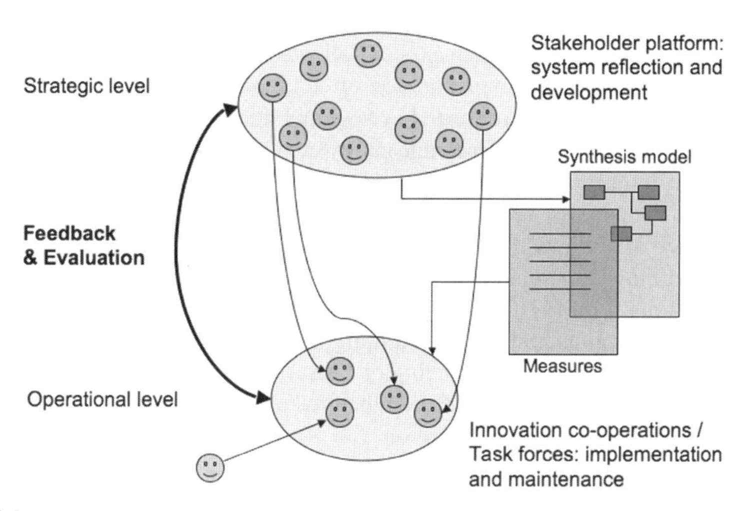


**Fig. 20.3** Organisational design of the platform 'Stotzigwald'. The development of a common understanding of the situation (system reflection) and the elaboration of common objectives and the appropriate measures (system development) occur at the strategic level of the stakeholder platform. The implementation of the suggested measures and the evaluation of their success occur at the operational level by innovation cooperations or task forces

knowledge during the process, whereas the scientists brought in their experience from former and current projects, e.g. NRP 48.

New scientific knowledge mainly contributed to the development of an evaluation strategy. This included the development of measurable indicators for assessing the changes in forest structure, tree regeneration, wildlife density and forage quality in response to the management measures taken. The scientists of the Swiss Federal Institute for Forest, Snow and Landscape Research (WSL) assisted with the establishment of indicator plots in the forest in order to observe changes in detail. A modelling study contributed new knowledge about the period over which the changes can be expected, and have to be monitored. In addition new methodological knowledge, of the assessment of wildlife densities and their relationship to browsing intensity, contributed to the selection of appropriate indicators for monitoring habitat development in Stotzigwald.

## 20.4 Recommendations

### *20.4.1 External Evaluation of the Platform Process*

The platform process and the resulting management strategy were evaluated by an external partner using criteria from political science. The materials used for the analysis were the integrated management concept, the protocols of the platform meetings and the results from two questionnaires, which were conducted during the platform process in 2002 (end of System Reflection Phase) and 2004 (end of System Development Phase). The following conclusions can be drawn with respect to the questions addressed by the project:

- Conflict solving: The platform process allowed for a common understanding of the situation and for the elaboration of common solutions (integrated action plan). All participants were treated as co-equal during the learning process and communicated factually. Knowledge gaps were detected and could be filled. The facilitation by a moderator and the neutral position of the scientists positively influenced the process. The only criticism concerned the high expenditure of time and the lack of obligation to implement the suggested measures.
- Mutual learning: The applied methods of collaborative learning and mental modelling on a platform for resource use negotiation proved of value. The exchange of different opinions and mutual learning between practitioners and scientists contributed to a better understanding of causal interrelations and to a better appreciation of the perceptions of the other stakeholders resulting in a generally broader understanding of the situation.

### *20.4.2 Successful Formation of a Stakeholder Platform*

The following preconditions should be fulfilled when initiating a platform process for resource use negotiation and conflict solving:

- There is an approved conflict situation.
- The concerned stakeholders have a strong interest in solving the existing conflict.
- The platform participants are legitimate representatives of their interest/stakeholder groups.
- The platform participants show willingness to participate in the platform over an extended time period.
- The platform participants agree on certain principles of cooperation and rules of communication.

### 20.4.3 Process Related Success Factors

The following aspects represent requisites for a successful platform process:

- The engagement of a locally accepted moderator creates a trustful atmosphere and allows for an open dialogue between the platform participants.
- The information and documentation of the system in question have to be sufficient to create a common knowledge basis.
- A common understanding of the system in question is the precondition to successfully formulating common goals and measures. Mental modelling is a valuable way to create a common understanding.
- Mutual learning between practitioners and scientists facilitates the conflict-solving process and amplifies the knowledge basis.
- Sufficient time has to be available to ensure a successful collaborative learning process.
- A transparent procedure and periodical evaluation of the platform process and its results motivate the platform participants to further cooperate, e.g. to implement the suggested measures.

**Acknowledgments** The projects 4048-064409 of Dr. Peter Brang "Required levels of tree regeneration in forests protecting against natural hazards: Model-based indicator development" and 4048-064439 Dr. Josef Senn "Silver fir and the mountain forest – ungulate conflict: Do browsing ungulates drive forest landscape changes in the Alps?" have been supported by the National Research Programme "Landscapes and Habitats of the Alps" (NRP 48) of the Swiss National Science Foundation.

## References

Baskent, E.Z. and Yolasigmaz, H A.: 1999, Forest landscape management revisited, *J Environ Manag* 24, 437–448.

Brändli, U.-B.: 1995, Zur Verjüngungs- und Wildschadensituation im Gebirgswald der Schweiz, *Schweiz Z Forstwesen* 146, 355–378.

Brang, P., Heinimann, H. and Dobbertin, M.: 2001, *Required levels of tree regeneration in forests protecting against natural hazards: Model-based indicator development*, Proposal to the National Research Programme *Landscapes and Habitats of the Alps*, Retrieved August 25, 2006, from http://www.nrp48.ch/projects/projects_detail.php?nfprojnum=9.

Buck, S.J.: 1999, Multiple-use commons, collective action, and platforms for resource use negotiation, *Agr Hum Val* 16, 237–39.

Checkland, P. and Scholes, J.: 1990, *Soft systems methodology in action*, John Wiley and Sons Ltd., Chichester, 329pp.

Flood, R.L. and Jackson, M.C.: 1991, *Creative problem solving: Total systems intervention*, John Wiley and Sons Ltd., Chichester, 268pp.

Heeb, J. and Roux, M.: 2002, Platforms and innovation cooperations for sustainable development of landscapes and regions. In: M. Flury and U. Geiser (eds.), *Local environmental management in a north–south perspective*, Vdf Hochschulverlag, Zürich, pp. 121–137.

Hindenlang, K.E., Heeb, J., Gerig, G. and Walker, J.: 2005, *Neue Wege im Wald- und Wildmanagement. Erfahrungsbericht der Plattform Stotzigwald – Wald- und Wildmanagement im Kanton Uri*, Retrieved August 25, 2006, from http://www.wsl.ch/programme/waldwild/downloads-de.

Kupferschmid, A.D.: 2005, Effects of microsites, logs and ungulate browsing on Picea abies regeneration in a mountain forests, *Forest Ecol Manag* 205, 251–265.

LeMaster, D.C. and Huebner, A.E.: 1997, Framework for federal and state forest planning. In: N.A. Moiseev, K. von Gadow and M. Krott (eds.), *Planning and decision making for forest management in the market economy*, Cuvillier, Goettingen.

Messmer, T.A.: 2000, The emergence of human-wildlife conflict management: Turning challenges into opportunities, *Int Biodeterior Biodegrad* 45, 97–102.

Roux, M. and Heeb, J.: 2002, *Gemeinsam Landschaft gestalten: Werkzeuge für gesellschaftliches Lernen*, LBL Verlag, Lindau, 160pp.

Selin, S.W., Schuett, M.A. and Carr, D.: 2000, Modeling stakeholder perceptions of collaborative initiative effectiveness, *Soc Nat Resour* 13, 735–745.

Senn, J. and Suter, W.: 2003, Ungulate browsing on silver fir (*Abies alba*) in the Swiss Alps: Beliefs in search of supporting data, *Forest Ecol Manag* 181, 151–164.

Senn, J., Suter, W., Hindenlang, K.E. and Roux, M.: 2001, *Silver fir and the mountain forest – ungulate conflict: Do browsing ungulates drive forest landscape changes in the Alps?* Proposal to the National Research Programme *Landscapes and Habitats of the Alps,* Retrieved August 25, 2006, from http://www.nrp48.ch/projects/projects_detail.php?nfprojnum=16.

Skutsch, M.M.: 2000, Conflict management and participation in community forestry, *Agroforest Syst* 48, 189–206.

Staehelin-Witt, E., Saner, R. and Wagner Pfeifer, B.: 2005, *Verhandlungen bei Umweltkonflikten. Ökonomische, soziologische und rechtliche Aspekte des Verhandlungsansatzes im alpinen Raum*, Research Report NFP 48, Vdf Hochschulverlag, Zürich, 169pp.

Steins, N.A. and Edwards, V.M.: 1999, Platforms for collective action in multiple-use commonpool resources, *Agr Hum Val* 16, 241–55.

Wehrli, A., Zingg, A., Bugmann, H. and Huth, A.: 2005, Using a forest patch model to predict the dynamics of stand structure in Swiss mountain forests, *Forest Ecol Manag* 205, 149–167.

Weisberg, P.J., Bonavia, F. and Bugmann, H.: 2005, Modeling the interaction effects of browsing and shading on mountain forest tree regeneration (*Picea abies*), *Ecol Model* 185, 213–230.

Wilson, K. and Morren, G.: 1990, *Systems approaches for improvements in agriculture and resource management*, MacMillan, New York, pp. 361.

# Chapter 21
# Retrofitting Postwar Suburbs: A Collaborative Design Process

**Carole Després, Andrée Fortin, Florent Joerin, Geneviève Vachon, Elise Gatti and GianPiero Moretti**

**Abstract** This chapter reports on the transdisciplinary process initiated, structured and orchestrated by the Interdisciplinary Research Group on Suburbs (GIRBa) between September 2002 and November 2003 with the goal of redefining the future of postwar neighbourhoods in the City of Québec.

The developed methodology combined scientific analysis, action research and participatory design in an effort to identify which urban design and planning solutions could have a positive impact on the vitality of aging suburbs. Its successive phases involved a diagnosis of the demographic, physical and social characteristics of postwar suburbs, the definition of general orientations and objectives for retrofitting these suburbs, and the elaboration of a strategic revitalisation plan along with an implementation strategy.

The first part of the chapter explains the context and motivations behind this transdisciplinary approach and briefly discusses the theoretical underpinnings that guided our work, namely Jürgen Habermas' *Theory of Communicative Action*. The next section describes the structure and execution of the 15-month collaborative process. The last part reports on the results of an Internet based survey that sought to evaluate participants' perceptions of the strengths and weaknesses of the collaborative exercise.

Preliminary findings indicate that the process led participants to reach and share a better understanding of postwar suburbs, as well as to increase their capacity for action as professionals. Numerous aspects of the reflection on the future of postwar suburbs were also integrated into government planning policies, from the municipal to the provincial levels.

**Keywords:** Urban design · Urban planning · Suburbs · Collaborative planning · Participatory design · Background

✉ C. Després
École d'Architecture, Université Laval, Québec, Canada
e-mail: Carole.Despres@arc.ulaval.ca

G. Hirsch Hadorn et al. (eds.), *Handbook of Transdisciplinary Research*,

## 21.1 Background

### *21.1.1 The Societal Context*

The first ring suburbs of Québec (City), the capital of Canada's French-speaking province of the same name, are facing a dilemma: they are both shrinking and aging, unable to attract the younger households and second-time homebuyers attracted to third and fourth ring suburbs, fuelling low-density greenfield expansion with its concomitant ecological, social and economic costs. The most recent census data show that several of these postwar districts have lost up to 10% of their population, a trend that is predicted to continue. Their residents have reached or are reaching retirement age, and a fair percentage can expect to lose their driver's licence in the future, consequently suffering a lack of autonomy and mobility.

This aging trend is not limited to the social sphere; it is also reflected in the built environment, constructed primarily between 1950 and 1975, and which consists largely of bungalow houses situated on wide lots and a hierarchical street pattern made up of commerce-laden arterials and looping residential streets. The single-family houses have been generally well maintained by their owners but are still in need of updating. Rental apartments have not fared so well and most are in need of considerable renovation. Street infrastructure, including sewer and water systems, pavement and lighting, also require attention, as do school buildings, several of which would be scarcely populated if not for numerous children brought in daily from other neighbourhoods at high costs. Recreational infrastructure, such as public ice rinks, swimming pools and park equipment also need to be upgraded while public transportation is difficult to operate due to low population density and the heavy use of the private automobile. The latter facilitates the use of shopping malls and warehouse retailing, making it difficult for neighbourhood services to survive.

It is in this context that the Interdisciplinary Research Group on Suburbs (GIRBa), Université Laval, initiated, structured and orchestrated a participative design process between September 2002 and November 2003 with the goal of redefining the future of postwar neighbourhoods in the City of Quebec.

### *21.1.2 Theoretical Underpinnings*

GIRBa's current approach to research and action is the result of a double conviction acquired over the years (Després, Brais and Avellan, 2004; Lawrence and Després, 2004). It is inspired by Jürgen Habermas's *Theory of Communicative Action* (1984, 1987) and echoes the works and writings of various collaborative planners and designers, most notably Bernardo Secchi and Paola Vigano (Istituto Universitario di Architettura di Venezia, Italy), Patsy Healy (University of Newcastle-upon-Tyne, UK), Judith Innes (University of California, Berkeley, USA), John Forester (Cornell University, USA) and Susan Fainstein (Columbia University, USA).

First, in order to solve complex planning and design problems, various types of knowledge are required (Joerin and Nembrini, 2005). GIRBa classifies knowledge into four categories: scientific, instrumental, ethical and aesthetic. Scientific knowledge corresponds to what is generally held as 'what is true' and is most often the result of empirical research. Researchers and consultants are its main spokespersons. Instrumental knowledge refers to practicality or to 'what is possible', the knowledge of how to go about things. Experienced professionals, technicians or workers are generally, although not exclusively, the main channels for this type of knowledge. Ethical knowledge corresponds to 'what is good' and is linked to customs, beliefs, values and past experiences that help people to determine what is wrong and what is right on a specific issue. Citizens and elected officials are primary sources of ethical knowledge. Finally, aesthetic knowledge, or 'what is beautiful', comprises images and refers to aesthetic judgment and experience, as well as to tastes, preferences and feelings about the built environment. Artists and designers as well as citizens are key providers of this type of knowledge. Identifying stakeholders belonging to these four types of knowledge means transcending not only the frontiers between scientific disciplines and their experts but also those between researchers and practitioners, between researchers and artists, and between researchers and citizens.

GIRBa's second conviction is that stakeholders of these four types of knowledge must have the opportunity for face-to-face communication in order for original reflective thinking to occur and consensual solutions to emerge. Following Habermas's (1984) cue, rationality in group decision-making is achieved through a process of language-based communication during which participants accept and contest various 'validity claims' in order to negotiate a consensus or mutual understanding of a situation. Language can be visual or image-based communication, a fundamental mode of communication in participatory design processes. During the process of confronting different types of knowledge through a series of dialogic encounters, a fifth type of knowledge progressively emerges, one which is more than the sum of the four others: it is what Habermas calls intersubjectivity. As Innes and Booher (1996) explain, this process of 'making sense together' assumes that participants are able and willing to engage in reflective thinking, characterised as the thoughtful consideration of the information and interpretations of others while allowing their own assumptions and interpretations to be questioned by other participants. Through this transformative process, incoherence of thought and arguments are revealed and collectively overcome.

## 21.2 The Collaborative Process Step-by-Step

In January 2001, the Government of Quebec introduced legislation aimed at consolidating urban jurisdictions across the province, thus requiring Quebec City to amalgamate with 13 neighbouring cities and towns, effective January 1, 2002. This was the beginning of the legal and symbolic reconstruction of Quebec City's enlarged territory. The new city was divided into eight 'arrondissements' or boroughs, several of which correspond to former suburban cities. Because the amalgamation called for important municipal and regional restructuring, as well as for the unification and revision of various existing planning policies, different authorities in the provincial, regional, metropolitan and municipal governments were given the mandate to specify their planning orientations for their respective territories within a 2004–2006 horizon. Aware of the opportunity to inform the mandates of several government authorities about Quebec City's new largely suburban territory, GIRBa gathered the heads of different planning services at a meeting during which they were invited to subscribe and engage in a participative planning and urban design process to redefine the future of postwar suburbs. In the months preceding the participative process, there was the timely publication of GIRBa's book *Suburbia Revisited* (In French, Fortin, Després and Vachon, 2002) in which research on Quebec City's suburbs was presented, along with an editorial series written by senior GIRBa members in the daily newspaper *Le Soleil*. GIRBa was also contacted by several radio, television and newspaper journalists for interviews about suburbs. This phase was important in raising the profile of the process.

The collaborative exercise was drafted in the summer of 2002 but the process was officially launched in September of the same year. The 15-month strategy involved stakeholders from three scales of territorial governance: macro, meso and micro.

- From the macro-scale were decision makers and planners representing regional, metropolitan and municipal governments whose decisions could impact on the future of first ring suburbs.
- The meso-scale corresponds to the borough scale. Participants were borough office directors and locally elected officials, as well as representatives of the local development centre, school board and local community services.
- The micro-scale refers to neighbourhood and community organisations, specific subgroup representatives, and the population at large.

Overall, more than 100 representatives from different sectors of government were involved at some point during the process, and close to 500 citizens were consulted through face-to-face interviews, focus groups and an Internet survey.

The process included three overlapping phases, each corresponding more or less to one academic semester (the exercise being associated with an urban design teaching programme, a dozen graduate students were involved at some point during the process). The first one, the diagnostic phase, aimed at reaching a shared understanding of the future challenges facing Quebec City's first ring suburbs. The second phase sought to collectively define orientations and objectives for retrofitting this portion of Quebec City's territory. The last phase consisted of elaborating strategic plans for these suburbs along with implementation strategies during an intensive participatory design session (traditionally called 'charrette' in the architectural and planning professions).

Throughout the process, GIRBa was responsible for planning and organising all meetings and for producing minutes (validated by participants) as well as easily understood summary maps. At the end of Phase I and Phase II, consolidated reports of all meetings (including orders of the day, minutes and PowerPoint presentations) were put together and handed out to all participants (GIRBa, 2003a,b). All documents were also made available on GIRBa's website (http://www.girba.crad.ulaval.ca/Francais/projets.htm). Figure 21.1 summarises the collaborative planning and urban design process.

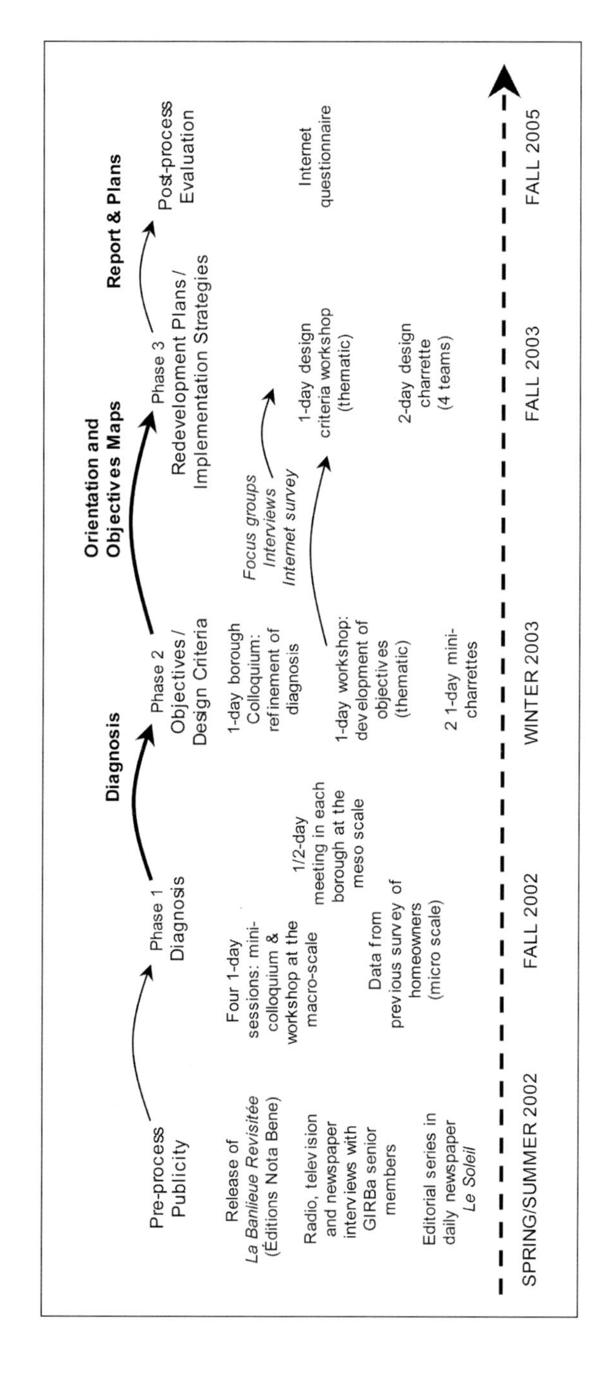


**Fig. 21.1** GIRBa's collaborative planning and urban design process

### 21.2.1 Phase I: Developing a Diagnosis

At the macro-scale, in fall 2002, GIRBa coordinated a monthly series of four one-day mini-colloquia and workshops during which 20–30 people (including GIRBa members) collectively identified the stakes and challenges involved in planning the future of first ring suburbs. Each gathering consisted of a half-day of presentations (from GIRBa and involved partners) in order to draw the most complete portrait of first ring suburbs. The goal of these presentations was, on the one hand, to disseminate scientific knowledge about the history, morphology, demography, uses and meanings of postwar suburbs, and on the other hand, to share instrumental knowledge provided by civil servants and administrators about suburban territorial planning and management policies and practices at different scales of government.

Each mini-colloquium was followed by a workshop where participants gathered in subgroups of 10–15 people to discuss and annotate or interpret maps of the city and the concerned suburban boroughs. This process favoured a back-and-forth exchange between theoretical/general presentations and concrete/precise spatial representations of the diagnosis, as well as the acquisition by participants of an in-depth knowledge of the concerned boroughs.

At the meso-scale, GIRBa met separately for half-day sessions with each borough council, as well as representatives from the school board, local centres for community services and local centre for development. The goal was to identify their main interests and challenges, using a map to point at specific buildings, streets or neighbourhoods. GIRBa reported these preliminary diagnoses to macro-scale representatives; those diagnoses were completed by key actors from each borough during a one-day colloquium at the beginning of Phase II.

At the micro-scale, the difficulty in identifying and mobilising key actors from various community groups within a population of close to 200,000 made it difficult for participants to be included in the diagnosis phase. At the beginning of the process, GIRBa acted as spokesperson of the residents, referring to its survey of 173 homeowners to integrate their concerns.

### 21.2.2 Phase II: Defining General Orientations and Objectives

The next phase aimed at validating general orientations for the future of first ring suburbs, as well as specific planning and urban design objectives. If a consensus for the diagnosis was relatively easy to reach, the second phase turned out to be more challenging. In other words, it is easier to agree on 'what is' than 'what should be'. This was accomplished through four one-day meetings between January and May 2003: one borough colloquium, one objectives-defining workshop and two mini-charrettes.

The borough colloquium was dedicated to identifying the stakes and challenges facing all four boroughs. Key actors from each district presented a refined diagnosis and engaged in a discussion with macro-scale representatives, which led to a better understanding of the interface between the boroughs themselves and the larger metropolitan area. The next workshop aimed at defining objectives for five areas of concern facing postwar suburbs (socio-demography, ecology/environment, functionality/infrastructure, finance/economy and culture/heritage), as well as exploring ideas as to how these objectives could be reached. Small subgroups of participants with diversified professional profiles worked on each of these themes and presented their reflections during a plenary session, leading to comments and discussion. Ten macro and 14 meso-scale representatives attended these two meetings.

The next meetings consisted of two mini-charrettes, each covering two boroughs. Their purpose was to test the relevance of the objectives with regards to the specificity of each borough. This was achieved through discussion, drawing, and design. The number of macro-scale stakeholders was cut in half during these activities in order to integrate key actors at the micro-scale or community level. Liberating local actors from their daily obligations turned out to be much more difficult than anticipated; only a few elected officials and members of local interest groups were able to join. To be more inclusive of citizens' perspectives in Phase III, GIRBa developed three additional knowledge bases. First, 20 focus group meetings were conducted with local associations of immigrants, teenagers, seniors and single-mothers, as well as two with small shopping mall managers. Second, 75 semi-structured face-to-face interviews were completed with tenants to complement an existing database on suburban homeowners. Third, an Internet survey was launched on GIRBa's website in summer 2003, to which 200 suburban residents answered within the period of one month.

### *21.2.3 Phase III: The Strategic Plan and its Implementation Strategy*

Phase III aimed at elaborating a strategic plan to orient the future of first ring suburbs, along with clear indications as to implementation. This was achieved with two distinct activities: the design criteria workshop and the final design charrette.

The design criteria workshop consisted of a one-day meeting with 11 macro-scale, 11 meso-scale, and 6 micro-scale stakeholders, along with a dozen GIRBa members, in which the objectives identified in Phase II were validated and potential solutions to be explored in the final plans prioritised. The results of the Internet survey, face-to-face interviews with tenants, as well as focus groups were presented by GIRBa members during the first half of that session as a way of integrating citizens' interests. During the afternoon, three subgroups mixing macro-, meso- and micro-scale participants were created and each presented with an exhaustive list of objectives and design criteria for two of the six following themes: society and demographics, environment and ecology, infrastructure and spatial planning, economy and finance, heritage and culture, and mobility and accessibility. For each theme, participants were asked to identify and prioritise five design criteria to be applied during the forthcoming participatory design session (charrette).

The strategic plan for the redevelopment of Quebec City's first ring suburbs was drawn during the charrette, which involved 23 key actors from the macro-, meso- and micro-scale territories, 12 GIRBa members and 10 graduate students in urban design. The two-day meeting took place in a large room divided into four sub-spaces to accommodate each of the borough teams. Each section of the room was equipped with pinboards displaying the Diagnosis Plan, the Borough Orientations and Objectives Plan, as well as an aerial photograph of the area. Each group also had 1: 5,000 and 1: 2,000 maps, as well as markers and tracing paper with which to work. The open-floor plan favoured the circulation of participants from one borough to another such that they could get information from key actors about shared territories or problems.

The first day started with a plenary session wherein the work programme was explained and after which the four teams retreated to their quarters to elaborate a Strategic Borough Plan. During the first evening, GIRBa's members as well as some voluntary participants redrew the sketches in preparation for a debriefing session at the start of the second day. This led to a refining of the borough plans and, in the afternoon, to developing plans for specific sub-sectors. On that second day, each borough delegated participants to work on either linking each Borough Strategic Plan with the larger city or identifying precise avenues for implementing the planning and urban design proposals. The charrette ended with the presentation of the amalgamated four borough strategic plans and detailed plans of the sub-sectors. In conclusion, participants were asked to freely comment on their experience throughout the collaborative process, which turned out to be very positive.

Following this activity, GIRBa started working on a report based on the recommendations and proposals defined in Phase III which would include a generic PowerPoint presentation, the final plans, as well as their related explanations and implementation measures. This final report has yet to be distributed. Although the work has been close to completion since summer 2004, a lack of time and resources has prevented it from being concluded.

## 21.3 Measuring the Outcomes

Two years later, with the objective of gaining lessons from this process, GIRBa posted an Internet survey on its website and emailed all key actors with an invitation to evaluate their perceptions of the strengths and weaknesses of the process, as well as of the relative success of its outcomes. The two year period proved to be ideal for a retrospective evaluation. Indeed, several planning exercises initiated by different government authorities following municipal amalgamation were completed or near completion at the time, which allowed participants to evaluate to what extent they benefited from the process in their own work and whether or not the knowledge issued from it has made its way into some orientation documents.

The Internet questionnaire included 79 items, 70 of which could be answered on a four-point Likert scale to measure the degree of agreement with a statement. Each item could be supplemented with open-ended comments. The final questionnaire was structured into four parts to evaluate participants' perceptions of: (1) their initial expectations, (2) the quality and efficiency of the organisation, (3) the quality of the activities themselves and of their associated production, and (4) the learning outcomes and transferability of knowledge toward ongoing planning and urban design exercises. Six additional items allowed participants to be categorised according to their professional profiles.

The survey was launched in November 2005, with participants given one month to respond. Out of 106 stakeholders who had taken part in the process at some point, 34 could not be contacted (change of address, retired, deceased, on leave, etc.), 13 did not feel they had been regular enough to provide valuable information. Out of the 54 who agreed to participate, 40 completed the questionnaire: 20 associated themselves with the macro-scale, 14 with the meso-scale and 6 with the micro-scale (among which 5 said they belong to more than one scale). Respondents were aged between 27 and 80 years old; the gender split was 14 F, 25 M and 1 unspecified. (In order to glean insights into our own learning, GIRBa members were also invited to participate in the evaluation. Their responses are excluded from this evaluation.)

The overall results suggest that participants have a very positive perception of the collaborative planning and urban design exercise. Indeed, at least three quarters of participants were 'in total agreement' or 'rather in agreement' with 54 out of the 70 positive affirmations about the process suggested to them.

Fourteen items were related to participants' perceptions of their initial expectations about the process, namely in terms of learning about postwar suburbs and about participatory methods, of networking and collaboration, as well as of their potential influence on the future of postwar suburbs and/or strategic orientation documents. The highest levels of agreement had to do with anticipated learning about participative processes, the least, although still high, with their anticipated influence on the process, and more precisely with the elaboration of the strategic plan, showing that participants' were more concerned with the process itself, rather than the results (Table 21.1).

Eighteen additional items were linked to participants' perceptions of the quality of the logistical and organisational aspects of the collaborative strategy. More precisely, these questions evaluated their perceptions of its coherence and clarity, of the appropriateness of the meeting types, durations and frequency, of the diversity and competencies of participants, and of the usefulness of the information provided. Here again, average ratings are high. The highest scores reflect an appreciation of the variety of meetings and communication tools used throughout the process. It is interesting to note that the participants not only greatly appreciated meetings where discussion was favoured, but also those sessions during which basic information on the suburbs was provided. Concerning the types of media used to convey information, traditional media (paper) were preferred to electronic ones (Internet site), even though most of the participants were professionals and likely to be comfortable with the latter. The lowest scores are related to some participants' irregular attendance as well as to the absence of certain stakeholders. Given the timeframe of the process as well as reorganisation of several government services after the municipal amalgamation, there was an inevitable turnover of participants. Furthermore, as some respondents specifically noted, there was the absence of representatives from the

**Table 21.1** Participants' perception of their initial expectations

| Expectations | **Affirmations used with Likert Scales in Internet Survey** 1 Full disagreement 2 Partial disagreement 3 Partial agreement 4 Full agreement | Average rating |
|---|---|---|
| Learning | To learn about the diversity of points of view concerning the future of suburbs | 3.67 |
| | To learn about suburbs through the amalgamation of a diversity of expertises | 3.64 |
| | To learn about participative processes in urban planning | 3.42 |
| | To collect ideas for our own strategic plans | 3.28 |
| | My objectives for participating in the process were met | 3.26 |
| Collaboration | To develop contacts useful in our work | 3.59 |
| | To work with representatives from the academic community | 3.31 |
| | To develop a partnership with one or more institutions | 3.18 |
| Influence | To contribute to the evolution of innovative methods of urban planning practices | 3.46 |
| | To participate in the construction of a consensus on planning suburbs | 3.36 |
| | To influence the evolution of suburbs | 3.21 |
| | To act against urban sprawl | 3.08 |
| | To promote one or more of my institution or organisation's projects | 2.92 |
| | To produce a plan for the suburbs | 2.66 |

Ministry of the Environment and the Ministry of Transportation, as well as a low number of representatives from the private land development sector, elected officials and citizens (Table 21.2).

Eighteen additional items were linked to participants' perception of the activities themselves: first, the quality of the exchanges among participants and of the exchanged information, of their contribution to the final proposals, and of the quality of the proposals themselves. The highest average ratings indicated that stakeholders felt that the conditions of speech were fair and their contributions welcome. The lowest scores in this category indicate that 8–11 participants did not really feel that they can claim authorship for the final products, that 10 fully or partially disagreed with the fact that the opinions of all participants were all weighed equally, and finally, that 6 thought that some information might have been hidden or manipulated by some participants (Table 21.3).

Finally, 20 items were related to participants' perception of the outcomes of the collaborative planning and urban design exercise to their work: more specifically in terms of their learning, networking benefits, and transfer of collaborative methods, knowledge and produced documents. These ratings were slightly lower. The highest scores concerned participants' learning about participative methods, postwar suburban planning issues, and the diversity and interests of participants. Seven out of ten participants felt the process somehow contributed to strengthening existing or developing new professional relationships. We ourselves noticed that after formal exchanges during the first meetings, participants seemed to truly

**Table 21.2** Participants' perception of the quality of the organization

| Organization | **Affirmations used with Likert Scales in Internet Survey** 1 Full disagreement 2 Partial disagreement 3 Partial agreement 4 Full agreement | Average rating |
|---|---|---|
| Process | The process was generally coherent and easy to follow | 3.26 |
| | From the very first session, the objectives of the process were made clear | 3.08 |
| Meeting | I appreciated the charrette (intensive design session) | 3.71 |
| | I appreciated the workshops with maps (discussion with maps in the afternoon) | 3.68 |
| | I appreciated the Mini-colloquia (PowerPoint presentation in the morning) | 3.47 |
| | I appreciated the small group discussions | 3.43 |
| | I appreciated the plenary discussions | 3.29 |
| | The monthly sessions of autumn 2002 and winter 2003 were – not – too close together | 3.03 |
| | The sessions were – not – too numerous | 2.97 |
| Participants | The competencies of different participants were complementary | 3.33 |
| | I – did not – had to convince my institution of the advantages of my participation | 3.09 |
| | Number of participants per session was – not – too high for fruitful exchanges | 2.79 |
| | The irregular presence of certain participants was –not– a problem | 2.73 |
| | Important stakeholders in this reflection on the future of suburbs were – not – missing | 2.68 |
| Information | Compendia: they were useful | 3.43 |
| | Meeting minutes: they were useful | 3.42 |
| | Meeting agenda: they were useful | 3.36 |
| | Website: it was useful | 2.83 |

work together and trust each other. The informal interaction between people with different disciplinary profiles is considered an important but less tangible outcome of interdisciplinary collaborative processes. Indeed, some participants acknowledged during the process that even within their own ministry or borough, they rarely had the opportunity to discuss with their colleagues from other divisions. Also of note, several participants felt they learned more from the process than they contributed to it, although the frequency of this perception and the high ratings of the quality of the process suggest that all participants made important contributions. This confirms our hypothesis that through intersubjectivity, a new type of knowledge is created, one that is more than the sum of individual contributions.

The lowest ratings for the survey were those related to the use by participants of the different produced plans, perhaps explained by the fact that GIRBa never made the plans available in their final form, and several professionals, from community health centres or school boards for instance, did not have to produce orientation

**Table 21.3** Participants' perception of the collaborative activities

| During the activities... | **Affirmations used with Likert Scales in Internet Survey** 1 Full disagreement 2 Partial disagreement 3 Partial agreement 4 Full agreement | Average rating |
|---|---|---|
| Speech | I felt free to express myself during these meetings | 3.73 |
| | All participants were given a fair hearing | 3.43 |
| | Access to speech was equitable | 3.38 |
| | I – did not – felt in conflict with one or more participants | 3.32 |
| | Everything was open to debate, including all assumptions and basic hypotheses (for example, urban sprawl is a problem, Quebec's population is decreasing) | 3.11 |
| | Discussions were terminated only after all participants' interests and questions were examined | 3.00 |
| | The opinions of certain participants was considered more credible or given more weight (due, for example, to their special status) | 2.71 |
| Information | I did not hide or manipulate information | 3.92 |
| | My standpoints during the process conformed those of my institution or organization | 3.45 |
| | Thanks to information provided through the presentations or the meeting minutes, all participants had a level of knowledge sufficient to participate in the process | 3.41 |
| | When there were doubts about the clarity, significance or reliability of information used, efforts were made to lift uncertainties (jargon, technical terms..., etc.) | 3.19 |
| | Information was not hidden or manipulated by participants | 3.04 |
| Contribution | I have the impression of having contributed to a collective reflection | 3.45 |
| | These planning proposals are generally close to my point of view | 3.14 |
| | GIRBa was a neutral orchestrator of the process | 3.11 |
| | I feel I can claim authorship of the planning proposals (diagnostics, schemas, plans) | 2.76 |
| Plans | Relative to conventional processes of decision-making, the participative nature of the process allowed for the production of better schemas and plans | 3.30 |
| | The redevelopment plan is a 'good' plan | 3.17 |

documents in their professional mandates and so did not require the plans as such. As for the lower degrees of agreement with items related to the use of similar or related participatory methods in their own work after the process, several respondents disagreed only because they said they were already using them. During the final debriefing that followed the design charrette, several participants indicated that they wished to replicate this type of collaborative process in their own work, even in contexts other than planning. A final important result is that several key actors (9–13 according to various items) acknowledged a transfer of knowledge. Some participants' comments indicated that general orientations, objectives and design criteria had made their way into several government agencies' orientation documents,

Table 21.4 Participants' perception of the outcomes of the collaborative process

| Outcomes | **Affirmations used with Likert Scales in Internet Survey**<br>1 Full disagreement 2 Partial disagreement 3 Partial agreement 4 Full agreement | Average rating |
|---|---|---|
| Learning | I learned about participative planning processes | 3.36 |
| | I am more cognizant of questions surrounding urban planning in relation to my work | 3.31 |
| | My vision of urban problems linked to the future of suburbs has widened | 3.28 |
| | I learned about the diversity of viewpoints and interests of other participants | 3,23 |
| | The process allowed me to discover relevant solutions in my professional activity | 3.12 |
| | I have the impression that the process allowed others to better understand my interests and points of view | 3.11 |
| | I noticed that the process changed the discourse on suburbs (among my colleagues, within my institution, in the media) | 2.82 |
| Collaboration | Today, I consider interdisciplinary and intersectorial collaborations in a more positive light | 3.24 |
| | The process solidified existing personal and professional relationships | 2.86 |
| | The process allowed me to develop new personal and professional relationships | 2.84 |
| | My participation in the process changed my appreciation of interdisciplinary and intersectorial collaborations | 2.74 |
| | Since the end of the process, I have organized discussion sessions with people from other services, organizations or with citizens | 2.70 |
| | Since the end of the process, I have undertaken similar approaches within the context of my institution (participative, collaborative, interdisciplinary processes, etc.) | 2.61 |
| | Since the end of the process, I have consulted people whom I met during the process | 2.56 |
| | The process allowed me to put in place one or more partnerships or collaborative projects between my institution/department and other institutions/departments | 2.39 |
| Transfer | I have often communicated the results of the process to my colleagues or members of my organization | 2.71 |
| | I use the plan created at the sub-districts scale | 2.17 |
| | I use the plan created at the boroughs scale | 2.03 |
| | I use the plan created at the city scale | 1.97 |
| | I have referred to the plans or to the process during a conference, colloquium, interview with the media, etc. | 1.92 |

something that GIRBa was able to verify in official documents and websites, and this despite the fact that the process documents were not made available following the process and that participants stated that they were more interested in the process than in the results (Table 21.4).

## 21.4 Recommendations

Overall, GIRBa's transdisciplinary research and action programme has produced a rich body of knowledge and allowed for a better understanding of the complexity of suburban settings, the challenges facing them and, most of all, of avenues for action. If GIRBa's initial intention was to share its extensive knowledge with decision and policy makers in the context of a reflection on the future of postwar suburbs, the survey, in addition to a cursory examination of the orientation documents of several involved partners, suggests that the outcomes of the collaborative planning exercise go far beyond what we anticipated. We believe that the main outcomes for participants has been a finer understanding of postwar suburbs as object as well as of the issues and challenges facing their future in the context of their physical and social aging; the development of a shared mission as well as of common objectives and proposals for retrofitting these suburbs; an increased capacity for action through building and sharing a knowledge basis on suburbs; and learning from transdisciplinary collaborative planning methods.

Furthermore, the very high number (95%) of participants agreeing that they met their original objectives for participation can essentially be attributed to two principal characteristics of the process. First, participants seemed to appreciate the substantial amount of information disseminated through the process, reinforcing the role of information sharing in participative events (Hanna, 2000; Joerin and Nembrini, 2005). Second, participants also demonstrated a high level of interest in the process itself, finding value in learning and experimenting with, a new communicative and interdisciplinary approach to decision-making.

On a general level, our experience has convinced us that: (1) scientific research is not performed in the same way when there is close and constant collaboration between researchers from different disciplines, (2) design research is a legitimate and autonomous way of producing knowledge on a given problem, one which accepts intuition and uncertainty, and (3) action research is an alternative mode of knowledge production that recognises practical reasoning, material and organisational constraints, and values public debate.

Architecture and planning seem to be fertile grounds for transdisciplinary contributions because of their very nature as multidisciplinary or 'undisciplined' disciplines, involving the natural and social sciences, as well as action-oriented practices aiming at transforming the built and natural environment. These circumstances correspond to a precondition for transdisciplinarity: disciplines capable of a constructive dialogue with other domains of knowledge, with a multidisciplinary functioning and project-oriented identity (Pinson, 2004). In this regard, GIRBa's programme of transdisciplinary research and action in Quebec City exemplifies the critical contribution of universities towards training professionals and researchers to work together in creating intersubjective knowledge, with the impacts being felt in all sectors of society.

**Acknowledgments** The reflections presented in this paper are the outcome of transdisciplinary and intersubjective work among GIRBa members. More specifically, the authors would like to thank Sergio Avellan, Nicole Brais, Mélanie Bédard, Martin Bussières, Mireille Campagna,

Alexandra Daris, Vickie Desjardins, Daniel Lacroix, Sébastien Lord, Nik Luka, Dominique Morin, David Paradis and Thierry Ramadier, all of whom contributed to bringing these ideas to maturity.

## References

Després, C., Brais, N. and Avellan, S.: 2004, Collaborative Planning for Retrofitting Suburbs: Transdisciplinarity and Intersubjectivity in Action. In: R.J. Lawrence and C. Després (eds.), Transdisciplinarity in theory and practice, *Futures* 36(4), Special issue on transdisciplinarity, pp. 471–486.

Fortin, A., Després, C. and Vachon, G. (eds.).: 2002, *La banlieue revisitée*, Nota Bene, Québec.

GIRBa.: 2003a, *Volet participatif. Recueil de comptes rendus et de présentations de l'automne 2002*, Université Laval, Québec.

GIRBa.: 2003b, *Volet participatif. Recueil de comptes rendus et de présentations de l'hiver 2003*, Université Laval, Québec.

Habermas, J.: 1984, *The Theory of Communicative Action, Volume 1. Reason and the Rationalisation of Society*, Beacon Press, Boston.

Habermas, J.: 1987, *The Theory of Communicative Action, Volume 2. Lifeworld and System. A Critique of Functionalist Reason*, Beacon Press, Boston.

Hanna, K.S.: 2000, The Paradox of Participation and the Hidden Role of Information, *J Am Plann Assoc* 66(4), 398–410.

Innes, J.E. and Booher, D.: 1996, Consensus Building and Complex Adaptative Aystems. A Framework for Evaluating Collaborative Planning, *J Am Plann Assoc*, 65(4), 412–423.

Joerin, F. and Nembrini, A.: 2005, Post-Evaluation of the Use of Geographic Information in a Public Participatory Processes, *URISA J* 17(1), 15–26.

Lawrence, R. and Després, C.: 2004, Futures of Transdisciplinarity. In: R.J. Lawrence and C. Després (eds.), Transdisciplinarity in theory and practice, *Futures* 36(4), Special issue on transdisciplinarity, pp. 397–405.

Pinson, D.: 2004, Urban Planning: An Undisciplined Discipline. In: R.J. Lawrence and C. Després (eds.), Transdisciplinarity in theory and practice, *Futures* 36(4), Special issue on transdisciplinarity, pp. 503–513.

# Part V
# Cross-cutting Issues

# Chapter 22
# Participation

**Aant Elzinga**

**Abstract** Participation is a core element of transdisciplinary research. A look at the different project descriptions reveals that participation is more prominent in the first and the third phase of the research process. Also, the intensity and the specific meaning of participation differ substantially between the projects. Transdisciplinary research could benefit from more reflexivity on questions such as who is empowered by participation, or on which criteria are used to decide who is in and who is out.

**Keywords:** Constructivism · Empowerment · Epistemology · Politics · Stakeholders

## 22.1 Introduction

The literature about public participation is largely dominated by case studies relating to technology assessment, risk analysis and the formation of science and technology policy. Participation theory is underdeveloped and evaluations of participation methods are often 'limited to ad hoc suggestions and criticisms about advantages and disadvantages of the various techniques, and the lack of a clear framework for criticism makes it difficult to compare and contrast their relative merits' (Rowe and Frewer, 2000; also see Rowe and Frewer, 2005 for the concept of 'public participation'). The present paper seeks to avoid such cataloguing of recipes distilled from the chapters in the present book. Instead, findings in a number of chapters will be laced with remarks on, and critical analysis of, the contexts and the concepts of participation that gives a broader perspective. I begin by situating the theme in an historical perspective focusing on the emergence and shifts of meaning of some associated concepts. Thereafter participation in the three 'phases' of transdisciplinary projects are discussed. Towards the end some findings from scholarship

✉ A. Elzinga
Department of History of Ideas and Theory of Science, University of Göteborg, Göteborg, Sweden
e-mail: vetae@hum.gu.se

in science and technology studies (STS) are brought in, and a couple of issues are raised for further research.

## 22.2 Interdisciplinarity as a Prelude

In the late 1980s, in various countries, a series of new research fields emerged. These were prompted by societal problems and pressure from user groups. Among such fields appeared the following: peace and conflict research, development studies (related to development aid to what were then called underdeveloped countries), systems ecology, human ecology, work life research, women's studies, research into higher education, social work research, nursing care research and police research. In some cases social movements provided the motivation – anti-imperialist, peace and environmental movements, women's liberation and labour movements. In other cases it was a combination of welfare state concerns and professional occupational groups that provided external relevance pressures and participated in the establishment of new academic teaching programmes and areas of scientific expertise.

Occupational groups such as social workers, midwives, nurses, and police, sought increased status and professional legitimacy through 'scientification' of their knowledge, which often rested on tacit know-how and personal command of practical skills. Scientification brought with it codification of certain parts of their knowledge, theoretical concepts and new standards of public certification. Social and cognitive legitimation strategies were used to argue the case for participation in academic knowledge production. Social legitimation strategies referred to specific societal problems that had to be addressed, while cognitive legitimation strategies often started out by pointing to a lack of scientific knowledge in relevant areas.

Considerable effort was devoted to integrating workplace tacit knowledge into research projects, influencing the formation of new concepts. In nursing care research for example, it was pointed out that while medical expertise included a lot of important knowledge about the human body, diseases and the like, academic medicine failed to address the aspects of 'caring' for patients. Concepts like empathy and coping were therefore given new theoretical connotations corresponding to the goals of patient care in the ward. In social work and life work research the concept of 'action research' became prominent in order to denote the process wherein researchers interacted with target groups of people (e.g. workers in the workplace) as a part of their field work; whereby data was accumulated and societal problems were translated into researchable scientific problems to be tackled by a discipline such as sociology. Inspiration for the development of new concepts also derived from interaction with user groups and the latter's representations or images of experienced reality. Furthermore, interaction between researchers and user groups also occurred during the validation of research results.

Later, when the new fields took on a life of their own and began to function as new disciplines or interdisciplinary specialities, the links between researchers and user groups became more routine, which meant that professional representatives of the user groups took over as the ombudspersons for user group interaction with research projects. In areas where social movements were strong a similar process occurred in the form of participation by professional activists belonging to one or other non-governmental organisation (NGO). In particular, the role of social movements in the setting of standards and regulating the design of new technologies, deserves study (Eyerman and Jamison, 1989; Hård and Jamison, 2005).

## 22.3 Entry of the Prefix 'Trans' and Related Concepts

It is interesting to note how in the first wave of discussions the terms interdisciplinarity and action research attended the co-production of new social and cognitive orders. Primacy was still given to the relative autonomy and sovereignty of academic disciplines, in as far as the terms inter- and multidisciplinarity had the question of interaction between members of different academic tribes (existing disciplines) as its centrepiece. Later the term transdisciplinarity became more prevalent, signalling the goal of transcending disciplinary boundaries as such. Of course Alvin Weinberg had already used the term trans-science back in the early 1970s, but then it was as part of his cognitive strategy to uphold boundaries around real science. Trans-science was taken to refer to an arena outside science proper where science based knowledge is employed in the public arena for advice in decision making and policy (Weinberg, 1972).

The discussions in the 1990s by contrast, in the context of neo-liberal, market oriented thinking that replaced welfare state and social movement oriented thinking, gave a new twist to the idea of participation. Now it was meant to highlight research or knowledge production in the actual context of application, so-called Mode 2 (Gibbons et al., 1994) as distinct from research in an academic setting (Mode 1). The new discourse emphasised the crossing, not only of boundaries between disciplines, but perhaps more so, the move beyond disciplines. Thence transcendence or 'transdisciplinarity' becomes an appropriate term (Elzinga, 2004). There is also talk of a 'new social contract for science' (Lubchenko, 1997; Gibbons, 1999). We should be careful however not to exaggerate the differences vis-a-vis the late 1960s. Nor, for that matter, should it be compared with the struggle of engineering in the early part of the 20th century, to become a recognised university based mode of higher educational training (with its own academic doctor's hat) and knowledge production; even though user groups and practitioners of various types participated in its establishment. Often, when they are introduced, new terms have ideological connotations.

In political discourses on democracy terms like 'citizen participation' in community change and 'participatory democracy' appeared in the late 1960s. Among others they entered into the discussion of: technology assessment (TA); risk assessment; science and technology policy in the mid-1970s; and OECD reports, where 20 years later the concept of 'governance' was in vogue. David Dickson, writing in 1984, noted how in the U.S. there had been a struggle about who should define the substantive content of participation. A chapter in his book, The New Politics of

Science, is devoted to delineating the tensions between technocracy and democratic control (Dickson, 1984). The technocratic approach is based on the imposition of solutions to problems relating to the social impact of science through consensus by experts. Public participation might be encouraged, but only in the process of reaching consensus about solutions usually expressed in technical terms. This was the 'rational' top–down approach, ignoring questions regarding the basic political structures through which solutions were to be put into effect. In contrast, a bottom–up democratic approach stressed the importance of procedures as much as goals, arguing that the rationality of solutions offered by experts is often illusory, and the best protection against this is a form of participation that simultaneously calls for a redistribution of political powers as well as the insight of technical expertise. It was an approach that had developed under the influence of the environmental movement and the morass of the Vietnam War, whence grassroots democracy was revitalised.

By the late 1970s the surge of the protest movements had subsided and the technocratic approach dominated once more, partly refined so as to include 'participation' in processes where the economic, political and scientific elites controlled the structure of the decision making agenda, e.g. by deciding which kinds of public participation would be admitted at what stage, laying down boundary conditions for participation, and determining what kinds of arguments would be considered by decision makers. Thereby, the substantive content of participation was narrowed and skewed towards legitimating established power politics. The political challenge to the technocratic approach had been successfully contained and the meaning of participation transformed and tamed to fit the power elite. Part of the Mode 2 discussion that emerged in the 1990s fits well into this mould.

On a more critical note some scholars involved in research into public policy have introduced further terms, such as 'mandated science' (Salter, 1988) and 'regulatory science' (Jasanoff, 1990) that came into our vocabulary in the early 1990s. These terms derived from studies relating to the interaction of research and policy in the area of environmental protection. It was found that differences in institutional and cultural dimensions of knowledge production and validation were significant and had to be taken into account. Salter was looking at procedures and the setting of standards for acceptable toxic levels in the work place environment, or in neighbourhoods adjacent to pollution prone industrial enterprises. She argued that in mandated science, policy considerations are closely integrated at every step in the production and use of knowledge. Secondary activities, such as evaluation, screening and meta-analysis, play an important role in such research. There is a significant component of knowledge synthesis.

Jasanoff has been particularly concerned with comparison of different countries' ways of managing risk related to modern biotechnologies. She, among others, pointed out how 'research science' (as distinct from 'regulatory science') places greater value on papers published in journals for academic peers, while science conducted for policy is rarely innovative and may not be subject to peer review. In as far as prediction enters into the picture, the transdisciplinary researcher is often asked to assign an estimate of the risk attached to different options when it comes to policy decisions or measures. Reporting channels may often be reports to public agencies or user groups. A characteristic of transdisciplinary research then, is that its results tend to have a dual audience – other researchers as well as practitioners and user groups in society at large.

Related terms that are still used today are post-normal science (Funtowicz and Ravetz, 1992; Chapter 23) and post-academic science (Ziman, 1996). The emphasis here is on the transcendance of academic disciplinary areas and the entry of external practitioner or user group participation in the validation of research results and their adaptation for implementation in policy and decision making in society at large. Thence we also have the terms 'extended peer communities' (Funtowicz and Ravetz, 1992) and 'regulatory peer review' (Jasanoff, 1990). The latter refers to peer review being imported into user oriented bureaucracies or agencies. The former involves entry of outside users into academic based societal relevance (and quality) review processes. Affiliated with these moves one can find new ideals of democracy in a knowledge society. Extended peer review refers to more and more groups being drawn into the process of evaluating the research results and reworking them, resulting in recommendations for political action to mitigate identified risks or threats (e.g. emission of greenhouse gases in global climate change scenarios). In this context experts' affiliation with one or other of the stakeholders (industry, groups of environmental activists, owners of forests, game hunters in an alpine forest area, etc.) may introduce a bias in the extended peer review process, something that may be compensated by consciously introducing a demand for transparency and articulation of different interests – in a word, 'reflexivity'.

The literature concerning the theory of participation identifies two main approaches: a liberal functionalist or pluralist one; and a theory of direct participation. The former emphasises group representation while the latter gives priority to individual citizen involvement as 'amateurs' who are supposed to become progressively more knowledgeable in the participation process. Both views, however, converge around a number of criteria, e.g. the participant should be independent, involved in the research process as early as possible, and be given resources to effectively influence decision making. There should be a clear task differentiation, structured decision making, transparency for all concerned and cost-effectiveness (Laird, 1993; Rowe and Frewer, 2000). However, greater clarity concerning these points is by itself not enough to induce a deeper reflexivity. For this we also need awareness of some basic structural and institutional dynamics of science-in-society.

As Peter Weingart (1999) has argued the two sides, science and politics, operate along different institutional codes or logics. The Mode 1/Mode 2 dichotomy fails to take this into account, and it tends to blur boundaries, losing sight of the paradoxical nature of the science–politics relationship. The greater the number of societal problems that become subject to scientification, the larger the number of areas in which science based controversies may figure, a process that is attended by a simultaneous diminishing of the authority of scientific expertise. Increased scientification under such circumstances, therefore, simultaneously leads to more, not less, politicisation. In order to sustain their authority, scientists, institutionally, tend to cling to the old linear model of 'truth speaks to power', while policy makers do the same by relying on existing advisory arrangements. Thus one gets an entrenchment of the traditional norms, values and perspectives, or 'logics' in both spheres, an intricate dialectic of paradoxes that needs to be taken into account when calling for greater reflexivity.

## 22.4 A Constructivist View

A constructivist view of scientific knowledge production takes as its point of departure the fact that scientific models of issues such as climate change do not by themselves always produce final outputs of interest to policy users. Therefore, one wants to look more closely at the further process of judgement and evaluation whereby decoupled modelling is introduced in order to reach a final output that will serve as the input-advice in the policy making arena. The strength of user group participation in such settings makes a difference in the context of implementation. A further point is that observational data against which a model (e.g. climate model) is tested or validated are themselves not free from theoretical commitments and assumptions (Jasanoff and Wynne, 1998).

With the foregoing in mind I now turn to the project descriptions in this volume. A guiding idea in the Handbook is that there are three distinguishable phases in projects: (a) problem identification and structuring, (b) probing aspects of a problem, concept and hypothesis formation, and analysis (problem analysis), (c) implementation, whence results from a project are integrated in a real world setting (bringing results to fruition). The activities classed here in three different categories tend to mesh in actual research projects, but the division into three phases is a convenient analytical device. I will use this framework to structure the discussion of the projects. This should however not be confused with the structure of the handbook: I will discuss participation in (a)–(c) independently of whether a project appears in Part 1, 2 or 3 of the handbook. A look at the different project descriptions reveals that, more often than not, participation figures are more prominent in the first and third phases, while focus on internal interaction, or collaboration between members of different disciplinary tribes, is greater in the second phase. Of course, such interaction is also very important in the first phase in order to establish rapport and develop a shared understanding of the object of study. Thus Baccini and Oswald refer to the need for researchers from different disciplines to sort out their 'cultural differences' early, before a more sophisticated conceptual framework can be developed. A common language has to be developed.

## 22.5 Participation in Problem Identification and Structuring

In the science studies literature one finds similar observations, sometimes expressed in terms of establishing a suitable 'boundary object'. This may at first hinge on a metaphor that is later transformed into a bridging concept, such as the redefinition of what is meant by 'mobility' (Chapter 6). In the project on mobility in a model city, external participants, at the outset, were planners from two model cities. Interests of the urban population and local government were taken into account by defining the 'societal' or 'practitioner' problem, with their help. This project highlights that the identification and definition of a societal problem are not enough. Such a problem may not be immediately amenable to research but has to be translated into a researchable problem or 'research problem'. This means that the problem must be conceptually transformed and incorporated into a scientific discourse where it stimulates researchers to play with different models. The transformed problem, i.e. the research problem, becomes the shared research object that forms the starting point. It is not essential that everyone concerned has exactly the same perception of the problem or leading concept. The role of the model as a boundary object is to be partly open, leaving room for interpretative flexibility, so that various researchers can meaningfully refer to it, have different disciplinary angles on it, but still collaborate around it. The model serves as an integrative object along the boundaries of different disciplines, socially and cognitively. It also permits the researchers to avoid monodisciplinary reductionism (a point emphasised in several of the project descriptions), i.e. the tendency to promote a single interpretation of a term or concept as the only correct one.

Schwaninger et al. deal with the societal problem of citizen behaviour relating to environmentally benign practices – solid waste separation in particular. Municipal authorities were consulted to obtain information regarding possible enabling and constraining effects of different types of incentive structures that were modelled with an eye to finding appropriate policy alternatives. An implementation feedback to the municipal authorities and other interested parties was envisaged, but in practice it apparently fell outside the scope of the project (Chapter 13).

Hubert et al. deal with developing a grazing menu for sheep and other livestock in the Mediterranean rangeland in southern France. Stakeholders include livestock farmers, foresters and local political and administrative authorities (Chapter 7).

The authors devote considerable attention to the process by which the societal problem became translated into a research problem. The farmers' knowledge, reflecting their daily lives, was tapped and systematised from various angles. Scientific concepts and tools were used to construct new 'objects' to constitute the problem qua eco-systemic research problem. The authors nicely trace the dialectical tacking motion from the immediately concrete farmers' perceptions and know-how to the scientifically reflected new 'objects' (in a new research discourse) and back again to the farming context, now with more appropriate suggestions for science based interventions. In a sense one sees here, without reduction, a rethinking of practitioner knowledge with the help of transdisciplinary research, focusing on the issue of a complex interconnected system of animal production and environmental maintenance. The farmers and shepherds thus have a central position in the partnership with researchers during the phases of issue identification, problem (re-)definition and structuring, influencing the setting of parameters for modelling a 'menu' for livestock meals. It is noted that modelling can also stimulate alternatives and probe further possibilities for change at a more general level.

In the climate change modelling project described by Held and Edenhofer (Chapter 12) the stakeholders are major global actors, governments, multinational corporations, global NGOs and especially the EC. This is primarily a project of analysis. A novel feature is its incorporation of future technological change as an endogenous factor in the analysis, and also the way conservationist and non-conservationist values are factored into the modelling in terms of a maximum allowable threshold value of global mean surface temperature change in the future (2°C). Thus the potential user group input is indirect (virtual participation), in part matched by direct informal consultation with representatives of two German federal governmental ministries (economic and environment), German energy suppliers and Greenpeace, in order to iteratively obtain acceptable options. In this case, the threshold value of the global temperature is a boundary constraint in the selection of realistic but relevant options for policy – relevant in the sense that the aim is for political realism.

In the regional forest and wildlife management project (Chapter 20), stakeholders were brought together (27 stakeholders representing forest owners, hunters and game wardens, foresters, nature conservationists, road planners and tourist officers) at the outset and continued to play a role as partners in a collaborative inquiry to generate science based knowledge about herbivore impact on forest ecosystems. In this case the researchers not only assembled data and a variety of mental models, but also went on to mediate in a process of negotiation with an eye to resolving a conflict. The 'synthesis model' again served as a boundary object in which different stakeholders could identify some of their own interests and input and therefore develop a sense of ownership regarding the bridging model.

Another important element the authors point out is the development of trust between researchers and users as a precondition for enrolling participants and networking. It is an ingredient that also appears to be important in many of the other projects, for facilitating interaction among researchers, and between the researchers and stakeholders and user groups (e.g. Chapter 17 concerning nomadic communities).

## 22.6 Participation in Learning and Analysis

The core of the project described by Hindenlang et al. (Chapter 20) consisted of a dialogue amongst practitioners and between researchers and practitioners. The process as described compares well with what scholars of scientific controversies (controversy studies) have found out about the evolution and dynamics of a public controversy and its termination (Beauchamp, 1987; McMullin, 1987; also see Nelkin 1995, Martin and Richards, 1995). These scholars sometimes distinguish between epistemic (i.e. knowledge related) and non-epistemic factors in a controversy. The latter refer to personality traits of those involved, institutional pressures, political influences, stakeholder interest and the like. Whereas positivist approaches to controversies focus on agreement reached on the basis of rational discussion and argument, regarding all non-epistemic factors as irrational 'noise', constructivist studies of controversies indicate that such rational termination of conflicts may never be reached, but that 'closure' of a controversy may still occur. When this happens it may be due to external intervention by courts or the adoption of procedural rules.

However, there is also 'negotiation closure' – settlement by intentionally arranged and morally unobjectionable resolution – acceptable to the principals in the controversy even if no party's ideal is reached. In participation theory the method is referred to as 'negotiated rule making' (Fiorino, 1990; Laird, 1993). In the case described by Hindenlang et al. researchers contributed to such a process by obtaining factual information, doing conceptual analysis, developing a common framework of principles, exposing inadequacies and unexpected consequences of various courses of action, raising examples and counter-examples to arrive at a set of acceptable measurable indicators to be used in forest and wildlife management.

The project described by Schelling et al. (Chapter 17) involves a broad range of stakeholder interests at various levels: local, regional, national and international. Stakeholders in Chad were continually consulted as resource persons in scientific studies devoted to developing health indicators and designing a single united human and veterinary medical and health delivery system. The cultural dimension is prominent, since treatment seeking behaviour on the part of nomadic men, women and families were found to be 'strongly influenced by cultural norms'. Thus socio-geographical surveys of ecological, economic and political dimensions of user groups' livelihoods were important. But user groups also had direct input into data and concept formation: interviewees alerted researchers that a vaccine used against anthrax was contaminated. The laboratory examination confirmed the concerns of the livestock owners. Here local knowledge was important for setting researchers on the right track. Participation implied a certain degree of empowerment of the nomadic communities involved.

Walter et al. (Chapter 14) describe a project that applies an interesting methodology for systematic interaction between researchers and stakeholder groups in all three phases. The importance of structure is emphasised as a precondition for flexibility. A significant anchor for participation is the project's co-constitution as a joint academic & cantonal authority endeavour – there was a co-determination of ownership of the problem from the outset. The links between a series of research groups focusing on an analytically differentiated set of facets on the one hand, and stakeholder groups involved in different sectors (e.g. agriculture, silviculture, textile industry, and regional political or administrative activities) on the other hand, allowed for continual reciprocity and a mutual learning curve on both sides. This mode of organising interaction provides a stable socio-epistemological baseline for deconstructing the societal problem analytically into different facets that may thereafter be reconfigured in a model to suggest various scenarios for landscape transformation.

The systemic model that evolved evidently served as a boundary object between the researchers and stakeholder groups, facilitating scenario construction of interest to both sides – users being encouraged to provide not only input for validation purposes but also, when needed, to reconfigure their own thinking about the problem prior to the implementation phase. One interesting outcome mentioned is a political decision relating to the future of tourism.

Rip's Chapter 9 contains a programme declaration for social science supported technology assessment focusing on nanotechnology. It is guided by a recognition of the societal mismatch whereby in practice the bulk of material and cultural resources mostly appear on the side of technology drivers, while exercises in anticipatory intelligence on the part of citizens who will be impacted (as potential beneficiaries and/or victims) by new and emerging technologies is minimal or sometimes non-existent. A summary review of Foresight practices as well as Technology Assessment readily confirms the situation that Rip and his colleagues seek to overcome with their programme for Constructive Technology Assessment (CTA). Their approach is one that is organised around citizen participation in conjunction with a social science based understanding of the dynamics and potential impacts of science and technology.

One of the practical goals of Rip's programme is to use CTA expertise to offer nanoscientists and technologists – as well as other actors, including NGOs – a 'support system' for reflection and strategy articulation. The chapter thus homes in on the important question of 'reflexivity' and on clarifying and counteracting the structural bias whereby actors constantly project linear futures defined by their own intentions and – one might add – tunnel vision (compare Weingart's point above). In the programme, CTA projects relating to technological trajectories are overlayed with studies of cross-cutting aspects like risks, images, ethics and governance, in other words a combination of technical, social, political and cultural aspects or dimensions. Workshops involving researchers and stakeholders of various categories will be used to consider socio-technical scenarios of possible futures to stimulate multi-actor deliberations and mutual learning. The implementation phase still lies ahead. It will be interesting to see what comes out of it; considering the elusively futuristic and abstruse character of the impact of nanotechnology, a new technology that seems to present exceptionally tricky challenges for 'participation', requiring, for example, recognition and identification of sources of nanotoxicity (Royal Society and The Royal Academy of Engineering, 2004; Service, 2005).

## 22.7 Participation in the Implementation Phase

Having hitherto concentrated on the first two phases of the projects reviewed, let me now come back to the role of participation in the implementation phase. As already indicated this aspect is more pronounced when the goal is to develop a forest and wildlife management strategy (Chapter 20) or community based action involving barefoot veterinaries and midwives in nomadic communities without immediate access to village based infrastructures for health and veterinary services (Chapter 17) In the latter case (four) national stakeholder workshops helped articulate community demands and prepare the way for new types of medical and veterinary extension services. In the former, communication and negotiation-interaction formed part of a series of (10 one-day) meetings that gradually led to the shaping of a management tool. Two evaluations helped trim the process and give it public visibility. Researchers in both cases reported their findings and experiences in scientific journals, as well as in reports to government and practitioners.

In the climate modelling case (Chapter 12) members of the public at large were not included in the participation process. Information on the options was primarily aimed at influencing established opinion leaders in economic and environmental spheres, with an eye to influencing current debates for the future of the Kyoto process, especially via the European Commission, a key actor on the global arena.

The CITY: *mobil* project (Chapter 6) seems to have ended up in an intermediate position. Discussion of the ecology of transportation and traffic genesis analysis using cost-benefit scenarios was highly relevant and interesting to scientific peers. At the same time a condensed outcome was fed back to societal practitioners in the form of a guide or handbook – a catalogue of strategic measures. It was hoped that practitioners would feel inclined to use these measures as instruments to help clarify policy options for decreasing the numbers of cars, improving the performance of the public transport system, promoting cycling, and in developing new planning tools for connecting the city's transportation planning and budget planning. Success here ultimately depends on the city authorities' ability to overcome barriers that maintain and continually reinforce the culture of sectoral differentiation of key functions and responsibilities in a city. To be effective in this situation transdisciplinarity has to allow trans-sectoral cooperation and integration by the users and official participating partners. With its primary focus on the development of a knowledge base for a new approach to urban mobility, enrolment of external partners to the cause seems to have become a secondary consideration.

## 22.8 The Need to Problematise the Concept of Participation and Introduce 'Reflexivity'

Members of the public and users may be invited to participate in research and deliberation for many reasons: to begin a process of cultural and institutional change; to observe and collect data; to complement expertise; to make or implement decisions; to attempt to 'educate' citizens about science; to gauge public opinion for the purpose of market research; to overcome public mistrust, stifle objections or defuse critique (for a review of some of the functions of public participation or engagement with science and technology, see Irwin, 1995, 2001, 2004; Sclove, 1998 emphasises the democratic imperative).

In order to dig deeper it is useful to take up a theme that is very much at the heart of participation, viz., 'public understanding of science' (PUS), a subject that has received much attention by science studies scholars during the past 20 years. In this literature it is now stated as common knowledge that the traditional mode of PUS is a science centred one. Therefore, it is referred to as the deficit model, and highlights its concern with scientific illiteracy and knowledge gaps on the part of the public. Since communication is not seen as a question of dialogue but rather as a uni-directional flow where scientists speak to the public, it is also called the linear model of science information (as distinct from an interactive model involving two-way communication). In such an idiom researchers are assumed to be the possessors of rational and expert knowledge. Participation then becomes a matter of the public, or practitioners, receiving and appreciating scientific knowledge in their daily concerns. If there is any problem it is perceived as the participants' misunderstanding and use of technical knowledge. Thus 'public', 'society', 'practitioner', or 'user' tends to be problematised while science is taken at face value. In the minds of scientists, users are seen as passive receivers of their goods.

In the wake of scientific controversies and a fear of science's loss of authority in society there emerged an interest in moving to a more interactive model of communication with the public (Levidow and Marris, 2001; symptomatic of this move is a recent EC journal special thematic issue on 'science dialogues' – European Commission, 2005). Research in STS since about the mid-1980s has contributed to this trend by helping to turn the question around, signalling the need to investigate how people actually define and experience 'science'. It has been found that, in the cultural appropriation of scientific concepts and new technologies, different meanings are attached to scientific information, since citizens or users differ in the way they incorporate science into their everyday life-worlds. Attention has also been directed to the way in which particular scientific constructions disseminated to the public incorporate closed models of social relationships. These reflect dominant power relationships in society at large and therefore should be subject to critical scrutiny and negotiation on the part of the receiver (Wynne, 1995). A constructivist perspective on PUS and the science-policy interface thus seeks to problematise science as well as the public and users (see Jasanoff and Wynne, 1998 for detailed review).

Furthermore, it has been found that researchers tend to view 'users' or layperson participants in science and technology in ways that help sustain the privileged position of the expert, protecting them from public scrutiny of the way commitments, rules of inference or methods of standardisation may contain a structural bias. Certain stakeholders are privileged, particularly those who possess material and cultural resources to act and change conditions in society. This question has been discussed, in the case of genetically modified foods (GM-food), in connection with the use of focus groups and layperson participation in deliberations regarding different policy options. Results from interviews may reflect what the researchers imagine the participants (the laypersons, consumers, environmental activists) believe rather than the participants' actual feelings. (Callon, 1999; Nowotny et al., 2001; Maranta et al., 2003). Focus groups and workshops moreover involve a process of collaborative learning whereby laypersons may appropriate the technically oriented, expert centred perspective of the researcher, leading to engagement in the latter's strategy. When outcomes from such consensus seeking consultative exercises contradict the interests of strong economic lobbies and political elites, the government may still decide to ignore the advice, as happened in the GM nation debate in the UK when the public refused to believe in the benefits of GM food technology (see European Commission, 2005, p. 25 in an article under the heading 'Two-way communication', an interview with Steve Miller at University College London).

The foregoing suggests that the conflict in the 1970s between the approaches of technocracy and democratic control is being replayed in new ways. In a strategy of containment of public dissent the word 'participation' takes on a patronising meaning. Its rhetoric hides the technocratic approach while steering clear of the alternative, the 'democratic control' approach as outlined by Dickson (1984, see above). Levidow and Marris (2001), in particular, have noted how the language of participation is mobilised to refurbish the old linear model, shifting the construal of the public as 'ignorant' to one of the public as 'distrustful'. Instead of disseminating information to overcome ignorance it is now a matter of defusing discontent. If the aim really is to 'relegitimise decision-making, government will need to 'unlearn' many institutional assumptions and to redefine the problem at stake. Rather than seeking ways to change the public, it is necessary to change the institutions responsible for promoting innovation and regulating risks. In particular they need to change their preconceptions of science, technology and public concerns. In such a process, public concerns offer a useful starting point and social resource for organisational learning' (Levidow and Marris, 2001, p. 357). Here we are back to what Dickson was talking about more than 20 years ago.

## 22.9 Who is Empowered by Participation?

The question in the subheading is a double one (Elzinga, 2000). Firstly, it is important to distinguish between effective participation and symbolic or token participation. The former leads to empowerment while the latter involves would-be participants going through the motions of being consulted without really having any bearing on the problem definition, analysis, or ultimate implementation of the results. Laypersons, or NGOs representing them, may also enter researchers' (simulation) models as virtual participants who determine boundary conditions, in which case they do make a difference. At other times they figure as imaginary 'users' in the minds of researchers when the latter discuss user consultations. Secondly, there is the question of who gets invited to participate and who simply gets left out. What criteria are used to define and target user groups as relevant participants, and who does the defining and targeting? Who gets empowered by transdisciplinary research and, the other side of the same coin, who gets marginalised?

Unfortunately, most of the project descriptions lack a degree of reflexivity; they do not provide sufficient information on this particular aspect of participation processes to warrant a cross-referential discussion. Baccini and Oswald implicitly touch upon the subject when they refer to 'generating cooperative majorities for shared target qualities' (Chapter 5) using participatory workshops. The question that arises is how do these 'cooperative majorities' differ from non-cooperative groups. Do they have specific stakes and do they possess superior material and cultural resources to make them worthy of inclusion while more marginal groups are excluded?

The issue of inclusion/exclusion mechanisms in participatory processes has captured the attention of philosophers, historians, sociologists and political scientists who point to a trend associated with globalisation. It is a trend where local governance and choice available to actors at the microlevel in local arenas, concern arrays of options that are predetermined by forces and structures beyond their control. A sense of individual participation is cultivated in a culture of individualism, with promises of radical freedom of self constitution of one's appearance, one's body, lifestyle, or local affairs, while decisive decisions concerning one's livelihood are made in centres where elites concentrate their power. The possibility of radically changing one's own situation and life condition presupposes access to material and cultural resources that many people do not possess. As Zygmunt Bauman (2001) argues, it may well be that we are living in an increasingly polarised society, where some have an opportunity to practice meaningful freedoms and some do not; new mechanisms of inclusion and exclusion prevail when it comes to defining who is a relevant participant and who is not.

## 22.10 Conclusion

In this chapter project descriptions have been used to highlight salient features of the notion of participation. This has been done by looking at the historical context in which the discussion of interdisciplinarity began in the late 1960s and early 1970s. Twenty-five years later this became transdisciplinarity. It is noted that participation enters into transdisciplinary research projects in varying degrees, depending on the focus and goals of each project, and on the particular phase one looks at.

Apart from some general lessons for improving dialogue – the building of mutual trust, commitment, clear delimitation of expected tasks, transparency, and reflexivity – there are also project-specific features that are contingent on the context: the way a project is initiated, who the stakeholders are and what roles they (can) play. A distinction may also be made between real physical and virtual participation of external users or practitioners. A further distinction is made between effective and token or symbolic participation. We are asked to be more sensitive to power relations that prevail in transdisciplinary endeavours, taking into account skews in material and cultural resources available to different actors.

By asking who gets empowered, and what potential users get left out, we plead for a greater degree of reflexivity and at the same time point to a problem that calls for further deliberation, research and innovative social experimentation.

## References

Bauman, Z.: 2001, *The Individualized Society*, Polity Press, Cambridge.

Beauchamp, T.L.: 1987, Ethical Theory and the Problem of Closure. In: H.T. Engelhardt and A.L. Caplan (eds), *Scientific Controversies*, Cambridge University Press, Cambridge, pp. 27–48.

Callon, M.: 1999, The Role of Lay People in the Production and Dissemination of Knowledge, *Sci Tech Hum Val* 4(1), 81–94.

Dickson, D.: 1984, *The New Politics of Science*, Pantheon Books, New York.

Elzinga, A.: 2000, What Participation? (Rev. of Jamison, A. and Ostby, P. (eds), Public Participation and Sustainable Development), *Sci Cult* 9(1), 121–128.

Elzinga, A.: 2004, The New Production of Reductionism in Models Relating to Research Policy. In: K. Grandin, N. Worms and S. Widmalm (eds), *The Science-Industry Nexus. History, Policy, Implications*, Science History Publications, Sagamore Beach, pp. 277–304.

European Commission.: 2005, *RTD info. Magazine on European Research*, special issue November 2005 devoted to Science dialogues.

Eyerman, R. and Jamison, A.: 1989, Environmental Knowledge as an Organizational Weapon: The Case of Greenpeace, *Soc Sci Inform* 28(1), 99–119.
Fiorino, D.J.: 1990, Citizen Participation and Environmental Risk, *Sci Tech Hum Val* 15(2), 226–243.
Funtowicz, S. and Ravetz, J.R.:1992, Science for the Post-Normal Age, *Futures* 24(10), 963–976.
Gibbons, M.: 1999, Science's New Contract with Society, *Nature* 402, C81–C84.
Gibbons, M., Limoges, C., Nowotny, H., Schwartzman, S., Scott, P. and Trow, M.: 1994, *The New Production of Knowledge*, Sage, London.
Hård, M. and Jamison, A.: 2005, *Hubris and Hybrids. A Cultural History of Technology and Science*, Routledge, London
Irwin, A.: 1995, *Citizen Science: A Study of People, Expertise, and Sustainable Development*, Routledge, London, New York.
Irwin, A.: 2001, Constructing the Scientific Citizen: Science and Democracy in the Biosciences, *Publ Understand Sci* 10(1), 1–18.
Irwin, A.: 2004, Expertise and Experience in the Governance of Science: What is Participation For? In: G. Edman (ed.), *Expertise in Law and Regulation*, Aldershot and Burlington, Ashgate, pp. 32–50.
Jasanoff, S.: 1990, *The Fifth Branch*, Harvard University Press, Cambridge.
Jasanoff, S. and Wynne, B.:1998, Science and Decisionmaking. In: S. Rayner and E.L Malone (eds.), *Human Choice & Climate Change*, Batelle Press, Columbus, pp. 1–87.
Laird, F.N.: 1993, Participatory Analysis, Democracy, and Technological Decision Making, *Sci Tech Hum Val* 18(3), 341–361.
Levidow, L. and Marris, C.: 2001, Science and Governance in Europe: Lessons from the Case of Agricultural Biotechnology, *Sci Publ Pol* 28(5), 345–360.
Lubchenko, J.: 1997, Entering the Century of Environment. A New Social Contract for Science, *Science* 279, 191–497.
McMullin, E.: 1987, Scientific Controversy and its Termination. In H.T. Engelhardt and A.L. Caplan (eds.), *Scientific Controversies*, Cambridge University Press, Cambridge, pp. 49–91.
Maranta, A., Guggenheim, M., Gisler, P. and Pohl, C.: 2003, The Reality of Experts and the Imagined Lay Person, *Acta Sociologica* 46(2), 150–165.
Martin, M. and Richards, E.: 1995, Scientific Knowledge, Controversy and Public Decision Making. In: *Handbook in Science and Technology Studies*, Sage, London, pp. 506–526.
Nelkin, D.: 1995, Scientific Controversies. In S. Jasanoff, G.E. Markle, J.C. Petersen and T. Pinch (eds.), *Handbook of Science and Technology Studies*, Sage, London, pp. 444–456.
Nowotny, H., Gibbons, M. and Scott, P.: 2001, *Re-Thinking Science*, Polity Press, Cambridge.
Rowe, G. and Frewer, L.J.: 2000, Public Participation Methods: A Framework of Evaluation, *Sci Tech Hum Val* 25(1), 3–29.
Rowe, G. and Frewer L.J.: 2005, A Typology of Public Engagement Mechanisms, *Sci Tech Hum Val* 30(2), 251–290.
Royal Society and The Royal Academy of Engineering.: 2004, *Nanoscience and Nanotechnologies*, RS & REA Report of July 2004, London.
Salter, L.: 1988, *Mandated Science*, Kluwer, Dordecht
Sclove, R.E.: 1998, Better Approaches to Science Policy Editorial, *Science* 279, 1283.
Service, R.F.: 2005, Calls for More Research on Toxicology of Nanomaterials, *Science* 310, 1609.
Weinberg, A.: 1972, Science and Trans-Science, *Minerva* 10, 209–222.
Weingart, P.: 1999, Scientific Expertise and Policy-Making: Paradoxes of Science in Politics, *Sci Publ Pol* 26(3), 151–161.
Wynne, B.: 1995, Public Understanding of Science. In: S. Jasanoff, G.E. Markle, J.C. Petersen and T. Pinch (eds.), *Handbook of Science and Technology Studies*, Sage, London, pp. 361–388.
Ziman, J.: 1996, Post Academic Science: Constructing Knowledge with Networks and Norms, *Sci Stud* 9(1), 67–80.

# Chapter 23
# Values and Uncertainties

**Silvio Funtowicz*** **and Jerome Ravetz**

**Abstract** Transdisciplinary research and post-normal science (PNS) are a complementary pair of approaches to the new understanding of science. Post-normal science concentrates on questions like 'what-about?' and 'what-if?' The issue of 'safety' is an exemplar of the post-normal approach, since it cannot be managed by a linear, reductionist analysis.

**Keywords:** Post-normal science · Dialogue · Uncertainty · What-if? · Safety

## 23.1 Introduction

In traditional scientific research, even when oriented towards policy, it was assumed that values were irrelevant and uncertainties could be tamed by statistical methods. Now the public, many scientists, and some policy makers, appreciate that certainty and objectivity are not to be attained in the products of such research. Something will have to change in our understanding of science and also in its conduct. Such a complex problem calls for a variety of perspectives; transdisciplinary research and post-normal science (PNS) are a complementary pair. In the former, experience of the new sorts of tasks for science led to this new synthesis. In the latter, the approach was more philosophical, considering how radical the changes in our conceptions of science would need to be. In practice, the two approaches have much in common. Focusing on post-normal science, we show in this chapter that it has a characteristic methodology, concentrating on questions like 'what-about?' and 'what-if?' The issue of 'safety', which perhaps defines the problematique of modern industrial civilisation, is an exemplar of the post-normal approach, since it cannot be managed by a linear, reductionist analysis.

✉ J. Ravetz
James Martin Institute for Science & Civilization, University of Oxford, Oxford, UK
e-mail: jerome-ravetz@tiscali.co.uk

*The views expressed are those of the author and do not represent necessarily those of the European Commission.

## 23.2 Three Sorts of Inquiry

We recall that PNS is based on a threefold distinction between different sorts of inquiry (Funtowicz and Ravetz, 1991, 1992, 1999). In the following diagram, which defines PNS, there are three circular strips in the quadrant. The ideas behind the names of the axes are familiar by now; 'systems uncertainties' and 'decision stakes'.

Down in the lower left hand corner, each is low. That segment is called 'applied science'. This is the 'normal science' where, in the policy relevant fields of science, simple puzzle-solving is effective. All the routine work of monitoring people and the environment, and of building databases about the effects of environmental factors on the behaviour of organisms, comes into this class. But we know that in many critical cases, especially involving our major environmental problems, that straightforward 'applied science' is not effective. What next?

What sort of work is involved in that next, intermediate category? We call it 'professional consultancy', in order to suggest the work that is done by a surgeon or a senior engineer. Someone doing those jobs has to be trained in the relevant science, but there is more to the job than just applying the science. Even if most cases are routine, the professional must always be prepared to cope with the unexpected. They must not merely be ready to change plans rapidly, perhaps improvising as they go along but more seriously, any mistake can have drastic consequences for their patient or client. The most extreme example of professional consultancy is that of the military commander. He knows that he must have a plan before going into battle. But he also knows that on contact with the enemy the plan is the first casualty!

As the problems of technology and the environment became more severe through the latter half of the twentieth century, the first reaction was to manage them professionally. For example, the design and regulation of the new technology of civil nuclear power called forth a new profession of 'probabilistic risk analysts'. Faced with a novel and complex technological system, its practitioners had no 'normal science' of existing safety practice to rely on. In response they developed sophisticated mathematical techniques for estimating the probability of various sorts of

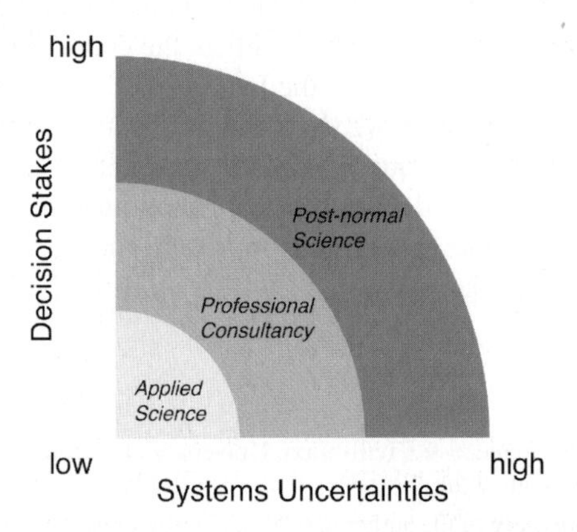


**Fig. 23.1** Post-normal science diagram (adapted from Funtowicz and Ravetz 1993, p. 745)

accidents, and the harm that would result from each of them. The general conclusion of the American risk analysts was that although no one could guarantee a zero risk, there was only a one-in-a-million chance of a serious nuclear accident in the dominant reactor type occurring in any given year. In policy terms this was taken to mean that the technology is safe. Then came the Three Mile Island incident, and the general realisation that a core meltdown had been averted more by luck than by design, and which led to the end of the construction of civil nuclear reactors in the USA. Now we may well ask what sorts of scientific professions are capable of managing the emerging problems of safety, health, environment and ethics within the new technologies of genomics, robotics, artificial intelligence, neuroscience and nanotechnology. That's why post-normal science is now on the agenda for all debate about science.

The essential insight of post-normal science is expressed concisely by the rainbow-quadrant diagram. It reminds us that there are hosts of urgent policy problems involving science, for which routine expertise is totally inadequate, and for which even the best professional knowledge and judgement are insufficient. This happens when, as in the outer strip, systems uncertainties and/or decision stakes are large. But if all the trained people can't tell us what to do, how are we ever to make good, correct decisions on these difficult and urgent issues? Such situations are also characteristic of transdisciplinary research. The essay by Hindenlang et al in this volume describes what might be a type-case: the impact of herbivores on tree regeneration. There are sharp conflicts of interest between hunters and conservationists, and so deep is the disagreement and confusion that 'stakeholders often fail to discriminate between scientific facts, value judgements and intention'.

## 23.3 The Need for Dialogue in Science

When we come to the situations where post-normal science is appropriate, where uncertainties and value-loadings cannot be denied, that old model of scientific demonstration is inappropriate. Instead we need dialogue. In this situation, everyone has something to learn from everyone else. Of course the experts will have a special command of the more technical issues. But others might know better how well, or how badly, the scientific categories fit in with the reality that they experience. These are the members of the 'extended peer community'. Many policy debates hinge on 'safe limits'. A person doesn't need a Ph.D. to be able to ask intelligent questions about safety tests, and to inquire whether they are truly realistic in relation to practice. We might query whether lab tests or surveys, even if performed quite properly using 'applied science' criteria, might turn out to be irrelevant or misleading if applied uncritically in a post-normal situation. Thus, we might need to know whether the sample populations in a study included (for example) children and pregnant women, or animals that breathe air close to the soil. Or we might need to know whether the specifications for safe use of equipment are likely to be respected in real industrial or agricultural situations (it is prudent to assume that they are not). People with local or practical knowledge can spot these sorts of flaws more effectively than experts who are trained in a standard doctrine.

There is an ever increasing stream of 'dialogues' and 'consultations' on science and technology policy. It is very easy to see which are genuinely post-normal, and which are some sort of exercise in manipulation. Is there a genuine extended peer community, in which all sides learn from each other? If not, then the exercise will be fairly low down on the classic 'ladder of participation' of the American Sherry Arnstein (Arnstein, 1969).

## 23.4 Science from the Post-normal Science Perspective

In order to facilitate that process, we will review some of the insights of the post-normal science approach. First, we recognise that we are now going through a revolution in our implicitly assumed conception of science. It is now common sense that science does not deliver certainty, nor is it free of values, personal judgements or institutional agendas. The faith in the truth and objectivity of science, established by Descartes and Galileo, is overthrown. In policy issues, the facts are frequently inconclusive. Their selection and shaping depends on the choice of problem, which in turn is influenced by the framing of policy issues; and this is a matter of politics. In many policy issues involving science, our ignorance is more important than our knowledge. After centuries of triumphalism, the rule of dogmatic science is at an end.

This loss of faith in science derives partly from the character of science based technology in the 'knowledge economy'. Those who prophesied the growth of hitherto unimaginable powers over nature never thought that those powers might run out of control. But these powers are now commanded partly by agencies working outside a political and moral consensus. And even those agencies themselves cannot prevent the occurrence of 'unintended consequences' of all degrees of severity. Any genuine understanding of a technology must now take into account malevolence; not only of those attacking the technology (such as hackers), but also of those applying it to anti-human ends.

These developments are well appreciated by the new public, which is required by the knowledge economy to be sophisticated in order to produce and to consume its products. (Who has not witnessed the effortless mastery of IT systems by the young?) Thanks largely to the environmental crisis, the sophistication of this new public now extends to assessing the whole interrelated system of production, consumption and governance; increasingly, 'consumers' become critical citizens. While the legitimacy of actions of the modern state depends strongly on public welfare, the authority for its policies depends, in the last resort, not on divine sanction or birth or wealth (as in the past), but on science. When official science is revealed as incompetent or corrupt, there is a recognised 'crisis of trust'. When people ask, 'Why should we begin to trust you now?' after scandals like BSE in the UK, the crisis of governance, focused on science but not restricted to science, is real.

## 23.5 A New Post-normal Methodology for Science

Under these new post-normal conditions, science is conducted in different ways. As we say, PNS applies when, typically: facts are uncertain, values in dispute, stakes high and decisions urgent. Quality assurance of both process and product can no longer be left to a community of disciplinary 'peers'; the task requires an 'extended peer community'. Their contribution is not restricted to putting critical questions to the experts. In their creative engagement, they will ask open-ended questions like 'what-about?' and 'what-if?'. Through their participation they will develop a sense of ownership of the problem, which is crucial for their motivation. This is well described by Held and Edenhofer in their essay in this volume (Chapter 12). Also, they will deploy their 'extended facts', which can include anecdotal evidence, confidential information, local knowledge and ethical commitments. A practical example and a theoretical context of these 'extended facts' is provided in the essay by Després et al in this volume (Chapter 21). The example they give is the aging suburbs of Québec, where an Interdisciplinary Research Group on Suburbs created a participative design process; and where, inspired by Habermas' Theory of Communicative Action, they classified knowledge in four categories: scientific, instrumental, ethical and aesthetic. The latter source of knowledge includes artists and designers, and citizens.

Under these new conditions, we move from the traditional dream of conclusive scientific demonstration to the new ideal of dialogue for reconciling real antagonisms. The truncated self-awareness of scientific expertise, necessary for its reductionist strategy, is being replaced by an enhanced self-awareness of all participants, which includes their own uncertainties and commitments. The extended peer communities contribute special insights and forms of creativity, so that previously insoluble problems become occasions for new forms of knowledge and social action.

This post-normal awareness includes the recognition of the inescapable value-loading of every inference; statistical or scientific. The choice of confidence-limit in every statistical test embodies the value-laden decision between the errors of excess selectivity and of excess sensitivity. When, as is now typical, the scientific evidence is inconclusive, there is a policy driven choice between principles of inference. At one extreme is the principle that absence of conclusive evidence of harm is equivalent to conclusive evidence of absence of harm. At the other extreme is the principle that absence of significant evidence of harm is equivalent to suggestive evidence of official concealment of harm. Such choices are at the heart of the post-normal condition; they cannot be reduced to hard facts or to political or scientific authority. Methodology is now politicised, and is itself part of the post-normal processes of debate.

## 23.6 'Safety' as a Theme for Post-normal Science

A new sort of science, that must to some extent arise in opposition to this troubled mainstream 'globalised' science, can be organised around a new theme: safety (Ravetz, 2003). The traditional goals of Western science, knowledge and power, are compromised under the new conditions of the knowledge economy. What industrial innovation, however beneficial in the long run, has not had its investment motivated by either Profit, Power or Privilege? The very achievements of science in improving the living conditions of at least the world's rich, have created the need for a new ideal. Its successes in producing safety, locally and in the short run, have in the long run, created new dangers for the whole planet. The attempted reduction of 'safety' to 'acceptable risk' will prove as unsuccessful as the nuclear power industry for which it was developed. By contrast, 'safety' is a complex, post-normal concept, which is at once pragmatic, recursive and ethical. It is full of paradoxes, as in the question, 'how safe is safe enough?'; and these remind us that it is a post-normal concept, not amenable to linear, reductionist reason.

The sciences of safety develop their own methodologies, which can be seen in their characteristic leading questions. In the search for knowledge, the leading questions are 'what/how?'; and in the development of devices, they are 'how/why?'. In the post-normal sciences of safety, the leading questions are 'what-about?' and 'what-if?' (Ravetz, 1997). Such new questions are calling into being the corresponding new techniques of inquiry and with them the appropriate social institutions and practices. They are the natural questions for members of the extended peer community, just as they are strange and unnatural for those trained in the myopic puzzle solving of 'normal science'.

## 23.7 Post-normal Methodologies

Post-normal science requires new methodologies, and is in the process of creating them. On the more narrow front of the technical management of uncertainty, there is the NUSAP family of methods of assessment of the qualitative aspects of quantitative information. In its original form, this provides five categories for characterising any quantitative statement: Numeral, Unit, Spread, Assessment and Pedigree (Funtowicz and Ravetz, 1990). This has now been developed into a very powerful tool by the collaboration of Jeroen van der Sluijs at Utrecht University with colleagues at the Netherlands Environmental Assessment Agency (RIVM/MNP) notably Arthur Petersen (see Jansenn and van der Sluijs, 2003). Through a set of check-lists, this provides practitioners in any field with qualitative aspects of their quantitative information. It also provides the users of such information with a perspective on its strengths, limitations and weaknesses. In that way, it contributes both to its more effective utilisation, and to strategies for its improvement. It enables a true dialogue between equals: between all the members of the extended peer community.

On the social side, there is a growing family of methodologies for structuring dialogue on the post-normal issues. They have a great variety of forms and political functions. In all of them, there is an attempt to integrating, indeed to reconcile, the very different sorts of knowledge characteristic of the scientific approach, including IT techniques, with the involvement of citizens in an extended peer community. On the social side, 'community research' and 'democratising technology' are vigorous new growths (The Institute for Community Research, 2003; Wakeford, 2004). In this volume, Schelling et al. show how the nomadic pastoralists of a very poor country can perform their own effective quality control on sophisticated medicines, and convince the scientists of the correctness of their assessment (in this instance, that the anthrax vaccine was contaminated). Indeed, all the projects discussed in this volume can be seen as examples of the post-normal science approach, enriching its insights through their own practice.

The practical achievement, in the creation of such new tools for the improvement of scientific work, is considerable. The philosophical contribution of this approach is equally important. It shows that the demise of the long-standing dogmatism of science need not lead to a postmodern anarchy of nihilism or of relativism about facts, values and reality. We can grant that scientific knowledge, like any other image of reality, simultaneously reveals, distorts and conceals. The great philosophical challenge of our time is to comprehend these apparently contradictory, but actually complementary, aspects of knowledge. With a holistic, systems conception of knowledge itself, it is possible to undertake the reconstruction of our philosophy of scientific knowledge, with its varied dimensions of knowledge, power and experience.

## 23.8 Conclusion

This new philosophy of science is not merely a matter of a better understanding of the world. As we contemplate the impending decline of American based globalisation as a hegemonic world order, along with the aggravating global ecological crisis, the task of constructing an appropriate new philosophical synthesis takes on great urgency. We need a conception of scientific knowledge that is not so much designed for the traditional societal goal of the attainment of power, as for the urgent task of the achievement of reconciliation.

We believe that the radical reform of the conceptions of science offered by post-normal science will enable all practitioners in this area, including those in trandisciplinary research, to be more imaginative and to produce a new synthesis of theory and practice that will enhance our lives and lead to a truly sustainable society.

## References

Arnstein, S.R.: 1969, A Ladder of Citizen Participation, *JAIP* 35(4), 216–224, Retrieved June 12, 2007, from http://lithgow-schmidt.dk/sherry-arnstein/ladder-of-citizen-participation.html.

Funtowicz, S.O. and Ravetz, J R.: 1990, *Uncertainty and Quality in Science for Policy*, Kluwer, Dordrecht.

Funtowicz, S.O. and Ravetz, J.R.: 1991, A New Scientific Methodology for Global Environmental Issues, *Ecol Econ*, Columbia University Press, New York, 137–15.

Funtowicz, S.O. and Ravetz, J.R.: 1992, Three Types of Risk Assessment and the Emergence of Post-Normal Science. In: D. Golding and S. Krimsky (eds.), *Social Theories of Risk*, Greenwood Press, New York, 251–273.

Funtowicz, S.O. and Ravetz, J.R.: 1993, Science for the Post-Normal Age. Futures 25 (September), 739–755.

Funtowicz, S.O. and Ravetz, J.R.: 1999, Post-Normal Science – An Insight Now Maturing, *Futures* 31(7), 641–646.

Jansenn, P. and van der Sluijs, J. (eds.).: 2003, *Guidance for Uncertainty Assessment and Communication*, RIVM and MNP, Bilthoven, Retrieved July 7, 2006, from www.rivm.nl and www.nusap.net.

Ravetz, J.R.: 1997, The Science of 'What-If?' *Futures* 29(6), 533–539.

Ravetz, J.R.: 2003, A Paradoxical Future for Safety in the Global Knowledge Economy, *Futures* 35(8), 811–826.

The Institute for Community Research.: 2003, Retrieved July 7, 2006, from http://www.incommunityresearch.org and The Loka Institute, Retrieved July 7, 2006, from http://www.loka.org.

Wakeford, T.: 2004, *Democratising Technology/Reclaiming Science for Sustainable Development*, Retrieved July 7, 2007, from www.itdg.org/docs/advocacy/democratising _technology_itdg.pdf.

# Chapter 24
# Learning from Case Studies

**Wolfgang Krohn**

**Abstract** The question of how transdisciplinary research contributes to scientific knowledge cannot be answered without calling into question a broadly accepted view of the nature of scientific knowledge. Transdisciplinary projects are mixtures of idiosyncratic and nomothetic knowledge structures and the strategies combine research, development, and implementation. The classification into four types of learning offers an analytical view without forcing the projects into epistemic boxes. Distinguishing different perspectives and models of collective learning lowers the burden of legitimising the 'scientific value' of case studies. This attempt to understand transdisciplinary research from an epistemological point of view employs categories developed in the 19th century for defining the differences between the humanities and the natural sciences. Transdisciplinary projects are equally committed to the disciplinary knowledge bases of the natural sciences and technologies and to the value-laden themes of the humanities, but also to procedural methods of the social sciences.

**Keywords:** Causal analysis · Ideographic knowledge · Nomothetic knowledge · Recursive learning · Types of learning

## 24.1 Introduction

This chapter addresses the question of how transdisciplinary research contributes to scientific knowledge. It cannot be answered without calling into question a broadly accepted view of the nature of scientific knowledge. One of the essential features that gives knowledge the status of being scientific is its generality: the less circumstantial and conditional an achieved piece of empirical knowledge is, the higher its scientific value. Even though the history of science abounds with rare discoveries and strange experiments of the highest importance, eventu-

✉ W. Krohn
Institute for Science and Technology Studies, Bielefeld University, Bielefeld, Germany
e-mail: wolfgang.krohn@uni-bielefeld.de

G. Hirsch Hadorn et al. (eds.), *Handbook of Transdisciplinary Research*,

ally their scientific relevance depends on their actual or potential contribution to theory. Admittedly, after the 'experimental turn' in the philosophy and sociology of science, beginning with Hacking's (1983) 'Representing and Intervening', theory can no longer claim the top position in the advancement of empirical knowledge. Still, it seems safe to state that knowledge production without the aim of contributing to theoretical knowledge can hardly be acknowledged as a scientific endeavour.

However, most cases presented in this volume are different. They exhibit three characteristics that are not in accordance with the standard view. They emphasise

- the regional and social specificity of the objects or fields under consideration: the study of *real world problems*;
- the participation of various groups with conflicting needs, interests, reasons and values: involvement of *heterogeneous actors*;
- a procedural orientation leading to problem definition, conflict resolution, and managerial implementation: deliberative *strategic planning*.

While theory oriented research aims at getting rid of contingent factors, in transdisciplinary research it is declared to be important to learn as much as possible about their specificity in order to include those factors in the projected models. One can argue that the move to specific conditions is typical of applied research where existing knowledge is moulded to meet certain parameters, to be commensurable with knowledge from other disciplines, to fit into stakeholder interests, and to conform to political conditions. Important as this may be to the success of innovative projects, most transdisciplinary research projects presented in this volume are more ambitious. They raise fundamental questions concerning:

- the relationships between the social and the natural,
- the conceptual integration of views kept apart by disciplinary perspectives,
- the methods of modelling complexities and picturing scenarios,
- the epistemological coordination of different types of knowledge, and
- the integration of values and knowledge.

The questions go far beyond the clarification of pragmatic problems of knowledge application, interdisciplinary communication and the negotiation of the expectations of politicians, organised lobbyists and concerned people. As already reported by the editors (Chapter 2) several observers have attempted to understand transdisciplinarity as a foundation for reconciling the prospects of scientific knowledge production and life-world concerns of people. If successful, it would contribute to changing the self-description of science and to constituting new institutional frameworks for coordinating research and reform. Is there a way out of the tension between the moderate development of a methodical toolkit for applying knowledge resources to local social and natural conditions and the ambitious new research agenda aiming at a better theoretical understanding of complex problems? In what follows I sketch an epistemic model which gives coordinates to the tension without dissolving it. However, it allows the question regarding the scientific value of transdisciplinary research to be answered.

## 24.2 The Ideographic versus Nomothetic Controversy

The distinction between locally adapted and theoretically generalised knowledge was already at the heart of a 19th century controversy about the respective merits of the humanities and the natural sciences (Chapter 2). The assumption at the time was that they were based on different ideals, called the nomothetic and the ideographic ideals: stating laws versus describing the characteristics. I do not use the terminology in order to return to the unfortunate division between disciplines, but because it seems helpful in developing an understanding of the epistemic specifics of transdisciplinary research projects. In the words of its inventor, philosopher Wilhelm Windelband, the terms are explained as follows: 'in seeking knowledge of what is real, the empirical sciences are looking for the general in terms of natural laws or for the singular in historically determined shape; they consider partially the steady form and partially the unique self-determined content of the real events. Those are sciences of natural laws, these are sciences of events' (Windelband, 1894). Transdisciplinary research projects cannot be positioned precisely. Usually these projects are case studies and the epistemological ambivalence seems to be an essential feature of such case studies. On the one hand, they heavily rely on, and expand, the nomothetic knowledge bases of the sciences. On the other hand, they focus on the case's peculiar, if not unique characteristics. To fully understand these is as demanding as any study of historical or cultural events. Nomothetically, a case is taken as an exemplar of a more general kind and its specific features will be neglected. Eventually it becomes an abstract instance of a general law (e.g. a falling body as an instance of gravitation). Ideographically, the case is taken as an object to be valued for its unique composition of highly specific features: it counts in its own rights (e.g. a falling Chinese vase). Windelband's disciple Heinrich Rickert made the neglect of or emphasis on values the discriminating criterion between the two explanation ideals: 'There is not only a necessary connection between the generalizing and the . . . value-free observation of objects, but also a necessary connection between the individualizing and the value-laden understanding of objects' (Rickert, 1924).

Windelband and Rickert wanted to appease the heated controversy between the natural scientists' reductionism and the culturalists' holism. The reductionist claim is that whatever can be explained by science is covered by general laws. The holistic claim is that an understanding of any historical, cultural and even natural event presupposes an acquaintance with its complex unity or totality. Whether explanation by laws, or understanding by insight is chosen, depends on interest and context. Transdisciplinary research, it seems, is somewhere in between. The projects either oscillate between both ideals, which is unsatisfactory, or they strive for a new epistemology, thus going beyond the alternative. This, however, is much easier proposed than done. In the following I describe and arrange the aspects of the epistemological problems without pretending to solve them.

To begin with, I consider how non-scientific actors bring in the ideographic features of a project. Most importantly, local actors care for their case, and not for any general knowledge. They force researchers to be as specific as possible and develop their models and scenarios close to the circumstantial conditions. To solve the problem 'in principle' would not be acceptable to the audience and the local actors who push the case. If the knowledge bases and methods of scientific experts cannot sufficiently treat the subtleties of the case, local experts and practitioners fill in with their experience based knowledge. Negotiating the acceptability of the research strategy with local authorities, concerned people, and lobbyists of organised interests, compels the experts to be as explicit as possible about potential risks and uncertainties. Each case has its individual value, because the involved actors are engaged in solving their specific case, not a general problem. If local actors were invited to use their case as an appropriate example for finding a general solution to related cases their motivation would decline. They do not want to use their case because, in an essential sense, they are part of the case. Eric Higgs in his influential 'Nature by Design' wrote: 'Thousands of restoration projects take place in North America each year. Many are community-based efforts that rely on volunteer support. The act of pulling weeds, planting, configuring a stream bank to match historical characteristics or participating in a prescribed fire that returns an old process to the land helps develop a ferocious dedication to the place. By investing labor one becomes part of that place' (Higgs, 2003). In short: stakeholders are driven by idiosyncratic concerns.

Scientists, in turn, are professionally trained to search for the more general features of a case, and they do so in several respects. Firstly, they try to see the case as an exemplar of similar cases so that knowledge applied successfully in one situation can be expected to work in another; and that knowledge achieved in one situation, can be tested elsewhere. Any such generalisation remains, of course, hypothetical; but not to work for it would be a rather unscientific attitude. Secondly, they are interested in developing methods and models that are consistent over more than one case. Thirdly, causal explanation of what happens when certain measures are taken, presupposes the need to distinguish between dependent and independent variables. This can be done only with a model that has a nomothetic structure. Fourthly, and perhaps most importantly, non-scientific actors (stakeholders and local authorities) wish to know as precisely as possible, what the state of the nomothetic art is and how experienced the experts are in handling similar cases. The non-scientific concerns that make a project highly specific have to be addressed by testimonies based on nomothetic reliability.

A further feature is added by the impact of the humanities, cultural studies and social sciences on at least some of the transdisciplinary projects. I have mentioned the idiosyncratic values of stakeholders. Though important, they are not exactly of the kind Rickert had in mind. Values that distinguish a case as being unique usually refer to categories that belong to the domains of these sciences. We speak of traditional, aesthetic, ethical, religious, and environmental values. When Rickert proposed values as responsible for perceiving and understanding the distinctive characteristic of something, he referred to the knowledgeable value judgement of scholars. Even if this may not directly apply to problem driven transdisciplinary projects, it can be said that relevant values attached to the problems have to be interpreted and ordered within the context of scholarly informed knowledge, e.g. in the context of urban architecture, ecological restoration, or landscape management. Rickert's philosophy of the humanities was formulated from the standpoint of hindsight as his paradigm discipline was history. The ideographic model as the name implies, does not call for action; but from the perspective of action oriented transdisciplinary research the challenge is to integrate ideographic features into strategic planning. They cannot and should not simply be absorbed as they are contingently present, but must be transformed by a project to fit the natural, social and cultural concerns.

In short: nomothetic and the ideographic ideals are perhaps still to find a place in leading reference disciplines, but transdisciplinary projects cut through the differences and search for methods, models and concepts to manage the coordination or even integrate the perspectives.

## 24.3 Ideographic and Nomothetic Features in Transdiciplinary Research

Before taking the next step a brief look at some of this volume's presentations will give evidence to the combined presence of ideographic and nomothetic features in transdisciplinary projects. The project, 'Constructing Regional Development Strategies', presents a method 'to tackle real world problems in multi-actor contexts' (Chapter 14). The method is in itself an example of nomothetic orientation applied 'from urban and rural cases to business cases, all dealing with complex, ill-defined problems concerning human–environment interactions'. The method provides for the involvement of concerned individuals and institutions, not only to include their 'specific expertise', but also to support their 'motivation' by developing a 'sense of ownership'. Here the value dimension becomes obvious. However, the case owned by the participants is also an 'example' for a variety of similar rural projects. The scientific stakes are made very clear: 'A case has to be selected from the viewpoint of science, which aims at deriving generally valid insights from a number of respresentative cases. . . . A case represents a general problem, but in its specific and unique shape'. This is a self-description, which carefully balances the weights of ideographic and nomothetic concerns.

Other cases shift emphasis towards the nomothetic. The Project 'Fischnetz' attempts to identify the causes for the shrinking of the brown trout population. Different locations have been chosen in order to include 'varying characteristics to represent the range of conditions' (Chapter 8). Still the project is considered to be transdisciplinary. It has integrated local knowledge for the various sites, turned stakeholders into co-researchers participating in as many as 77 subprojects. From a scientific point of view, it was clear from the beginning that to find a generally valid causal mechanism for the decline would be of higher value than tinkering independently with each case. Even if the different sites are considered unique, and 'owned' by the stakeholders, the vanishing of the trout is not.

Interesting with respect to nomothetic and ideographic tension is the case of CITY:*mobil* (Chapter 6). From the very beginning the project aimed at perceiving and solving a general problem pertinent not only to locality but to modern life style. The objective was to develop a general concept of sustainable mobility, which could be broken down into local strategies. Still, the project was transdisciplinary in its attempt to integrate heterogeneous actors and their 'contradicting normative, scientific, and political claims'. However, conceptual work and implementation were separated so that the ideographic constraints did not impact on the learning process.

Almost opposite are the projects that were clearly performed for the purpose of developing local management strategies. Cases in point are the project of Schelling et al. (Chapter 17) in the service of nomadic pastoralists and the project of Hindelang et al. aimed at understanding and minimising the damage caused by hoofed animals in forests (Chapter 20). Both projects were absolutely local. But long-term observation and monitoring have been established to satisfy the demand for badly needed nomothetic knowledge.

The confluence of value concerns and nomothetical theory orientation is especially visible in Baccini and Oswald's project (Chapter 5), 'Designing the Urban'. The report nicely describes the discipline based cognitive barriers between the architects/urban planners and the natural scientists/ecologists. In the internal terminology the former were the morphologists (who formed the 'Gestalt'), the latter the physiologists who care for nomothetic functions. Their cooperation aimed at developing a completely new view of the general issue, the so called urban sprawl. Their theoretical model of the structures and functions of the typical modern urban system is a rare example of interdisciplinary theory formation. On the other hand, they leave the attempt to define the 'good city' (i.e. the problem solution) aside, because 'the best method for defining the 'good city' is based on participatory, transdisciplinary work. We can only learn what a 'good city' is in the relevant social context'. Here again, it can be observed how the value dimension is related to the ideographic case. Only those who experience city life are entitled to evaluate the quality of city life.

In general, transdisciplinary research projects combine ideographic concerns about problem solutions with nomothetic expectations of generalised knowledge. They may put more weight on one side or the other, but both orientations are present. Typically, non-scientific representatives emphasise the local robustness of a solution. Scientists or professional experts expect contributions to nomothetic knowledge. For local actors the project is not a 'cool case'. They are attached to the values of the case, because it is part of their life-world. The Rickert principle, involving the values and the holistic shape of an ideographic case, holds true. Although scientists doing research in the realm of ecological issues can become attached to a case and treat it for its own sake, as scientists they want to learn some lessons for the next case and they hope to add some piece of knowledge to their research field. It is an important observation that in all projects methods are put forward which help to integrate conflicting interests, divergent views, disciplinary knowledge bases, and local experiences. These methods aim at consistency in the problem definition and at political effectiveness for implementing solutions. They methods assist in generating a valid ideographic model. Several project reports state that methodological work using such integrative methods is an important part of transdisciplinary research. One can say that the methodology of transdisciplinary research is one of the fields where generalisations are expected, since they are needed in all projects where the object demands ideographic scrutiny. However, these generalisations are not given without the proviso that their application to another case needs to be carefully adaptated to that case's peculiarities.

## 24.4 Learning from Case Studies

The guiding question 'what can be learned from transdisciplinary research projects?' is not yet answered. Are the cases telling us that some nomothetic lessons can be learned *despite* their situational conditions, or that lessons can be learned *because* they are embedded in real world contexts? There is no easy answer. Apparently the *despite* version is closer to the accepted methodology of the natural sciences (see for an elaboration of this view Shrader-Frechette and McCoy, 1994). But the *because* version is perhaps more appropriate, since – if the apparent semantic contradiction is allowed – contingency is essential in transdisciplinary research. If this is so, it is tempting to introduce a type of learning capable of coordinating the two ideals by finding general laws and understanding through knowing the specific shape. In the search for this type of learning, I propose to look at what may be called *expert learning*. Experts expand their knowledge base, not only by generalising experience, but also by becoming more and more experienced in seeing the specifics of a new case. Experts become steadily better at knowing more about their field and in interpreting various, seemingly contingent details (including surprises). Experts exist in numerous fields, in which being sensible to the case and having access to a knowledge base are related. Among the best known are doctors who are highly esteemed because they can relate seemingly accidental symptoms to general features of a disease to form a diagnosis. Other examples of experts are mountain guides, software maintenance specialists, management consultants, lawyers, or safety engineers. Just like the doctor they all are especially helpful if something goes wrong. Their ability to identify problems and suggest solutions does not seem to be governed by methods of induction or deduction, but by making use of a knowledge base that combines the singular and the general by means of the typical. The nomothetic part of the knowledge base consists of more or less codified professional knowledge, which can be learned, but cannot be reliably applied without practical training. Its ideographic components result from the specialists' memorising highly specific cases they have been involved in. However, what makes an expert better is not the pure sum of well documented cases, but the ability to take features of the cases to be typical. Since every new case is different, it is the ability to operate with similarities and dissimilarities across various cases. Expert work combines professional care for a given case and learning from that case, thus expanding the knowledge base, which in turn assists the more precise understanding of the next case. Generalisation is not helpful, because the knowledge base never becomes independent of the cases that structure the field; it cannot deductively be applied.

If one wishes to give a term to this intermediate learning between nomothetic and ideographic knowledge a reference to the Kantian term of reflexive judgement can be helpful. Experts expand their capacity to properly judge the next case. They become more prudent. I consider enhancement of judgement and prudence a form of learning that shares the broadening of the knowledge base with the ideal of generalisation; and shares the enhanced ability in understanding a case in its peculiarity with the ideographic ideal. The categories point towards a better understanding of the kind of learning achieved in transdisciplinary research projects. Anyway, the tension between ideographic and nomothetic explanation ideals should be accepted and used to develop a new model of learning which combines in situ innovation and scientific research.

The most obvious objection to calling transdisciplinary researchers experts and seeing research results as expertise is that it unduly downplays the scientific efforts. However, expertise does not say anything about the amount of research needed to make it reliable. The transformation of ill-defined complex problems to manageable innovative strategies can be extremely demanding. Or, in other cases, it can easily be done by the 'experienced' expert. Seen from the perspective of non-scientific actors, expert advice drawn from a reliable knowledge base and without the need to immerse all actors into the opaque waters of open questions would presumably be highly valued. Too many open questions that are out of reach of reliable theories and technologies certainly do not add to the trust between scientific and non-scientific actors. On the other hand, the scientific value of projects increases if solutions to difficult and complex problems are presented. If a transdisciplinary research project predominantly relies on adapting approved knowledge to a given case, the research does not aim at generating new knowledge. If the project ventures too far into new fields, the increasing research risks affect the innovation process. Expertise as a concept of knowledge that integrates some features of both the nomothetic and ideographic ideals is worth considering in more detail. This would include looking more closely at the results of transdisciplinary projects and observing how knowledge production relates, on the one hand, to options, strategies and decisions, and on the other, to the improvement of methods and models.

## 24.5 Types of Transdisciplinary Learning

So far I have made two major points with respect to the scientific nature of transdisciplinary research results. First, the knowledge produced relates to objects that combine, in varying proportions, features of culturally valued units and theoretically determined objects. Second, expert knowledge adds scientific value to projects: this knowledge is located somewhere between generalised theoretical knowledge and practical case solution. In the following, I assemble both these aspects in a rather simple two dimensional schema (Fig. 24.1). It maps on one axis the relevance of situational factors – high in ideographic, low in nomothetic cases – and on the other, the degree of research invested in a project. If orthogonally crossed, the dimensions provide four types of transdisciplinary research–learning, to which I have given provisional names. The horizontal dimension refers to the content of a project and maps the ideals of nomothetic explanation and ideographic understanding. The vertical dimension refers to the kind of research activity stretching from applying tried and tested routine knowledge, to generating completely new theoretical, methodical, or empirical knowledge. The first dimension can also be called ontological or content related, the second epistemic or knowledge related.

A short description of these types: the upper two types are research intensive with respect to understanding and solving the problems. If the essential features of a case can easily be isolated from its environment, and the parameters are controllable, research can move towards causal analysis or 'laboratory research'. Non-scientific actors can strongly participate in setting the agenda determining boundary conditions; however, their local experience is less important. If, on the contrary, the essential features of the case are ideographic, then situated knowledge, local concerns and values count. Cases are shaped by carefully integrating various perspectives. This can rarely be done in one step, but by recursive intervention. Real world experiments (Groß et al., 2005) are prominent examples. The lower two types

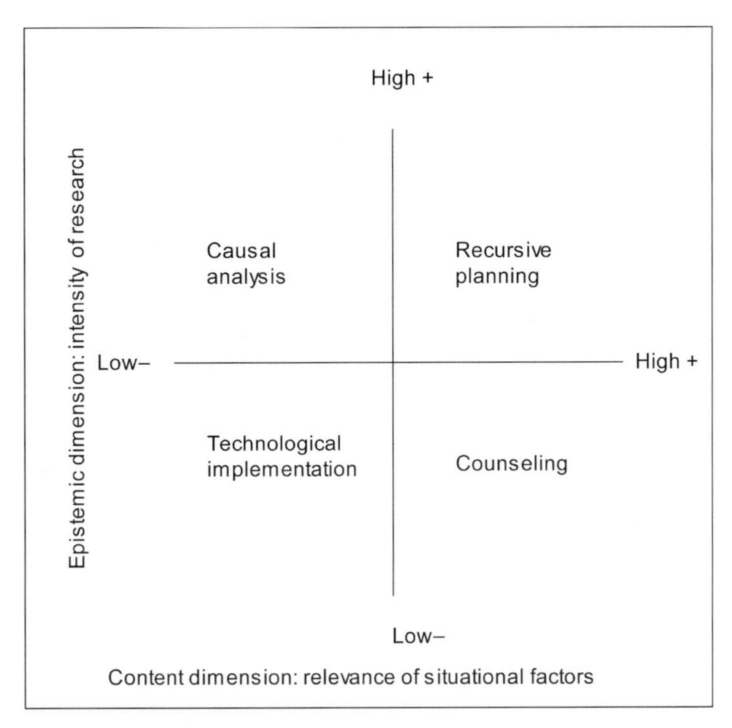


**Fig. 24.1** Types of transdisciplinary learning

pose fewer research risks, because either research results are already available, or they are not expected to be obtained. Technological implementation usually uses high amounts of nomothetic knowledge and is innovative only in combining it in a new design. If the case does not allow for much knowledge production and includes ideographic peculiarities, the best way to proceed is to organise participatory counselling and hope to find a broadly accepted solution. The diagram is a variant of a schema developed by Groß et al. (2005) for the analysis of ecological real world experimentation. It also bears some similarities to the diagram proposed by Funtowicz and Ravetz, but it would need some efforts to compare the parameters and types (Chapter 23).

As we have seen, transdisciplinary research projects are not ideal types but transgress the boundaries. Even if some projects predominantly fit into one quadrant, some components will belong in other quandrants. Furthermore, the time trajectory may lead them through several of the fields. Most projects are difficult to classify in such a schema – not because they don't fall into one of the boxes, but because they belong in more than one. A fair mapping would not locate the projects as points but give them a certain space and shape spreading them over the quadrant's fields. The schema is introduced here in order to provide a more analytical approach for understanding the learning effects of transdisciplinary research projects.

### *24.5.1 Causal Analysis*

If in a project the ideographic factors are low and there is an intent to generalise the findings, the advancement of nomothetic knowledge is predominant. Many examples of this kind relate to problems of health, long-term ecological hazards and other risks. They aim at determining and controlling the causes of risks either by statistical analysis or by identifying causal mechanisms. As many potential risks cannot be studied in the laboratory and frequently need collaboration from practitioners the design of studies will be transdisciplinary, whether or not this term is used, e.g. medical release trials usually include practitioners and patients. Field experiments with genetically modified organisms need the cooperation of farmers, regulatory boards, stakeholders and firms. As scientific knowledge production in a real world setting is the predominant purpose, measures are taken to arrange and control the conditions as accurately as possible. Causal analysis of this kind is an extension of laboratory research beyond the institutional limits of science. It is socially accepted, politically supported, and legally admitted because the expected knowledge is needed in a general context, e.g. the health care system or environmental policy. It may well be argued that the complete set of ideas forming this ideal type – controlled observation, inference of causes, predicition and control – is no longer applicable when the complexity of the real world counts. Basically, the argument is sound. However, it cannot be doubted that singling out certain cause-effect relations can lead to breakthroughs in many problem areas. I have already mentioned domains where the avoidance or at least the control of risks is essential. It is interesting to see, how the public debate here re-establishes the almost forgotten myth of the Baconian experimentum crucis. Are genetically modified organisms harmful? Is low radiation a hazard? Is a new drug free from side-effects? Only transdisciplinary research, including field experiments, epidemiological research on users of technologies and medical release tests, can give answers. The answers are rarely precise enough to settle the case. Commenting on the British field experiments, designed and politically established in 1999 to measure the effects of herbicide tolerant genetically modified crops, Lezaun and Millo (2004) wrote: 'Experiments may not dispel uncertainty or produce the consensus necessary to chart a broad regulatory discourse, but they help to define and legitimise specific decisions at important crossroads' (Lezaun and Millo, 2004). This is a politically moderated version of the decision experiment. It would be strange to refrain from performing causal analysis because experiments and statistics do not lead to clear cut answers when matters are complex. Even if only a few transdisciplinary research projects can be seen as representing an ideal example of causal analysis, its methodical virtues are present in almost all cases.

### *24.5.2 Recursive Planning*

If the expectations of new knowledge are high but its validity is limited to the solution of a particular case with highly idiosyncratic social and natural conditions, a completely different methodological approach is needed. Examples are environmental projects: natural resource management, ecosystem management, ecological restoration and the like. The objects of concern are valued because they are special, rare, and endangered. The leading methodological idea is that all action directed at shaping a system unavoidably becomes a component of it. The predominant idea is not to find causal links between single inputs and effects but to study the highly unpredictable changes in the system. Some years ago computer simulation games were invented by psychologists and ecologists in order to train players to improve the development of poor cities, communities or countries. The games simulated the recursive dynamics of systems, the unforeseeable effects of intervention, and the difficulties of understanding complex systems (Dörner, 1997). As in reality, every intervention becomes part of the system's operational dynamics and cannot be undone. Reality does not have a reset button. The term 'recursive learning' denotes the ongoing interplay between intervening and observing, which either leads to nothing, to destruction, or to an adapted management of the system's development. The case of planning by means of recursive learning is epistemologically interesting. Most of the systems – rivers, regions, traffic, landscapes – are so complex that their course of change triggered by interventions cannot reliably be predicted. This means in turn, that even a perfect scientific model of the system – whatever that is – would not really help. Trialling, monitoring, accounting for surprises and adapting action is the cycle of learning, which can hardly be overcome by better theoretical knowledge. The best available knowledge explaining complex systems states that they are intrinsically unpredictable. In other words: nomothetic knowledge about the behaviour of complex systems exists only at the metalevel. The 'laws' of complexity explain that a system's reaction to interventions cannot be predicted. One of the great achievements of classical science was the combination of explanation and prediction. However, complex systems in real world contexts can be described as unpredictable: it can be known, that future states cannot be known. The nomothetic background of this statement is the theory of non-linear dynamic systems if, and insofar as it is applicable to transdisciplinary research projects, it gives weight to research that takes a single case and tries to follow its unique development and understand its specific reactions to interventions. A well developed epistemology of this kind of research was put forward by Lee (1999) on behalf of the 'Resilience Alliance', which builds on the ideas of the adaptive management approach to ecosystems (Holling, 1978). It is a strategy for improving policies and practices by learning from practice, as well as from the general research programme, in order to reach a new understanding of resource management and sustainable development.

### *24.5.3 Technological Implementation*

In many cases transdisciplinary projects are also involved in technological implementation. Modern technologies can be considered as linking mechanisms between science based technological knowledge and social innovation. Even if the design of a new technology is much more capable of negotiating alternative options, it should be kept in mind that a large amount of reliable technological knowledge is available. Technological knowledge is nomothetic knowledge made applicable for use and made reliable by standards and norms. Technological implementation, understood as defining and putting into practice a technological solution to a complex problem, presupposes a lot of integrative work in combining the thousands of components. Construction of high buildings, bridges, airplanes or dams are cases in point. Transition towards recursive learning is becoming more frequent, since new technologies in the context of ecological innovation need step to step adjustments. Surprises and failures have become a routine experience of modern technology and can only be eliminated over time. Also, the distinction between implementation and theory oriented research can be transgressed if new technological solutions lead into unexpected open questions. Still, it is meaningful to distinguish between a type of innovation that, by and large, can rely on technological know-how that is already applied and approved, and is not intended to drift into open research. Transdisciplinary research projects of this kind apply features of industrial innovation to solving problems of ecological modernisation. As far as it goes, it is of great relevance, especially with respect to non-scientific actors. If technological experts can offer an action plan at the heart of which there is a feasible technology, the uncertainties of research and learning can be reduced, planning can be based on firm grounds, and the conditions of acceptability specified. In this sense Jahn and Keil correctly state: ‘There are numerous problems, which can be treated within the engineering approach of goal oriented multidisciplinarity’ (Becker and Jahn, 2006).

Admittedly, there are rather narrow limits for this type of project. Usually the problems are ill-defined and complex and don’t allow for a technological fix.Furthermore, technological experts tend to overestimate its usefulness and underestimate open questions of feasibility and side-effects. However, every transdisciplinary project contains modules, which come close to this type and are essential to the project’s success. With respect to learning, every new case adds to the stock of knowledge applicable to similar problem constellations. Its value is ‘reliability’ based on theoretically founded disciplinary knowledge, routines of methodical application, and case by case experience.

### *24.5.4 Consulting/Counselling*

If the transdisciplinary problems are highly idiosyncratic and not well understood by the sciences, experts and researchers can still contribute advice, offer strategies and develop scenarios. However, unlike in technological implementation type learning, in consulting/counselling the risks are almost completely shifted to the non-scientific actors. The politics of the projects, in terms of getting stakeholders involved, finding the common denominator of interests, and negotiating the distribution of costs and benefits, dominates the procedures. Carriers of scientific knowledge cannot contribute much; expectations of quick research results are illusionary. Specialists for organising, managing, and monitoring the social dimensions of the project can play an important role and guide it to success or failure. Respectively, scientific learning refers predominantly to improving the tools for performing transdisciplinary research projects. It is a field where social scientists, who study the adequacy, fairness and efficiency of various models of participatory procedures, are especially needed. In a certain sense, the models can be called social technologies of deliberative democracy. They can be tested and improved, as can any technology, and there are experts to evaluate their appropriateness and manage their application. Even if the body of knowledge with respect to the objective at stake is thin, the knowledge about procedures for getting people and parties involved, negotiating interests, ordering values, forming consensus, and monitoring progression and failure, has accumulated considerably (Renn et al., 1995; Abels and Bora, 2005; Chapter 20; Chapter 21).

## 24.6 Conclusion

As already mentioned, a precise subclassification of transdisciplinary research projects into the fields of the schema would not be adequate. Usually they consist of modules that conform more or less to the characteristics of one of the fields. In the terminology adopted here transdisciplinary projects are mixtures of idiosyncratic and nomothetic knowledge structures and the strategies combine research, development, and implementation. The classification into four types of learning offers an analytical view without forcing the projects into epistemic boxes. Distinguishing different perspectives and models of collective learning lowers the burden of legitimising the 'scientific value' of projects. If they are strong from one perspective they need not attempt to exhibit their strength in others. If a project is excellent in solving a highly ideographic case this is a welcome scientific contribution to local politics. Its value is not downgraded if there is no considerable contribution to theory. If, in turn, causal analysis promises a theoretical breakthrough without immediate payoff for practical solutions this is also acceptable. Most transdisciplinary projects aim at all four fields and classification may help to understand what weight is put on the different types of learning.

This attempt to understand transdisciplinary research from an epistemological point of view has employed categories developed in the 19th century for defining the differences between the humanities and the natural sciences. Transdisciplinary projects bridge these differences. The projects are equally committed to the disciplinary knowledge bases of the natural sciences and technologies and to the value-laden themes of the humanities and – as we would add today – to procedural methods of the social sciences. There is no environmental concern without seeing nature in its cultural dimensions; there is no solution to any problem without integrating human actors in its definition. However, if the metaphor of 'bridging differences' is taken seriously, the conceptual coordination between generalising knowledge to formulate laws and specifying knowledge to address unique events leads to confusion. This essay does not pretend to have dissolved confusion. It has aimed at making visible, how epistemologically exciting the new approaches of transdisciplinary research can be.

## References


Abels, G. and Bora, A.: 2005, *Demokratische Technikbewertung*, Transcript, Bielefeld.

Becker, E. and Jahn, Th.: 2006, *Soziale Ökologie. Grundzüge einer Wissenschaft von den gesellschaftlichen Naturverhältnissen*, Campus, Frankfurt a. M.

Dörner, D.: 1997, *Die Logik des Mißlingens. Strategisches Denken in komplexen Situationen*, Rohwolt, Hamburg.

Groß, M., Hoffmann-Riem, H. and Krohn, W.: 2005, *Realexperimente, Ökologische Gestaltungsprozesse in der Wissensgesellschaft*, Transcript, Bielefeld.

Hacking, I.: 1983, *Representing and Intervening*, Cambridge UP, Cambridge.

Higgs, E.: 2003, *Nature by Design. People, Natural Process, and Ecological Restoration,* MIT Press, Cambridge.

Holling, C.S. (ed): 1978, *Adaptive Environment Assessment and Management*, Wiley, New York.

Krohn, W. and van den Daele, W.: 1998, Experimental Implementation as a Linking Mechanism in the Process of Innovation, *Res Pol* 27, 853–868.

Lee, K.N.: 1999, Appraising Adaptive Management. *Conservat Ecol* 3(2), Retrieved June 11, 2007, from http://www.consecol.org/vol3/iss2/art3.

Lezaun, J. and Millo, Y.: 2004, Testing Times. In: ESRC (ed.), *Risk & Regulation*, Summer, pp. 8–9.

Renn, O., Webler, Th. and Wiedemann, P.: 1995, *Fairness and Competence in Citizen Participation*, Kluwer, Dordrecht.

Rickert, H: 1924, *Die Probleme der Geschichtsphilosophie*, Winter, Heidelberg.

Shrader-Frechette, K. and McCoy, E.: 1994, Applied Ecology and the Logic of Case Studies, *Philos Sci* 61, 2, 229–249

Windelband, W.: 1894, Geschichte und Naturwissenschaft. Straßburger Rektoratsrede. In: W. Windelband (ed.), *Präludien. Aufsätze und Reden zur Philosophie und ihrer Geschichte*, J.C.B. Mohr, Tübingen, pp. 136–160.

# Chapter 25
# Management

**Kirsten Hollaender, Marie Céline Loibl and Arnold Wilts**

**Abstract** This introductory chapter discusses management problems associated with transdisciplinary (TD) research. The chapter gives an outline of the relevance and challenges of TD project management. The means and skills necessary for the effective coordination of transdisciplinary research are discussed. The two main tasks facing TD management are identified as: facilitating mutual learning, and enabling shared goal definition in heterogeneous teams. A number of recommendations for the management of transdisciplinary research are identified based on this discussion and on the empirical chapters in this volume.

**Keywords:** Research teams · Project management · Skills development · Management tools · Adaptive approaches

## 25.1 Introduction

Transdisciplinary (TD) research transgresses both the boundaries between scientific disciplines and the boundaries between academic research and practice. TD project teams consist of researchers trained in different academic traditions who often use a variety of disciplinary theories, methods and research techniques. In addition, a number of different stakeholders are involved in most, sometimes all, stages of TD research. TD project teams, therefore, are characterised by heterogeneity and complexity (Hirsch Hadorn et al., 2006). Both these factors are double faceted. On the one hand, heterogeneity is a core strength of transdisciplinary teams. Through their heterogeneous composition they are regarded as generating more differentiated views and having access to a wide and differentiated knowledge base from a multitude of knowledge sources. Thus, a transdisciplinary team is potentially better

✉ K. Hollaender
Department of Sociology, University of Groningen, Groningen, The Netherlands
e-mail: k.m.hollaender@rug.nl

G. Hirsch Hadorn et al. (eds.), *Handbook of Transdisciplinary Research*,

equipped to achieve a 'fuller picture' – a more complex representation of a problem (Klein, 2004). On the other hand, this heterogeneity poses the main challenge for transdisciplinary teams. It leads, among other things, to a greater chance of disagreement about the validity of the knowledge and can result in conflict. Thus, whilst their heterogeneity is the core characteristic of transdisciplinary teams, it simultaneously represents the biggest threat; for conflict and destabilisation of the team. This can seriously undermine TD progress.

A potential source of conflict is disciplinary differentiation between team members in TD projects. Disciplines differ in the ways they assess and frame problems, and select appropriate methods. The different disciplinary backgrounds of researchers can lead to disagreement among team members about the nature and direction of TD projects. In addition, the input of NGOs and social interest groups may threaten the efficiency and progress of TD research. Cooperation, with practice, adds another area of knowledge and other forms of reasoning, which may have implications for the conduct of research. At the same time, the active involvement of researchers with different disciplinary backgrounds, as well as various non-academic stakeholders, is fundamental to effective TD research. Without that involvement there cannot – by definition – be any transdiscipinary research (Balsiger, 2004). Therefore, in TD research there exists a principal tension between heterogeneity and effectiveness. This is the transdisciplinarity paradox.

This paradox puts a particular strain on the management of transdisciplinary research. The main challenge for management of transdisciplinary teams becomes how to manage complexity and heterogeneity. Whilst heterogeneous transdisciplinary teams are better equipped than homogeneous teams to address highly complex problems, they are at the same time less well equipped to deal effectively with the complexity they create. Consensus within TD research teams is difficult to achieve at times because members may apply different criteria of judgment and relevance. Yet, effective ways of dealing with complexity are needed in all stages of TD research. Transdisciplinary research is complex in terms of its organisation and management. It generates additional complexity because it transgresses boundaries between disciplines (Nowotny, 2005). Also, the implementation of TD research solutions may lead to new problem definitions and research questions. At the same time, however, transdisciplinary research teams must find recursive ways to reduce this level of complexity (Pohl and Hirsch Hadorn, 2007).

TD research cuts across disciplinary divides and as a consequence its goals are generally broadly defined – more broadly than in disciplinary research projects that typically have a more narrow focus relative to transdisciplinary research (Lawrence and Després, 2004). TD research promises a better understanding of complicated practical problems, precisely because it integrates disciplinary perspectives and involves stakeholders such as end users or local experts at a very early stage of problem definition and solution. Therefore, TD research goals, typically include an explicit reference to practical relevance and societal value. It follows that the question of how TD research projects can be managed in such a way that critical problems can be avoided becomes particularly salient. What are the opportunities and challenges of TD research management? What are the means and skills that are necessary for effective TD research management?

## 25.2 Crossing Boundaries

Active management is very important for the success of transdisciplinary teams (Hollaender, 2006). A laissez-faire type of leadership, which hopes that the different parts of the work of transdisciplinary teams will grow together organically has not proven successful. Rather, the different parts have the tendency to diverge, making integration at a later point in time even more difficult. Thus, integration has to be part of the project from the start. However, management should not take the opposite position and try to determine the outcomes in too much detail because projects must be flexible and allow for some dynamic development.

Research capabilities of TD management are relevant, but successful management of transdisciplinary research also depends on whether research managers take on an active role in conflict resolution within the team, facilitating moderation between the different members' viewpoints and motivation of the team members. Effective management contributes to integration and consensus within teams. It promotes the development of a feeling of unity and a joint identity with the project. The promotion of the development of joint goals is an especially important management task. Teams that are managed and supported successfully are characterised by equality of rights, mutual acceptance, trust and mutual openness (see Defila et al., 2006 for practical recommendations).

Ongoing TD research often involves advanced forms of knowledge management (Schophaus et al., 2004). Improving communication and knowledge exchange between TD team members is a central task in the management of transdisciplinary research projects. TD research by its very nature crosses boundaries between scientific disciplines and between science and practice. For that reason, the aim of project management in transdisciplinary research is to ensure that boundaries do not become barriers.

TD research is always of an interdisciplinary nature. More than merely adding disciplinary perspectives, interdisciplinary research involves substantive integration and cooperation of theories and methods. This form of collaborative research aims to leverage cognitive resources in order to come up with concrete solutions to complex empirical problems such as land development or environmental pollution (Pohl, 2005). Because of its problem driven nature, this form of research cooperation differs from case to case, depending on situational factors and circumstances. These factors and circumstances become important determinants of the success or failure of interdisciplinary research projects. This means that it is very difficult to identify instrumental rules to support the management of collaborative interdisciplinary and, a fortiori, transdisciplinary research.

Regarding transdisciplinarity, in contrast to interdisciplinarity, most authors agree that specific problems have to be resolved to achieve a cognitive integration that is more than merely an addition of different perspectives and partial problem solutions (Laudel, 1999; Loibl, 2005; Hollaender, 2006). A transdisciplinary perspective aiming at integration requires substantial cooperation and management effort at the very core of the research project.

Different authors have observed that collaborative research that crosses disciplinary boundaries and involves non-academic participants is becoming increasingly important for both the development of new scientific knowledge and the successful practical application of knowledge. For instance, Gibbons et al. (1994) have characterised these interlocking processes of scientific development and practical application in terms of the emergence of a new mode of knowledge production. Arguably, this new mode of knowledge production distinguishes itself by the transdisciplinary nature of research practices. Although criticised by some (Weingart, 1997; also see Horlick-Jones and Sime, 2004), the so-called Mode 2 thesis has found widespread approval (Chapter 27).

Recently, management of sustainability research has received growing attention. For instance, Rabelt (2004) views sustainability research as a social process and presents instruments for enhancing research and for process support. Rabelt et al. (2006) promise that these instruments will contribute to supporting new forms of research practice required by the orientation towards sustainability. Schramm (2004) analyses potentials, restrictions and systematic problems of practically oriented research for sustainable development and derives recommendations for project management. Reflecting many of the characteristics of Mode 2 research, TD projects depend on teamwork. This is particularly apparent in research involving the natural science disciplines, such as environmental chemistry, eco-toxicology, and population biology (Chapter 8). Although at times less apparent, the same holds for research involving the humanities and social sciences (Chapter 15 and Chapter 16).

In all cases, effective cooperation between scientific disciplines is fundamental to the success of TD research. Since disciplinary structure is reflected in institutional structure, transdisciplinary research almost assumes collaboration across different disciplinary institutes. This means that TD research depends on the cooperation of researchers from different institutional backgrounds and working environments. TD research typically materialises as interinstitutional projects in which different research organisations are involved. Joint projects in which representatives of various disciplines participate, then, necessarily involve the cooperation of different academic institutions, often across academic disciplines (Chapter 5 – urban design, Chapter 12 – climate protection and economic growth, and Chapter 14 – institutional cooperation). Increasingly, these projects are organised across national borders (e.g. Chapter 11, on international cooperation in handling toxic chemicals, and Chapter 4 on international partnerships in Kenya).

In addition, representatives of interest groups and other experts, such as local administrators or farmers, are often included in TD projects teams. Of course, these different non-academic institutes and groups, follow their own logic and goals (Chapter 17). Therefore, management has to take into account that the actual members of research teams come from different systems or social and professional backgrounds, and have different orientations and interests (as illustrated in Chapter 20 – forest and wildlife strategy). In most cases, the composition of the transdisciplinary team reflects the assumed structure of the specific practical problem and hence, is characterised by multistakeholder processes (Hemmati, 2002). Thus, to a large extent, team composition mirrors the structure of a real world problem.

## 25.3 Creating Synergy

In TD research different frames of reference, both scientific and practical, come together. Transdisciplinarity, therefore, is a much discussed topic in fields that study problems with a clear societal relevance of impact, e.g. ecological issues (Attwater et al., 2005; Max-Neef, 2005), environmental problems (Pohl, 2005; Daschkeit, 2006), questions of urban design and development (Ramadier, 2004), and sustainability issues (Hirsch Hadorn et al., 2006).

TD research necessarily involves collaboration between heterogeneous actors who have different ideas about the purpose of their joint efforts. These actors operate within various frames of reference when it comes to defining research problems, identifying research tasks, and assessing research outcomes. This is problematic, not only because of theoretical incompatibilities between different approaches, theories, and research techniques of academic researchers, but because the different viewpoints of interest groups, such as state representatives and NGOs, or the local knowledge and observations of involved actors in a case study are critical to the success of TD projects as well. TD research management needs to be aware of this important point. TD management involves the planning of research as well as the stimulation of the cooperation between participants. In addition, transdisciplinarity has consequences for the construction of the specific problem, the research process itself and the eventual synthesis of its results. TD management can only create synergy when it is successful in differentiating and integrating different viewpoints and interests. Regarding the cognitive dimension of TD research, creating synergy can be achieved by applying an integrative methodology. With regard to the organisational aspects of TD research, synergy largely depends on effective forms of conflict resolution.

TD management is thus both a cognitive and an organisational task – and in both dimensions management may bring together the heterogeneous resources embodied by project teams to realise better research outcomes (Klein, 2004). Classifying colleagues in transdisciplinary teams as typical representatives of their discipline or their societal institution can often be observed in TD projects (Hollaender, 2006). This tendency should be dealt with by management at an early stage of the project (Hirsch Hadorn et al., 2006). Members should reflect on their different backgrounds, expectations and orientations, e.g. on how to deal with policy implications, how to deal with the necessity of young members to gain academic qualifications, and how to deal with differing priorities within the project – application vs. more fundamental research.

Establishing rules for the collaboration, agreed upon by all members, is another important management task. These are especially important because they reduce the number of situations where members need to be sanctioned for not performing well, and informal social control is reduced from the level it generally is in disciplinary projects. Equally important, are clear expectations about what are the duties of, and benefits to, members who contribute to joint results. Based on a case study of a project in the Swiss Priority Program Environment, SPPU, Mogalle (2001) develops and employs the concept of the research project as a firm where researchers are seen as entrepreneurs. Hence, it becomes possible to derive management recommendations for the operational and methodological levels of each research project.

The success of TD research depends largely on the effective organisation of research cooperation. However, given its heterogeneous and complex nature, the coordination of TD projects can only be conditional in nature. This means that project management must try to create conditions in which participants may learn from each other and must facilitate the processes by which participants can establish common research problems and shared visions about appropriate problem solutions. However, the management of larger projects, involving representatives from many different academic disciplines and social interest groups, can encounter serious difficulties.

The main challenges to the effective management of TD research, concerns firstly, the formation of shared goals among researchers, and between them and non-academic stakeholders and experts; and secondly, the effective flow of knowledge to enable mutual learning. These challenges have a social, communicative, and cognitive aspect (Hollaender et al., 2004). In order to meet these challenges TD research management must develop: social skills to build up good working relations with team members; communication skills to stimulate information flow and knowledge exchange within the research team; and cognitive skills to understand the theoretical and methodological issues surrounding transdisciplinary research questions.

The complexity of TD research is illustrated by the often very elaborate project structures (e.g. Chapter 6 and Chapter 14). The risk is that the organisational structures of TD projects may become too complex, hindering rather than helping mutual learning and goal adjustment. In addition, constructivist approaches are popular in TD project descriptions – in line with the inherently reflexive nature of Mode 2 research. For example, Deprés et al. (Chapter 21) draw on Habermas' theory of communicative action and discuss the intersubjectivity of the research process. In Chapter 7, Hubert et al. draw explicitly on the work of Gibbons and Nowotny when discussing the reflexivity inherent to TD research. However, an emphasis on social constructivism is not always compatible with the often very elaborate organisational structures of actual projects. Due to their scale and complexity, these structures generally need written agreements in which responsibilities, the division of managerial tasks, and decision making routines are formalised. This requires a more formal and at times bureaucratic form of research management.

## 25.4 Developing Skills

Although potentially a complicating factor, the heterogeneity of TD project teams is a core strength. This heterogeneity allows for the various viewpoints and types of knowledge of participants with different working styles and professional backgrounds to be incorporated into the design of the actual research project (Chapter 7). The teams working on such projects can therefore benefit from the wider range of expertise, skills, and professional competencies. Also, requirements for a sustainable application of research results can be integrated into TD research projects from the very first stages of problem recognition through to the last stages of problem solution and transfer of results (Klein, 2004).

The management needs to develop new skills to enable effective learning processes and to stimulate the formation of shared goals – this is the single most important task of TD management. This is stressed by Bergmann and Jahn (Chapter 6) who conclude that the effectiveness of TD research depends on the formulation of common research objects.

The integration of research results is an iterative process, which can be described as controlled confrontation. Scientists in interdisciplinary teams have a tendency to organise a project so as to avoid conflicting overlap, in a form of negative coordination. Typically, they will not 'intrude' on each others' claimed areas of knowledge. However, in order to make full use of a transdisciplinary team, management must organise some form of controlled confrontation of partial research results, diverging views on progress of the project, and opinions concerning the joint results; so that the results, checked from all disciplinary and stakeholder perspectives, can be accepted as joint results. This is particularly difficult because researchers may be confronted with different notions of inter- or transdisciplinarity in their home institutions (Aram, 2004). TD research management thus involves the complex coordination of a multistakeholder dialogue.

The many different ways to structure such a controlled confrontation range from getting results to enhance each other, to building a hierarchical and formalised form of integration into a systems model. Effective integration of the transdisciplinary team can, for instance, be achieved by concentrating on a joint social problem. Quite often, integration is achieved by studying joint objects or a joint spatial area. However, a difficulty with this approach is that the various disciplines may have different study priorities.

An additional factor complicating project management is that transdisciplinary research is typically public-good research (Pohl and Hirsch Hadorn, 2007). This is something that research managers must take into account when supervising TD projects. For instance, TD management differs from commercial research and development where management controls the resources, such as the financial means and the decision making authority, with regard to the distribution of research outcomes. Thus, conflicts about the orientation of research are minimised. All of the members of an R&D project can be assumed to be oriented towards the same objectives. However, transdisciplinary teams are characterised by different motivations. Whereas young researchers are often interested in their academic qualification, members of free institutes may be interested in the practical benefits, which would recommend them to potential clients and so on (Wilts, 2000). TD research cannot be controlled by management in ways similar to commercial R&D or traditional, disciplinary academic research. In contrast, the management of TD projects only partly controls research incentives. Frames of reference governing relevance and the reward structures differ not only between scientific disciplines but also between scientists and stakeholders. For example, it is often observed that stakeholders need to have visible results more quickly than the scientists involved.

A further factor that should be taken into account is that membership in TD project teams is transient. Members of the group still remain members of their respective 'systems' and may have other criteria of relevance to those of the TD team. For example, this is especially relevant for younger academics who are faced with institutional pressures of qualifying themselves through publications in disciplinary journals. Participating in a transdisciplinary project may put extra pressure on their need for qualification and academic recognition, because they have to meet project demands. In practical terms, recruitment is a particularly important management task in TD research.

A transdisciplinary project team reflects the differences between academic disciplines, and between academic research and social practice and the attitudes and research preferences of the team members. This is illustrated by Rip (Chapter 9), who shows that the reflexivity of TD research increases when interactions between 'scientists, technologists, entrepreneurs, insurance companies, regulators, [and] NGOs' are effectively organised. Coordinating joint research efforts by heterogeneously composed teams always involves the management of differences. This is even more likely when non-academic participants such as the representatives of interest groups, local experts and NGOs take part in TD projects (Ibid.).

## 25.5 Implementing Tools

In the implementation of tools for TD research management it is important to distinguish between issues of project leadership and project management. Project leadership and project management constitute two different forms of research support and require different inputs and coordination efforts. Since transdisciplinary projects typically are transient and interinstitutional, they include a combination of people who do not work together on a daily basis. Participants are often working together for the first time and, first of all, have to establish rules for their collaboration. Since members remain within their home institution, communication has to be actively taken care of – there is no 'joint coffee machine' or canteen where colleagues can meet each other. So, in addition to implementing tools to facilitate knowledge exchange, TD project management must aim to integrate project participants and stimulate communication between them. TD project management thus has cognitive, social, and communicative dimensions, at the same time (Hollaender et al., 2004).

The implication is that TD project management occupies itself with organising the communication between project partners – such as organising and facilitating regular workshops and project meetings where research results can be integrated (e.g. Chapter 8 – the management of the 'Fischnetz' project). In order to enable effective cooperation in this form of research management it is necessary for all parties involved to be in general compliance with the rules set up at the initiation of the project. Therefore, to an important extent, TD project management entails addressing issues of task division and task control. This means that a fundamental objective of TD project management involves identifying reasonable time frames for the completion of specified tasks, such as feasible deadlines for the publication of research results and their dissemination to non-academic stakeholders.

Management can achieve this through different means, depending on the extent to which project participants share goals and the extent to which information flows reach all those involved (Fig. 25.1). If information flows are problematic and goals are not shared among participants, management must take a more instrumental, perhaps interventionist role, to try and control conditions for effective TD research. If both knowledge flows and the extent to which goals are shared are optimal, it may be sufficient for management to monitor events, coordinating the TD process where necessary. Otherwise, management must focus either on improving information flows to enable mutual learning or help participants to recognise common goals and interests.

TD project management includes securing progress through the definition of shared goals and the exchange of knowledge and information among project members. It does not necessarily mean, however, that the project leader is personally performing daily management tasks. Management can be delegated or partly distributed among the team members, provided that the goals and project deadlines, as well as the distribution of resources and the identification of decision models, are no longer points of contention within the larger project team.

However, there are three aspects of project leadership that cannot be delegated and which should remain with the project leader. These are: 1. the overall

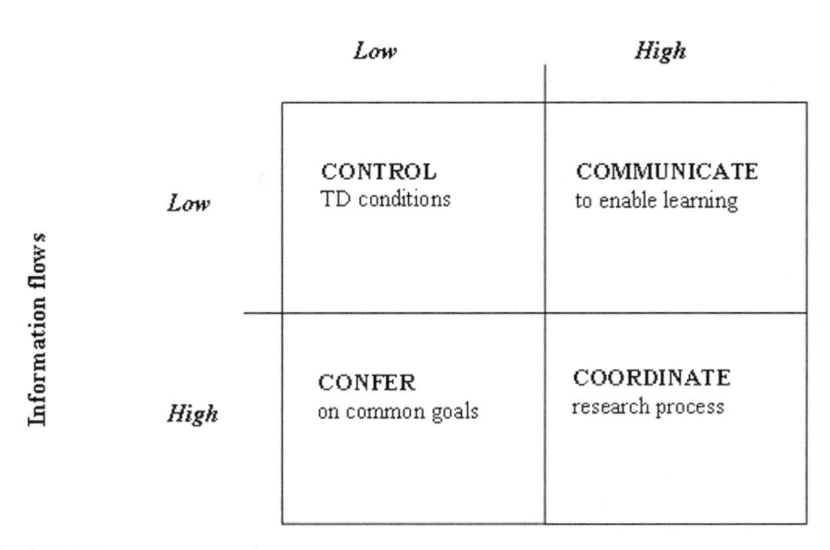


**Fig. 25.1** TD management tasks

responsibility for the implementation of shared working methods, partnerships, and decision rules; 2. represention of the project to external stakeholder contacts; and 3. the responsibility for overcoming the periods of instability and deadlocks that may emerge during various project phases (see e.g. Chapter 13 for recommendations).

## 25.6 Conclusion

The chapters in this volume present a variety of answers to the question of what can be expected from TD research; which characteristic challenges must be handled in such projects; and what the specific social and scientific relevance of the results can be. This introductory chapter has identified three key issues that are important for TD management: (1) facilitating mutual learning through the exchange of knowledge and information is a major factor in the success of transdisciplinary research; (2) creating synergy through the integration of interests and the formulation of shared goals is another important task of TD research management; and (3) stimulating mutual adjustment and compromise among project participants – both in cognitive and social terms – is a condition for the emergence of collaborative effort and a constitutive element of transdisciplinary research.

The process of mutual learning requires self-reflexivity of different theoretical and methodological positions as well as competent guidance for the process of linking these positions in a coherent approach (Pohl, 2005). Exploring and reflecting on this process within the team helps to reduce tensions resulting from the diversity of research objectives and quality criteria. Managers of TD projects should design and facilitate this process of reflection (Schophaus et al., 2004; Blankenburg et al., 2005; Defila et al., 2006).

An effective approach for providing such assistance is to discuss conflicts and complementary functions of institutional project environments, to reflect on the systems of power regulation within these institutions and to apply this strategy of fostering plurality rather than consistency, reflecting on the performance of the team itself and understanding the internal diversity of research interests and working cultures (Böhm, 2006). In fact it is this effort that will create common metacriteria for defining the nature and the relevance of differences in theoretical and methodological approaches, and in doing so, for reaching the analytical metalevel required to connect them and to stimulate mutual adjustment of positions (Loibl, 2006).

Transdisciplinary research projects should implement flexible modes of project steering because interim findings or changes in project environments often necessitate adjustments of the initial research plan or of the original methodological approach. In order to accomplish the research objectives, it may be necessary to adapt time schedules or modify dissemination strategies towards certain addressees. In order to decide on the need for such adaptations, it is important to define clear rules for this procedure at the beginning of a project. TD research management should implement and apply these rules and provide a set of adequate criteria for the continuous quality control within the project (Bergmann et al., 2006). According to Guggenheim (2006) transdisciplinary research strengthens organisational aspects of knowledge production while increasing attention to procedural forms of quality control.

Transgressing boundaries between disciplines and boundaries between research and practice demands attention to the nature, the backgrounds and the implications of such boundaries. We recommend working with adaptive management approaches based on regular self-reflection and systematic adjustment of the research process. Appropriate criteria and decision models will help TD teams to find out which modifications make sense in view of the research objectives and to handle the challenge of integrating methodological approaches and research findings. Therefore, it is recommended that TD project management analyse the project environment and the different backgrounds of the participants, as these are likely to be reflected in the preferences of the team members. An analysis of various evaluation criteria that are relevant for team members from different disciplines and institutions will show that 'quality' can be something very different for each of the team members – depending on methodological backgrounds and institutional requirements. This means that quality management of transdisciplinarity has to take into account the diversity of project goals and evaluation criteria relevant to the team members' disciplinary and institutional backgrounds.

Communication in TD projects does not simply happen. Managing communication has to be acknowledged as a separate task. In a TD setting, communication must be intense and ongoing if it is to achieve integration or exchange of different viewpoints, interests, convictions and scientific paradigms brought in by the various team members. Team members should work on the clarification of the ideas about teamwork and leadership held by participants and try to reach agreement upon decision models in order to prevent scientific or organisational conflicts. TD management thus includes the development and continuous adaptation of an internal and external communication concept.

Finally, an important management task is the management of knowledge. TD research is complex because any possible contribution to the solution of real world problems depends on the integration of scientific questions and questions regarding the application of results. The outcomes of a team's efforts have to be worked out in repeated integration cycles. The management of TD projects can greatly facilitate the cognitive integration of the contribution made by various participants. Management can, for instance, help team members to reach agreement on the definition of joint products or research results that the group wants to achieve. Management can also try to provide the resources that group members need for realising the adjustment and integration of results. The need to find a common framework for the interpretation of research makes it important that participants reach agreement on the nature and form of the integration mechanisms that are used to adjust the projects' outcomes. The management of TD projects can make a difference by identifying methods or concepts of integration.

## References


Aram, J.D.: 2004, Concepts of interdisciplinarity: configurations of knowledge and action, *Hum Relat* 57, 379–412.

Attwater, R., Booth, S. and Guthrie, A.: 2005, The role of contestable concepts in transdisciplinary management of water in the landscape, *Syst Res Behav Sci* 22, 185–192.

Balsiger, P.W.: 2004, Supradisciplinary research practices: history, objectives and rationale, *Futures* 36, 407–421.

Bergmann, M., Brohmann, B., Hoffmann, E., Loibl, M.C., Rehaag, R., Schramm, E. and Voß, J.-P.: 2006, *Quality Criteria of Transdisciplinary Research. A Guide for the Formative Evaluation of Research Projects*, ISOE Studientexte Nr. 13, ISOE, Frankfurt am Main.

Blankenburg, C., Böhm, B., Dienel, H.L. and Legewie, H.: 2005, *Leitfaden für interdisziplinäre Forschergruppen: Projekte initiieren – Zusammenarbeit gestalten. Schriftenreihe des Zentrums Technik und Gesellschaft der TU Berlin* (H.-L. Dienel and S. Schön, eds.), Franz Steiner Verlag, Stuttgart.

Böhm, B.: 2006, *Vertrauensvolle Verständigung – Basis interdisziplinärer Zusammenarbeit. Blickwechsel. Schriftenreihe des Zentrums Technik und Gesellschaft der TU Berlin* (H.-L. Dienel and S. Schön, eds.), Franz Steiner Verlag, Stuttgart.

Daschkeit, A.: 2006, From scientific environmental research to sustainability science? *GAIA – Ecol Perspect Sci Soc* 15, 37–43.

Defila, R., Di Giulio, A. and Scheuermann, M.: 2006, *Forschungsverbundmanagement. Handbuch für die Gestaltung inter- und transdisziplinärer Projekte*, vdf Hochschulverlag AG, ETH Zürich.

Gibbons, M., Limoges, C., Nowotny, H., Schwartzman, S., Scott, P. and Trow, M.: 1994, *The New Production of Knowledge: The Dynamics of Science and Research in Contemporary Societies*, Sage, London.

Guggenheim, M.: 2006, Undisciplined research: the proceduralisation of qualitycontrol in transdisciplinary projects, *Sci Publ Pol* 33(6), 411–421.

Hemmati, M.: 2002, *Multi-Stakeholder Processes for Governance and Sustainability – Beyond Deadlock and Conflict*, Earthscan, London.

Hirsch Hadorn, G., Bradley, D., Pohl, C., Rist, S. and Wiesmann, U.: 2006, Implications of transdisciplinarity for sustainability research, *Ecol Econ* 60, 119–128.

Hollaender, K.: 2006, *Interdisziplinäre Forschung: Merkmale, Einflußfaktoren und Effekte*, Ph.D. Thesis, University of Cologne, Cologne.

Hollaender, K.M., Loibl, M.C. and Wilts, A.: 2004, Management of transdisciplinary research. In: G. Hirsch (ed.), *Unity of Knowledge in Transdisciplinary Research for Sustainability*, Encyclopedia of Life Support Systems, EOLSS Publishers Co., Oxford, UK, Retrieved June 11, 2007, from http://www.eolss.net/.

Horlick-Jones, T. and Sime, J.: 2004, Living on the border: knowledge, risk and transdisciplinarity, *Futures* 36, 441–456.

Klein, J.T.: 2004, Prospects for transdisciplinarity, *Futures* 36, 515–526.

Laudel, G.: 1999, *Interdisciplinary Research Collaboration: Promoting Conditions of the Institution, Collaborative Research Centre'*, Edition Sigma, Berlin.

Lawrence, R.J. and Després, C.: 2004, Housing and health: from interdisciplinary principles to transdisciplinary research and practice, *Futures* 36, 487–502.

Loibl., M.C.: 2005, *Spannungen in Forschungsteams: Hintergründe und Methoden zum konstruktiven Abbau von Konflikten in inter- und transdisziplinären Projekten*, Verlag für Systemische Forschung, Heidelberg.

Loibl, M.C.: 2006, Integrating perspectives in the practice of transdisciplinary research. In: J.-P. Voß, D. Bauknecht and R. Kemp (eds.), *Reflexive Governance for Sustainable Development*, Edward Elgar, Cheltenham.

Max-Neef, M.A.: 2005, Foundations of transdisciplinarity, *Ecol Econ* 53, 5–16.

Mogalle, M.: 2001, *Management transdisziplinärer Forschungsprozesse*, Birkhäuser, Basel.

Nowotny, H.: 2005, The increase of complexity and its reduction – Emergent interfaces between the natural sciences, humanities and social sciences, *Theor Cult Soc* 22, 15–31.
Rabelt, V.: 2004, Nachhaltigkeitsforschung als sozialer Prozess – Begleitsintrumente für transdisziplinäre Forschung, *Ökologisches Wirtschaften* 6.
Rabelt, V., Büttner, T. and Simon, K.-H.: 2006, *Neue Wege in der Forschungspraxis*, oekom Verlag, München.
Ramadier, T.: 2004, Transdisciplinarity and its challenges: the case of urban studies, *Futures* 36, 423–439.
Schophaus, M., Schön, S. and Dienel, H.-L. (eds.).: 2004, *Transdisziplinäres Kooperationsmanagement. Neue Wege in der Zusammenarbeit zwischen Wissenschaft und Gesellschaft,* oekom Verlag, München.
Schramm, E.: 2004, Praxisorientierte Forschung für nachhaltiges Wirtschaften: Restriktionen, Optionen, Handlungsempfehlungen, *ISOE-Materialien Soziale Ökologie* 23.
Pohl, C.: 2005, Transdisciplinary collaboration in environmental research, *Futures* 37, 1159–1178.
Pohl, C. and Hirsch Hadorn, G.: 2007, *Principles for Designing Transdisciplinary Research, Proosed by the Swiss Academies of Arts and Sciences,* oekom, München.
Weingart, P.: 1997, From “Finalization” to “Mode 2”: old wine in new bottles? *Soc Sci Inform* 36(4), 591–613.
Wilts, A.: 2000, Forms of research organisation and their responsiveness to external goal setting, *Res Pol* 29(6), 767–781.

# Chapter 26
# Education

**Julie Thompson Klein**

**Abstract** When Roderick Lawrence and Carole Després introduced a special issue of the journal Futures on transdisciplinarity in 2004, they called it a word 'à la mode' (Lawrence and Després, 2004). More attention has been paid in the literature to research practice. From the beginning, however, the concept was linked with the goal of changing higher education and its relationships to society. This chapter presents an overview of theoretical and conceptual frameworks for transdisciplinary (TD) education, curriculum models, in situ modes of learning in professional practice and community settings, and a culminating reflection on transdisciplinary skills.

**Keywords:** Frameworks for education · Curriculum models · Learning · Skills · Competencies

## 26.1 Introduction

The term transdisciplinarity is traced conventionally to the first international conference on interdisciplinarity, held in France in 1970 and co-sponsored by the Organization for Economic Cooperation and Development (OECD). At the time, higher education was being pressed worldwide by calls for reform. The linear structure of universities was faulted for emphasising training in one-track careers and failing to serve the needs of society. In order to create a more purposeful form of education, Erich Jantsch (1972) proposed a hierarchical system of science, education, and innovation that moved from empirical, pragmatic, and normative to purposive levels. Disciplines and interdisciplinary linkages were still needed. However, isolated divisions of knowledge and common axiomatics were not enough to foster self-renewal and judgment in complex and dynamically changing situations. Jantsch's vision for a new universitas was driven by social systems design and the organising languages of general systems theory and organisation theory. Its institutional

✉ J. Thompson Klein
Department of Interdisciplinary Studies, Wayne State University, Detroit, MI, USA
e-mail: julietklein@comcast.net

structure was based on feedback among three types of units: systems design laboratories, function oriented departments, and discipline oriented departments with a focus on interdisciplinary potential. Some students, Jantsch envisioned, would be located in discipline and function oriented departments. Others would go through all three types of units with increasing emphasis on systems design laboratories, 'useful work', and self-education.

Transdisciplinarity was also conceptualised as an epistemological construct. In the OECD conference Jean Piaget treated it as a higher stage in the epistemology of interdisciplinary relationships. He believed that maturation of general structures and fundamental patterns of thought across fields would lead to a general theory of systems or structures. Piaget (1972) did not prescribe a programme of education. Nor did philosopher Joseph Kockelmans. However, in a 1979 collection of essays on interdisciplinarity published in the USA, Kockelmans situated transdisciplinarity in the philosophical and educational dimensions of sciences. He aligned the concept with the work of a group of scientists intent on systematically determining how negative effects of specialisation can be overcome to make both education and research more socially relevant. Discussion might also focus on concrete problems. In either case, though, the aim was to develop an overarching framework for addressing similar problems and transforming all forms of learning from simply 'training' into 'genuine' education. Kockelmans did acknowledge competing approaches to the task, including unification of sciences, a theoretical framework for empirical research, a unified worldview anchored in a common conceptualisation of culture, and renewed philosophical reflection on the presuppositions and unity of theoretical knowledge in all disciplines. For groups that emphasise social relevance, he added, the first objective is to re-establish contact between the university and society by building new interdisciplines and integrating new existing sciences.

## 26.2 Expanding Frameworks

As the concept of transdisciplinarity evolved, TD education appeared in a widening array of contexts. Some disciplines already had a transdisciplinary outlook. Philosophy, with its mandate for epistemological reflection on all forms of knowledge, is the oldest example. Their broad scope also conferred synoptic identity on the disciplines of literature, history, anthropology, geography, and religion, and the field of area studies. Some fields, such as women's studies, aim explicitly to transcend existing disciplines by forging new overriding paradigms. The TD label appeared in other educational settings as well, including peace and security studies, ecological and environment studies, arts, and engineering. In the closing decades of the last century, a new connotation arose in the USA – 'transdisciplinary science'. TD science is a form of 'transcendent interdisciplinary research' that fosters systematic theoretical frameworks for defining and analysing social, economic, political, environmental, and institutional factors in human health and well-being (Rosenfield, 1992). In areas such as cancer research it has led to new training programmes and career development outcomes.

Efforts to systematically integrate knowledge and to promote holistic education have continued but in a new context. The International University Reforms Observatory (ORUS) is a network of European, Latin American, and South African academics that promotes discussions of university reform with a TD and complex perspective being advanced in local National Reforms Observatories (http://www.orus-int.org/). The Centre International de Recherches et Études Transdisciplinaire (CIRET) is developing a new universality of thought and type of education informed by the worldview of complexity in science. This initiative, Basarab Nicolescu explains in an essay about 'The Transdisciplinary Evolution of Learning' on the CIRET website, enacts the four pillars of learning that Jacques Delors articulated at UNESCO's International Commission on Education for the 21st century: learning to know, learning to do, learning to live together, and learning to be. Knowledge of complexity, Edgar Morin adds, demands a politics of civilisation and reform of the university. The 'Declaration and Recommendations' adopted at the CIRET-UNESCO international congress in Locarno, Switzerland in 1997 included devoting 10% of teaching time in each discipline to transdisciplinarity, an itinerant UNESCO Chair to organise lectures involving the entire community with TD doctoral theses and the support of an international Internet site, and courses at all levels to sensitise students to transdisciplinarity (http://perso.club-internet.fr/nicol/ciret).

A new connotation focused on trans-sector TD problem solving has also led to new models of education and training. This imperative is not new. In 1982, the OECD announced that in order for the university to perform its full social mission, exchanges with the community must multiply, placing greater weight on 'exogenous' than 'endogenous' forms of interdisciplinarity and the traditional quest for unity of science (University and the Community 130). In 1990, Robert Constanza proposed making TD problem solving the primary function of academics, a goal requiring creation of permanent colleges, departments, or programmes and new inter- and transdisciplinary fields of study or 'meta-disciplines'. The major catalyst for a new TD discourse, though, evolved from European and North–South partnerships for sustainability. The distinguishing feature of the discourse, which forms the backdrop for this Handbook, is the externality of complex problems and the participation of a wider range of stakeholders. Problem domains vary, as the case studies in this volume demonstrate. Some collaborations focus on innovative technology and product development, while others focus on controversial social issues involving members of communities who are affected by planning and policy decisions.

## 26.3 Curriculum Models

Early models of curriculum were more ideal than real. In 1972, Jantsch found no links bridging the gap from normative to purposive levels, though he cited colleges of 'normative interdisciplinarity' in agricultural and environmental sciences, human biology, community sciences, creative communities, and engineering and applied sciences. Costanza (1990) noted interdisciplinary units in environmental fields, qualitative studies, and environmental and estuarine studies. By and large, though, explicit models of TD education are more recent. Generally speaking, academic programmes tend to be located within discipline-dominated institutions rather than autonomous institutions. Beyond the university, TD education also occurs in situ, in the workplace and in projects with community stakeholders. There is no systematic compilation of models and practices, but the literature is growing.

Publications emanating from conferences are an important source. The book Interdisciplinary and Transdisciplinary Landscape Studies emanated from a seminar held in 2002 at Alterra Green World Research in Wageningen, Netherlands. In the section on training professionals in research and policy, Ute Bohnsack describes the MSc in Ecosystem Conservation and Landscape Management, a programme run jointly by universities in Ireland and the Netherlands. In addition to learning about pertinent disciplines, students are placed in state agencies and private consultancies, and they do field and project work in multiple countries. At the graduate level, Gary Frey (2003) advises, TD education is often hindered by limited availability of supervisors in pertinent methods and by problems associated with students' lack of strong disciplinary identity. Frey urges more teamwork, a wider range of courses and seminars, and opportunities to mix with students of different knowledge cultures. In this volume, Baccini and Oswald also recommend framing Ph.D. topics in a way that permits time-limited projects of several years (Chapter 5). The average time of the thesis period in the urban design project they described was extended by about 20%, but doctoral students considered the experience to be a valuable supplementary step in their education.

CIRET's website (http://perso.club-internet.fr/nicol/ciret/) is also a valuable source of models. The list of innovative experiences cited at CIRET's 1997 Locarno congress includes the British Open University, the Academy of Architecture of Ticino, the American Renaissance in Science Education, the University of Basel, the Observatoire pour l'Etude de l'Université du Futur in collaboration with the Ecole Polytechnique Fédérale of Lausanne, the Maison des Cultures du Monde, and transcultural experiences in Catalonia. The most complex challenge, Nicolescu stresses, is the teaching of teachers and developing appropriate pedagogy. These imperatives are the focus of the March 2005 issue of the Bulletin Interactif du CIRET (#18), available on the CIRET website. In addition to conceptual articles and discussions of pedagogy, didactics, and professional formation, the issue presents experiences in Europe, Brazil, Mozambique, and Romania. A member of the CIRET network, the University of São Paulo's Centro de Educação Transdisciplinar (CETRANS), has also fostered curriculum, research, study sessions, and an educating educators project (de Mello, 2000).

Additional models and reflections may be located at <td-net> (http://www.transdisciplinarity.ch/). The site offers an overview of Graduate Studies and Continuing Education in TD projects in Switzerland. Individual sections of Publications in the Bibliography may be searched using the keywords 'education', 'teaching', 'learning', or 'curriculum'. The sublink on Journals also leads to periodicals in areas of action learning, science and technology education, higher education studies, urban studies, and a host of sustainability interests. Moreover, <td-net> affords access to many examples in preliminary workbooks for the 2000 International Transdisciplinary Conference, which are not readily available to the international audience. Two examples from the 2000 Conference illustrate the central role that project work and mutual learning play in TD education.

At the University of Basel a specialised master's degree in sustainable development can be earned. Contributing to sustainable development is a contemporary challenge that must be approached in an interdisciplinary manner and knowledge has to be achieved in transdisciplinary project work. Thus, in addition to specific disciplinary competences, the master's programme draws on aspects of sustainability that are generated by the natural, social and economic sciences. Apart from the opportunity to carry out a practical training outside the university, there are thematic modules, such as agglomeration and ecosystems, conservation and utilisation of natural resources, environment, values, societal transformation and health, as well as environmental problems in globalised world. After deepening the disciplinary and complementary knowledge, and before preparing the master's thesis, an interdisciplinary project has to be carried out. Students learn to approach real world problems, often in collaboration with experts from outside university, by identifying problems, defining aims, analysing the status-quo, developing and implementing strategies and instruments in order to achieve the interdisciplinary project task. Methodological aspects of the curriculum include the analytical and integrative skills and knowledge necessary to work on complex questions relevant to sustainability, together with reflexive skills and proficiency in dealing with instruments. Considerable importance is additionally placed on competences in communication, team development and project management in order to facilitate constructive work in transdisciplinary settings (Burger et al., 2000).

The ETH-UNS Case Studies form a compulsory part of environmental science education at the Federal Institute of Technology in Zurich. The programme has been conducting one case study per year since 1994. In Chapter 14, Walter et al. describe how teaching, research, and application are combined within a single project. Mutual learning is intrinsic to TD Integrated Planning and Synthesis (TIPS). Students develop knowledge and methods of science and collaborative skills by working in study teams and with case agents on complex problems in areas such as sustainability, agriculture, reintegration of industrial sites, urban and regional development, and the eco-efficiency of environmental interventions. The process is supported by a variety of case study methods, including formative scenario analysis, modelling systems dynamics, integrated risk model, future workshops, and life-cycle assessment.

## 26.4 In Situ Professional and Stakeholder Education

The complex problems of practice that confront traditional professions have reinforced the need for TD education. A new vision of engineering is being implemented at the Institute for Transdisciplinary Education and Research, housed in the

Department of Mechanical Engineering at Texas Tech University (USA). Courses in Transdisciplinary Programs in Design, Process and Systems are taught via on-site visits and distance learning (http://www.me.ttu.edu/home/research). Examples abound in health sciences as well. In this volume, Piko and Kopp describe efforts to apply social and behavioural sciences within medicine and health sciences in Hungary, introducing the biopsychosocial paradigm across levels of the higher education system (Chapter 19). At the University of Western Ontario in Canada, the Ecosystem Health Program has brought specialists together with the lay community in an undergraduate programme aimed at helping medical students understand the context of their patients' lives and social and environmental risks (McMurtry, 2000). Located at Brown University (USA), the Centers for Behavioral and Preventive Medicine provides information on transdisciplinary post-doctoral, new faculty, and internship training programmes. The programmes are aimed at preparing the next generation of physicians, psychologists, public health, and behavioural/social scientists to work jointly on complex questions and to produce data relevant to human health, behaviour, and social systems (http://www.lifespan.org/behavmed/trainingpostdoctoral.htm).

Professionals need training in the workplace as well. The Wageningen Initiative for Strategic Innovation, Jelleke De Nooy-van Tol (2003) reports, has provided professionals with the extra time, facilities, skills, and moral support to cope with complexity and value systems in their daily surroundings. Project leaders learn in the context of an actual project, following criteria that define the characteristics and preconditions for transdisciplinary research. A variety of learning formats has been used, including a formal course in communication skills and a pilot course in systems approaches, an atelier in the form of creative sessions with all stakeholders, support for using scenario-casting techniques and system approaches, and workshops for scientists involved in organic agriculture.

Several chapters in this volume call attention to the role of workshops. In addition to the chapters by Patricia Burkhardt-Holm (Chapter 8), Bergmann & Jahn (Chapter 6), and Baccini & Owald (Chapter 5), Esther Schelling and colleagues describe a series of national stakeholder workshops in Chad aimed at providing adapted health services to nomadic pastoralists and their animals (Chapter 17). The workshops and related focus groups set the aim of the programme, updated results, addressed barriers, readjusted priorities and interventions, and dealt with policy issues. Research and action activities, in turn, triggered community education initiatives. Després and colleagues' case study of retrofitting post-war suburbs in Quebec (Canada) depicts an intensive participatory design process led by the Interdisciplinary Research Group on Suburbs (GIRBa) at Université Laval (Chapter 21). In Phase I, at the macro-scale of decision makers and planners, the stakes and challenges were identified in a monthly series of four one-day mini-colloquia and workshops. Each mini-colloquium was followed by a workshop where participants discussed and annotated or interpreted maps of the city and suburban boroughs. At the meso-scale, meetings with each borough council and community representatives yielded a fuller picture of primary interests and challenges, followed by a colloquium involving key actors. In Phase II four one-day meetings, a borough colloquium, an objective-defining workshop, and two mini-charrette participatory design sessions helped to define general orientations and objectives. In Phase III, the strategic plan and implementation strategy were refined using a design criteria workshop and final design charrette. Of added note, the GIRBa also plays an incubator role for graduate students in architecture and urban planning projects.

Cyberspace is a rapidly expanding forum for academic coursework, professional training, and resources for TD education. The Holistic Education Network of Tasmania, Australia provides materials and links to organisations with interests in transdisciplinary inquiry, holistic learning, and transformative learning (http://www.hent.org/). Dora Marinova and Natalie McGrath define the elements of transdisciplinary education based on experiences in the Institute for Sustainability and Technology Policy at Murdoch University (See especially 'A Transdisciplinary Approach to Teaching and Learning Sustainability: A Pedagogy for Life'; http://lsn.curtin.edu.au/tlf/tlf2004/marinova.html). In the Mutual Learning Session on Education at the 2000 International Transdisciplinarity Conference, Abou-Khaled et al. (2001) and Reichel (2001) presented additional models. FACILE is the acronym for FACILitated distance learning environment for continuing engineering education. A project of the Technical University of Vienna, FACILE enhances the knowledge and skills of professionals in their daily work. The University of Hagen in Germany offers Internet-based distance courses and cooperates with national and international partner institutions. So does the British Open University. The NDIT/FPIT virtual university provides postgraduate studies and functions as a training and research partnership between universities and private companies in a consortium of over forty institutions (Nachdiplomausbildungé in Informatik und Telekommunikation/Formation Postgrade en Informatique et Télécommunications).

## 26.5 Cultivating Transdisciplinary Skills

Two recent reports furnish insights into the kinds of skills that are needed for TD research and the means of cultivating them. The 2004 report of the US National Academies of Science Committee on Facilitating Interdisciplinary Research did not address transector TD research. Yet, many strategies are generic to both ID and TD contexts. The top recommendations for students are to cross boundaries between disciplines and to take a broad range of courses while developing a solid background in one discipline. Undergraduates are urged to seek courses at the interfaces of traditional disciplines that address basic research problems, courses that study social problems, research experiences that span more than one traditional discipline, and opportunities to work with faculty who have expertise in both their disciplines and the interdisciplinary process. Graduate students are encouraged to broaden their experience by gaining knowledge and multiple skills in one or more field beyond their primary area, by doing theses or dissertations involving multiple advisers from different disciplines, by participating in conferences outside their primary fields, and, for all students, by working with mentors from more than one discipline. The top recommendations for educators are to develop curricula incorporating ID concepts, to offer more interdisciplinary studies, to take part in teacher-development courses on ID topics and methods of teaching non-majors, and to provide opportunities that relate foundation courses, data gathering and analysis, and research activities to other fields of study and to society at large. The report also recommends more training in ID research techniques, team management skills, and summer immersion experiences for learning new disciplinary languages and cultures (Centre for Educational Research and Innovation, 1982).

The 2005 report of the Academy of Finland Integrative Research (AFIR) team on Promoting Interdisciplinary Research (Bruun et al., 2005) did include transector TD. The AFIR team analysed a multiyear pool of applications for research grants across all funding categories of the Academy. The team did not address the topic of education in detail. Yet, their findings yield insights into how research education can overcome barriers to collaboration and integration in seven major areas: organisational structure, sufficient knowledge, cultural characteristics of fields, epistemological differences, methodological styles, psychological investments, and reception by varied audiences. A survey designed to gather information on the benefits of an interdisciplinary approach included breadth of preparation, networking with experts from different fields, and learning to combine differing knowledge, methods, and views. Most projects reported having, at least partly, an interdisciplinary approach in their particular areas. The largest group (41%) used a combined strategy. Research students were primarily educated in their home departments, but they were also given opportunities to learn about ID research. Another large group of projects (39%) exhibited more interdisciplinary than disciplinary traits in research education, achieved either by emphasising an interdisciplinary approach regardless of students' individual backgrounds or by making the combination of knowledge a central theme. Only 20% of projects delegated all research education to students' departments. Of particular significance, projects that organised research education in an interdisciplinary way experienced more synergies between research education and project design, development of interdisciplinary skills, and benefits for networking and creativity.

## 26.6 Conclusion

Three overriding lessons about the skills needed for TD research emerge from these reports and the frameworks, models, and practices surveyed in this chapter. The first lesson is the new quadrangulation of disciplinary depth, multidisciplinary breadth, interdisciplinary integration, and transdisciplinary competencies. In most visions of TD education, disciplines do not disappear. They provide essential building blocks. All students, though, need a contemporary view of disciplines that bridges traditional knowledge and the new developments and skills needed for complex problem solving. Breadth of exposure to multiple disciplines is also required and, Bohnsack (2003) adds, students in multidisciplinary courses need to be able to articulate the specific knowledge and experience gained in their own disciplines. The increased role of interdisciplinary fields must be taken into account as well. The most frequently cited skills in the literature on interdisciplinary education, Klein (2002) found, are the ability to locate and work with pertinent information, to compare and contrast different methods and approaches, to clarify how differences and similarities relate to a designated task, and to generate a synthesis, integrative framework, or more holistic understanding for a particular theme, question, or problem. The integrative process is inherently constructive, she adds, engaging students in acts of interrogation, decision making, and problem solving.

Finally, transdisciplinary competencies must be cultivated, leading to the remaining two lessons. The second overriding lesson is reconceptualisation of education as a dialogue of content and process. In establishing criteria for training in the MGU programme at the University of Basel, Ruth Förster (2000) and her colleagues were faced with the challenge of determining not only what skills and knowledge students need but also what training concepts and criteria teachers should employ. The underlying philosophy and method that emerged recognises the dynamic, reciprocal relationship of content and process. Content designates the knowledge, principles, and methods of different disciplines as well as inter- or transdisciplinary approaches, the ability to analyse complex problems, and familiarity with problem solving strategies. Process designates knowing how to organise and participate in inter- or transdisciplinary processes and projects and knowing how to communicate across academic disciplines and with external stakeholders. There is no one best method, Förster advises, rather a 'basket of options'.

Many of the elements of a transdisciplinary toolkit were identified in this volume, including modelling, scenario analysis, systems approaches, integrated risk assessment, group facilitation, and participatory models that foster joint decision making. New design tools were also reported, including the Netzstadt method for establishing a TD platform and procedure, the Synoikos method for participatory workshops, a computer learning-model of cities, and the information system of Least-Cost Transportation Planning. One approach – mutual learning – appeared repeatedly. Mutual learning requires skills of collaboration and negotiation. Knowledge is not simply exchanged but constructed and activated as individuals with differing views and stakes work together. Hindenlang et al. have also described the technique and skills of collaborative learning in their case study on regional forest and wildlife management (Chapter 20). In their chapter on urban design, Baccini and Oswald emphasised another recurring theme (Chapter 5). Understanding the language of others and perceptions hidden in use of the same words is essential. Even the most basic terms such as 'landscape', 'urban', or 'project' have different connotations.

The third and final lesson is the intertwined relationship of transdisciplinary competencies. Sharon Derry and Gerhard Fischer (2005) have conceptualised a set of overlapping categories of TD competencies for graduate STEM education that are applicable well beyond the STEM domains of science, technology, engineering, and mathematics.

- The first category is the ability to participate productively in reflective transdisciplinary communities, based on a deep understanding of the nature of communities, effective communication, metacognitive skills, and mindsets for engaging in TD work and study.
- The second category is the mindsets and metacognitive skills that enable lifelong learning, including critical thinking skills, learning on demand, and self directed learning.
- The third is the ability to understand, exploit, and design innovative sociotechnical environments, requiring fluency in using digital media.
- The fourth is the ability to develop, fund, and guide knowledge building communities as contexts of teaching and learning.
- The fifth is concern about real world needs, manifested in a willingness to become an engaged citizen.

The international team of Pohl, van Kerkhoff, Hirsch Hadorn, and Bammer offer a closing perspective on the skills needed to lead transdisciplinary and integrative projects (Chapter 27). The underlying theme is cognitive flexibility, manifested in a willingness to see beyond one's own discipline. Building on co-author Gabriele Bammer's work on the emerging specialisation of integration and implementation sciences, they designate the following needs in their contribution to this volume:

- the ability to appropriately scope problems and issues to determine how an integrative approach can best be applied, ensuring both multidisciplinary and multi sector involvement
- knowledge of and the ability to apply integrative methods and processes, such as modeling and group facilitation
- appreciation of different research epistemologies and the ability to draw on their strengths, tailoring them to each other and a common task
- understanding of policy, practice and product development and how these can be influenced by research
- the ability to foster research collaboration.

No transdisciplinary researcher, the authors advise, will be expert in all of these areas. Other team and project members can be brought in to fill skill gaps. Developing a larger framework for skills and allowing concentration on a subset of them can also provide the core of undergraduate and graduate curricula. The business of education has traditionally been the transmission of knowledge. The emergence of transdisciplinary orientations in the knowledge society, editors of the Handbook exhorted (Chapter 2), is not only curiosity driven: it is also committed to improving the lives of people in problem fields characterised by complexity and the interests of the common good. Yet, there is a crippling disconnection between local efforts and the abundant information and insights that have emerged. 'What has been learned on the job', the editors lamented, 'is seldom passed on to others for capacity building'. Handbooks and networks are crucial forums for banking and disseminating the wisdom of both theory and practice.

## References

Abou-Khaled, O., Coray, G., Delley, A. and Flückiger, F.: 2001, New Concepts for Continuing Education: A Challenge to the Entire Education System. In: *Transdisciplinarity: Joint Problem-Solving Among Science, Technology and Society, Workbook II: Mutual Learning Sessions*, Haffmans Sachbuch Verlag AG, Zürich, pp. 130–134.

Bohnsack, U.: 2003, The New Multi-University Msc in Applied Science 'Ecosystem Conservation and Landscape Management' (Ireland, Netherlands, Finland). In: B. Tress, G. Tress, A. van der Valk and G. Fry (eds), *Interdisciplinary and Transdisciplinary Landscape Studies: Potential and Limitations*, Wageningen, DELTA Series 2, pp. 124–138.

Bruun, H., Hukkinen, J., Huutoniemi, K. and Klein, J.T.: 2005, *Promoting Interdisciplinary Research: The Case of the Academy of Finland, Publications of the Academy of Finland*, Series #8/05, Academy of Finland, Helsinki.

Burger, P., Förster, R. and Jenni, L.: 2000, Transdisciplinary Training and Research. In: R. Haeberli, R.W. Scholz, A. Bill and M. Welti (eds.), *Transdisciplinarity: Joint Problem Solving among Science, Technology, and Society. Workbook I: Dialogue Sessions and Idea Market*, Haffmanns Sachbuch Verlag AG, Zürich, p. 84.

Centre for Educational Research and Innovation.: 1982, *The University and the Community: The Problems of Changing Relationships*, OECD, Paris.

Committee on Facilitating Interdisciplinary Research.: 2004, *Facilitating Interdisciplinary Research*, National Academies Press, Washington.

Costanza, R.: 1990, Escaping the Overspecialization Trap. In: M.E. Clark and S.A. Wawrytko (eds.), *Rethinking the Curriculum: Toward an Integrated Interdisciplinary College Education*, Greenwood, New York, pp. 95–106.

de Mello, M.F.: 2000, Transdisciplinary Evolution in Education. In: R.W. Scholz, R. Haeberli, A. Bill and M. Welti (eds.), *Transdisciplinarity: Joint Problem-Solving Among Science, Technology and Society, Workbook II: Mutual Learning Sessions*, Haffmans Sachbuch Verlag AG, Zurich, pp. 380–383; Retrieved August 3, 2006, from http://www.cetrans.futuro.usp.br.

De Nooy-van Tol, J.: 2003, Needs for Training of Professionals. In: B. Tress, G. Tress, A. van der Valk and G. Fry (eds.), *Interdisciplinary and Transdisciplinary Landscape Studies: Potential and Limitations*, Wageningen, DELTA Series 2, pp. 129–135.

Fischer, G. and Derry, S.: 2005, *Toward a Model and Theory for Transdisciplinary Graduate Education*, Paper presented at 2005 AERA Annual Meeting, Retrieved August 3, 2006, from http://l3d.cs.colorado.edu/~gerhard/papers/.

Förster, R.: 2000, Criteria for Training in Transdisciplinary Practice. In: R. Haeberli, R.W. Scholz, A. Bill and M. Welti (eds.), *Transdisciplinarity: Joint Problem Solving A Science, Technology, and Society. Workbook I: Dialogue Sessions and Idea Market*, Haffmanns Sachbuch Verlag AG, Zürich, pp. 93–97.

Frey, G.: 2003, Training Needs for Interdisciplinary Research. In: B. Tress, G. Tress, A. van der Valk and G. Fry (eds.), *Interdisciplinary and Transdisciplinary Landscape Studies: Potential and Limitations*, Wageningen, DELTA Series 2, pp. 118–123.

Jantsch, E.: 1972, Towards Interdisciplinarity and Transdisciplinarity in Education and Innovation. In: *Interdisciplinarity: Problems of Teaching and Research in Universities*, OECD, Paris, pp. 97–121.

Klein, J.T.: 2002, Introduction: Interdisciplinarity Today: Who? What? and How?. In: J.T. Klein, *Interdisciplinary Education in K-12 and College: A Foundation for K-16 Dialogue*, The College Board, New York, pp. 1–17.

Kockelmans, J.: 1979, Why Interdisciplinarity? In. J. Kockelmans, *Interdisciplinarity and Higher Education*, University Park, Pennsylvania State University Press, pp. 122–160.

Lawrence, R. and Després, C.: 2004, Introduction: Futures of Transdisciplinarity, *Futures* 36, 403–404.

McMurtry, R.: 2000, Reflections on Transdisciplinarity. In: M. Somerville and D. Rapport, *Transdisciplinarity: Recreating Integrated Knowledge*, EOLSS, Oxford, UK, pp. 179–184.

Piaget, J.: 1972, The Epistemology of Interdisciplinary Relationships. In: *Interdisciplinarity: Problems of Teaching and Research in Universities*, OECD, Paris, pp. 127–139.

Reichel, F.: 2001, Facilitated Open Distance Learning for Continuing Engineering Education. In: R.W. Scholz, R. Haeberli, A. Bill and M. Welti (eds.), *Transdisciplinarity: Joint Problem-Solving Science, Technology and Society, Workbook II: Mutual Learning Sessions*, Haffmans Sachbuch Verlag AG, Zürich, pp. 138–142.

Rosenfield, P.L.: 1992, The Potential of Transdisciplinary Research for Sustaining and Extending Linkages Between the Health and Social Sciences, *Soc Sci Med* 35(11), 1343–1357.

# Chapter 27
# Integration

**Christian Pohl, Lorrae van Kerkhoff, Gertrude Hirsch Hadorn and Gabriele Bammer**

**Abstract** Integration is a core feature of transdisciplinary research. The challenge we address is that there has been no systematic development of integration concepts and methods. Historically, crossing boundaries between disciplines and between research and practice became a particular feature of 20th century research. Three cognitive objectives influenced this development – (1) the ideal of a unity of all sciences and other disciplines, (2) solving problems in basic research by innovation, and (3) responding to the knowledge demands of the knowledge society.

We present key entry points for a more systematic discussion of integration methods. We develop a matrix with three basic types of collaboration – common group learning, deliberation among experts, and integration by a subgroup or individual – and four methods of integration – mutual understanding, theoretical concepts, models and products. After focusing 'inwards' on research methodology, we look 'outwards' at institutional support and constraints, and explore integration within and beyond science and integration across activities and across structures. Finally we explore the relationship between funding, capacity and demonstrated success, and argue that we are at a cross-roads which will determine whether this relationship becomes a vicious or virtuous cycle. We suggest that promoting a virtuous cycle requires the development of integrative methodology and a college of peers and outline steps towards this.

**Keywords:** Evaluation · College of peers · Institutional constraints and opportunities · Forms of collaboration · Means of integration

✉ C. Pohl
td-net, Swiss Academies of Arts and Sciences, Bern, Switzerland; Department of Environmental Sciences, ETH Zurich, Zurich, Switzerland
e-mail: pohl@scnat.ch

## 27.1 Introduction

In transdisciplinary research, integration is important throughout the research process: in problem identification and structuring, in problem analysis as well as in implementation. When taking plural relevant perspectives in the knowledge society on complex practical issues into account, linking and restructuring concepts, methods and results from heterogeneous bodies of knowledge becomes inevitable. Achieving this requires collaboration within a mixed team of people with different experiences and from varied institutional contexts: in science and other disciplines, public bodies, business and civil society. Integration is not a value in itself, but a necessary methodology for researchers who try to respond adequately to societal knowledge demands (Chapter 1; Bammer 2005; Pohl and Hirsch Hadorn, 2007).

The integration challenge is that, to date, relevant concepts and methods have been developed on an ad hoc basis. While excellent advances have been made, there have been few attempts to synthesise what we know into a coherent package. This makes it difficult to both transmit and build on available methodologies between projects and between research groups or to start thinking about developing rigorous high quality curricula. In this chapter we start by identifying lessons from history regarding what can be learnt from experiences with integration in basic and applied research. Then we address the 'who and how' of integration in transdisciplinary research by describing forms of collaboration and various means of integration. Many of these have been used in the research projects that are described in the Handbook. We then examine institutional constraints and opportunities. Transdisciplinary research needs both special skills and favorable institutions to enable, support and value efforts in integration. We describe how these two kinds of requirements can reinforce each other in a virtuous cycle. We conclude with an outline of high priority strategies for improving integration in transdisciplinary research.

## 27.2 Lessons from History and the Current State of Transdisciplinary Research

During the 20th century, crossing boundaries between academic disciplines in the sciences and humanities became a popular idea in research, higher education and related institutions. This was accompanied by a coining of terms such as interdisciplinarity, crossdisciplinarity, transdisciplinarity, holism, scientific integration, unity of knowledge, boundary crossing, mutual learning, and scientific nomadism. Those crossing disciplinary boundaries soon realise that this is not easy (Klein, 1990; Weingart and Stehr, 2000). Members of a discipline build a scientific community or college of peers by sharing cognitive and material elements such as concepts, theories and methods, which are elaborated and developed through research and handed over to the next generation through education. Disciplines have sophisticated means of communication. Institutional features, such as standard forms of publication, specialised curricula, as well as professional roles in academia and society, control membership and the development of disciplinary fields.

The cognitive and institutional functions of disciplines stabilise disciplinary differentiation as the basic unit of structure formation in academia (Stichweh, 2003). Kuhn (1962), in his developmental model of the natural sciences, points to the strong interactions among social and cognitive factors, which together constitute what he calls the 'paradigm' of any scientific community. If research does not match with the cognitive features of a paradigm – which includes concepts, theories and methods, standards and shared examples of good problems and their solutions – peers are unable to recognise and value the respective results. As a consequence, integrative research efforts are often contested as unsound or simply speculative, while researchers are likely to be devalued within the traditional scientific community. To gain acceptance requires the transformation and restructuring of the paradigm and of the scientific community, a process that Kuhn calls 'scientific revolution'. In a scientific revolution, an emerging community develops new common ground to determine meaningful research questions and accepted methods for the justification of results, together with the definition of membership and rules within the community. Within the Kuhnian picture, successful integration of heterogeneous cognitive elements into a new paradigm is closely linked with social transformations of the scientific community. Interdisciplinary research studies, starting in the 1980s (Chubin et al., 1979; Bechtel, 1986), fit nicely into this picture.

Three important cognitive objectives for crossing disciplinary boundaries that have influenced the development of the sciences in the 20th century are (1) the ideal of a unity of all sciences and other disciplines, (2) solving problems in basic research by innovation, and (3) responding to the knowledge demands of the knowledge society. These objectives can be combined and they may cross-fertilise each other. For instance, the idea that all sciences could in principle be translated and thus reduced into a physical super-theory was put forward by the logical positivists in the first half of the 20th century. This attracted eminent physicists to the field of biology, which gave rise to the development of molecular biology and to theoretical conceptions of ecosystem development. This idea has been revived by Wilson (1998). Another means of unification is the use of formal abstraction to develop structural conceptions. This can be applied in a variety of fields. One example is systems theory and analysis, which has stimulated innovations in various disciplines and is of major importance for structuring complex societal issues in transdisciplinary research (see Chapter 1). But in searching for innovation in basic research or responding to issues in society, unitary conceptions are not always the best way to go. The most promising kind of integrative approach depends on the specific questions to be answered – for instance, the ways different research approaches can be merged or innovative methods and concepts developed.

A paradigm is critical for integration to improve its scientific profile (Rosenblum, 1997). Paradigms do not simply emerge by themselves, but through systematisation and critical investigation of research practices. Integrative research also requires social and institutional factors such as available resources, opportunities for advancement in career or recognition. This prompted the United States National Academies, comprising the National Academy of Sciences, National Academy of Engineering and Institute of Medicine in their joint publication 'Facilitating Interdisciplinary Research' (National Academies, 2005) to describe and analyse in more detail funding structures, evaluation procedures and criteria, institutional settings, organisation and management, education, publication strategies and media in order to come up with recommendations for science policy.

Encouragingly, research practices that embody the idea of transdisciplinarity are plentiful, as the chapters in this Handbook show. As noted earlier, the knowledge society demands integration of knowledge as a means to improve societal problem-solving. A starting point for transdisciplinary research is given when knowledge about a socially relevant problem field is uncertain, when the concrete nature of a problem is disputed, and when there is a great deal at stake for those concerned by a problem and involved in investigating it. As a consequence, identifying and structuring problems and the corresponding research questions has to: (1) draw upon bodies of knowledge about the problem field from a variety of academic disciplines as well as from relevant sectors and stakeholders outside academia and (2) relate data, hypotheses, conceptions and interpretations across those domains.

In the Introduction (Chapter 1) it is argued that when the goal of transdisciplinary research is to respond to societal knowledge demands, three kinds of knowledge are required: knowledge about the origins and development of problems, including their interpretation in the 'life-world' (systems knowledge); knowledge about needs for change, desired goals and better ways of acting (target knowledge); and knowledge about technical, social, legal, cultural and other means of transforming existing ways of acting in desired directions (transformation knowledge). Meeting these knowledge demands requires grasping the relevant complexity of the problems, taking into account the diversity of scientific and societal views of the problems, linking abstract scientific reflection with relevant case-specific knowledge, and constituting knowledge with a focus on problem solving for what is perceived to be the common good (Pohl and Hirsch Hadorn, 2007).

It is a core task of transdisciplinary research to integrate the diverse scientific and societal views of the problems recursively during the research process. The first step for such integration is to acknowledge, to respect and to explore the diversity of perspectives. As Loibl (2006) states, this diversity is not a handicap to be overcome, but an invitation for creative interaction: '[T]eam performance is not necessarily combined with a high degree of accordance amongst team members but [. . .] it rather seems closely connected to the conscious handling of team heterogeneity and to a very open and down-to-earth analysis of divergences'. Based on such a positioning, the perspectives can begin to interact in a second step, which Bechtel (1986) describes as: 'to press each [approach] against the other, in the hope that the dialectical interaction might advance the understanding in both enterprises'. Giri (2002) advances this further by claiming that transdisciplinary research is what evolves within these interactions: 'In transdisciplinary striving, relationship rather than our separate disciplinary Being is the ground of our identity'. There are, however, various approaches and practices for achieving such integration in transdisciplinary research. We propose a typology in the next section.

## 27.3 Conceptual and Practical Challenges of Integration in Transdisciplinary Research

Researchers have developed a range of approaches to integration, depending on their perspectives and backgrounds, on the problem they address and on the particular context within which a project evolves. Yet across this diversity we can start to identify a number of forms of collaboration and means of integration that appear more or less consistently in transdisciplinary research.

We suggest that there are three basic ways a transdisciplinary research team may organise its collaboration in order to reach integration: common group learning, deliberation among experts, and integration by a subgroup or individual (Rossini and Porter, 1979). Common group learning means that integration takes place as a learning process of the whole group. Several case studies in the Handbook describe such a learning process among researchers and social stakeholders, including group model-building of solid waste management (Chapter 13), collaborative learning on forest and wildlife management (Chapter 20) and 'making sense together' in order to retrofit Québec's post-war suburbs (Chapter 21).

In deliberation among experts, team members with relevant expertise in the components of the problem amalgamate their views. Integration takes place during one or more rounds of exchange among the experts. The notion of expert is not restricted to academics but also includes experts in the 'life-world'. The case study that documents the development of one medicine for nomadic pastoralists and their animals (Chapter 17) is an example of such a deliberation among different disciplinary, nomadic and government experts.

In the third form of collaboration a specific sub-group or individual undertakes the integration. This is the case in the 'Fischnetz' example (Chapter 8) where the project leader was responsible for integration, and in the project-based teaching study (Chapter 14), where a particular sub-group took care of integration.

These three forms of collaboration are 'ideal' types and simplifications. In practice it may be more difficult to distinguish them. Further, different phases of the research process may ask for different forms of collaboration. However, they offer a starting point from which we can consider different approaches to collaboration in transdisciplinary research.

Moving on now to means of integration, we identify four primary classes of 'tools': mutual understanding, theoretical concepts, models and products. An ubiquitous complaint of those engaged in integration is 'not speaking the same language'. Problems of mutual understanding arise when someone does not know the meaning of other disciplines' terms, when someone is not aware how a word changes meaning depending on the context, or when someone is not familiar with specialised scientific definitions. If these problems are seen as problems of communication, they can be effectively addressed by making explicit what important terms mean (Chapter 15). Thus the first group of tools for integration facilitate mutual understanding through effective communication. One way to do this is by deliberately using everyday language and avoiding scientific terms. The transdisciplinary research programme 'Austrian Landscape Research' (http://www.klf.at/)

expected participants to use everyday language in collaboration and publications and instructed researchers in its proper use (Nicolini, 2001). In addition, everyday language is important for integrating stakeholders into the research process as participants, a feature of most transdisciplinary projects. However, everyday language is not a universal panacea, as it is also ambiguous and contextualised. Hence other methods have also been developed, including formal and informal glossaries to enhance mutual understanding by explicitly fixing the meaning of key terms.

Often, the challenges in integration are about creating or restructuring the meaning of theoretical and conceptual terms to capture what is regarded as relevant in problem identification and framing. Therefore, a second group of integration 'tools' comprises theoretical notions, which can be developed by (1) transferring concepts between fields, (2) mutually adapting disciplinary concepts and their operationalisation to relate them to each other, or (3) creating new joint bridge concepts that merge disciplinary perspectives. Integration of perspectives by transferring concepts from one perspective into another is what the syndrome approach is based on, within which environmental problems all over the world are analysed, classified and treated as if they were medical diseases that can be characterised and recognised by a particular constellation of symptoms (Lüdeke et al., 2004). Simoni et al. (Chapter 16) describe the mutual interconnection of concepts of science, law, psychology and sociology in the common definition of research questions as well as their operationalisation in problem analysis to arrive at an integrated understanding of the welfare of children of divorced parents. Examples of bridging concepts are provided in this Handbook, including 'three dimensional mobility' (Chapter 6), 'the Netzstadt' (Chapter 5) or the 'new objects of research' (Chapter 7).

A third means of integration are models, which can be situated on a continuum between purely quantitative (mathematical) to purely qualitative (descriptive). (Semi-) Qualitative system dynamics models are often developed in a collaborative learning process among researchers and other stakeholders, aiming at a shared understanding of the system, its elements and their interactions (e.g. Chapters 13, 14 and 20). Quantitative models are often used by researchers to integrate knowledge from different disciplines, e.g. by integrating a climate model with an economic model. Integration then involves connecting the quantitative descriptions of natural or social systems in order to learn about how they influence each other, as in integrated assessment (Chapters 11 and 12) or in biopsychosocial modelling (Chapter 19). The models are also used to project scenarios of the future in order to support decision-making (Chapters 3 and 14).

The fourth means of integration are products. Such products can be medical treatments (Chapter 17), a development plan (Chapter 4), technical devices, regulations (Chapter 10), exhibitions, and so on. In this case it is the intended result that defines the component sub-parts and that stimulates integration. When the primary purpose of the 'product' is not an end in itself, but to join the diverse interests of those involved, one may also use the term 'boundary objects' (Star and Griesemer, 1989) to describe this particular approach to integration.

**Table 27.1** Forms of collaboration and means of integration (based on Pohl and Hirsch Hadorn, 2007)

| Means of integration | Forms of collaboration | | |
|---|---|---|---|
| | Common group learning | Deliberation among experts | Integration by a subgroup or individual |
| Mutual understanding (Everyday language, glossary . . .) | | | |
| Theoretical concept (Transfer of concepts, mutual adaptation of concepts, bridge concepts . . .) | | | |
| Model (Qualitative model, quantitative model, scenarios . . .) | | | |
| Product (Technical device, database, regulation, exhibition . . .) | | | |

The four means of integration are again 'ideal' types and simplifications that present each of these means as optional tools for the purposes of clarity and comparison. In practice any mix of means may be used, since, for example, a model, a bridge concept and a product can be combined. In addition, efforts to ensure mutual understanding through effective communication will probably be involved in any attempt at integration.

Table 27.1 combines the three forms of collaboration and four means of integration in a matrix that demonstrates twelve primary ways of integrating. While the project design for some research might fit neatly into one cell, much of the time the skilful handling of integration will require that researchers change and adapt the forms of collaboration and the means of integration during the research process.

## 27.4 Institutional Challenges of Integration

In the previous section we focused on forms of collaboration and means of transdisciplinary integration that are founded on a specific organisational unit, such as a project, a programme or a research centre. This perspective looks 'inwards' towards ways of thinking that researchers involved in transdisciplinary integration can use to structure their research. Yet it is also useful to look 'outwards' from the project, to consider the institutions that support or inhibit integration. We use the term institution here to describe the various sets of rules and conventions, both formal and informal, that structure the ways in which research is funded, organised, conducted and evaluated. These institutions cannot be controlled by individual researchers in the same way a research design or methodology can be, but institutional arrangements can help or hinder the practice of integration by expanding or limiting the range of choices available to researchers.

Institutional factors, such as funding rules, evaluation criteria and organisational structures that favor disciplines are often cited as barriers to transdisciplinarity. In this section we focus on academia as the main institution of interest, but features of other institutions (such as government, business or civil society) can also affect the practice of transdisciplinary research. Indeed, despite the innovation and successes of many transdisciplinary projects, it nonetheless remains the case that these successes have often emerged from a struggle with formal and informal institutions that remain rooted in notions of disciplines, and a tension between the role of research in producing knowledge and the role of research in producing action and transformation. These issues emerge in many of the research projects reported in this volume. Yet reflections on how researchers can better capitalise on existing institutional opportunities are far less common. In this section we will briefly explore some of the institutional factors that researchers commonly identify as supporting or constraining integration in transdisciplinary research. We then ask: what are some of the connections between the 'inward' and 'outward' perspectives that can help us develop institutions that support integration in transdisciplinary research?

Institutional factors that attempt to support integration can be roughly categorised according to whether they focus on integration within or beyond the scientific setting; or whether they focus on research structures or activities (Figure 27.1; van Kerkhoff, 2005a). This can be a useful framework for considering institutional constraints on, and opportunities affecting, the ways in which learning proceeds.

*Institutional connections across activities, within academia.* Cross- or multi-campus collaborations in research or teaching are common vehicles for transdisciplinary integration. In research in particular, participants in these typically large, complex projects, often describe inflexible budgeting and administrative processes, and conventional approaches to research evaluation that do not 'count'

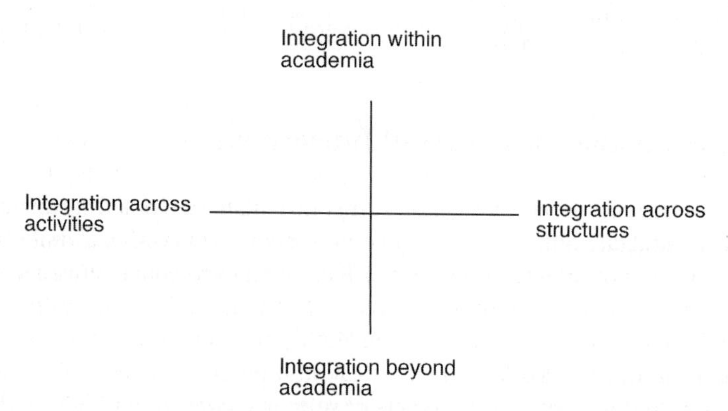


**Fig. 27.1** Four approaches to transdisciplinary integration (adapted from van Kerkhoff, 2005a)

less conventional research products, as institutional barriers. Bergmann and Jahn (Chapter 6) discuss the compromise of publishing under a group name, where all members of the group were included, but no individual could gain professional recognition for the work. A fair authorship and citation system that is able to cope with several authors of equal importance cannot be expressed in the conventional way authors are listed and papers are cited in science: this is one of the tasks to be addressed. In the meantime other criteria may be used, which have nothing to do with the paper's content. In our case the authors, who all contributed equally to this chapter, are listed in order of descending height. Short time frames for academic evaluation and shortage of academics with the capacity to evaluate transdisciplinary projects were also cited as barriers (Chapter 5). Some of these, such as administration and attributing authorship are common to any large, complex project, whether transdisciplinary or not. Inability to gain recognition for non-academic impacts such as policy change, and lack of qualified evaluators are core concerns for transdisciplinarity; they direct attention to research evaluation processes that do not take into account some of the main aims of transdisciplinary research.

*Institutional connections across structures, within academia.* New organisational structures, including cross- or multi-campus initiatives or Centers of Excellence, can often support transdisciplinary research more readily than discipline based departments. As Kiteme and Wiesmann (Chapter 4) have noted, such structures can also become a training ground for researchers and other staff in skills required to lead transdisciplinary projects, such as negotiation and facilitation. Yet these efforts tend to be ad hoc, bound to projects rather than to a broader community of transdisciplinary scholars. As mentioned earlier, transdisciplinary researchers do not have a college of peers that, in conventional disciplines, generates robust evaluation criteria, develops training curricula and fosters career paths. This is less a problem of the existing institutions than it is a problem of the currently scattered and dispersed transdisciplinary research community. Research funding programmes that support and explicitly demand integrated approaches, such as the European Commission's 6th Framework Program (Directorate General for Research, 2002), are central to encouraging researchers and research institutions to support training and career development in transdisciplinary research, but do not necessarily foster the development of an organised college of peers.

*Institutional connections within activities, beyond academia.* Research activities that extend beyond academia include projects or programmes where practitioners are involved and join with researchers to establish research questions, contribute to data, and/or interpret findings. Several authors pointed to the benefits of flexible project structures or management (e.g. Chapters 4 and 25), for accommodating the complexities that emerge from interactions and negotiations within transdisciplinary projects. Evaluating transdisciplinary projects that aim to generate real-world outcomes as well as scientific findings is also difficult, as successes in policy change, conflict solving or mutual learning (Chapter 20) are usually not acknowledged in academic institutions. This leads to a widespread perception that transdisciplinary integration is time-consuming and inefficient in comparison with disciplinary research (Chapter 14). A recent report to the US National Academy of Sciences on successful efforts to link research and practice (Clark and Holliday, 2006) highlighted the importance of practitioner-driven dialogue and problem-setting, but also pointed to the constraints posed by risk-averse political or funding environments, which are exacerbated by inconsistent (or sometimes non-existent) approaches to comprehensive evaluation.

*Institutional connections across structures, beyond academia.* It is widely held that conventional research funding sources are reluctant to support transdisciplinary integrative research. Yet recent years have seen the emergence of alternative funding sources that prioritise transdisciplinarity. In addition, co-funding models offer a useful middle road for integrative projects that seek to meet both scientific and active transformation goals (see Schelling's description of the co-funded National Centre of Competence in Research North–South venture, Chapter 17) and are offering new opportunities to researchers prepared to engage meaningfully with practitioners. In Australia, the Cooperative Research Centres Program has been a good model for offering incentives for researchers and practitioners to work together, with co-funding and co-governance requirements encouraging active partnerships (van Kerkhoff, 2005b). In Germany, the programme 'Social Ecological Research' (http://www.sozial-oekologische-forschung.org/en/index.php) of the Federal Ministry of Education and Research (BMBF) is a framework that encourages transdisciplinary research for sustainable development. Internationally, the United Nations Environment Program has proposed a multi-faceted approach to integrating global environmental monitoring activities that includes a conceptual framework, information network, capacity building programme, toolbox, and an 'assessment compact' connecting many environmental assessment bodies around the world in a commitment to institutional and process-based harmonisation (UNEP, 2006). Institutional arrangements that facilitate such models are an essential part of a transdisciplinary research infrastructure. Yet these models are generally not described or evaluated as transdisciplinary research. Dual accountability to scientific and practice organisations has been identified by others as an important element of institutions that support effective 'boundary crossing' between science and practice (Guston, 2001). Quality assessment and evaluation procedures that can encompass dual, and possibly multiple, goals are needed, but seldom forthcoming.

Paying close attention to the ways in which our institutions support or hinder integration is a key part of the future of transdisciplinary research. Yet given the early stage of development of transdisciplinarity as a field of research, it is often not clear whether these barriers are solely the result of clashes between an academic system founded on disciplines and the integrative underpinnings of transdisciplinary research, or are side-effects of the newness and fragmentation of transdisciplinarity itself. As yet there is no recognised college of peers, which is fundamental to developing shared methodologies, appropriate academic evaluation frameworks, training and career development. This affects not only individual projects or programmes, but also new institutional experiments like Cooperative Research Centres that are rarely evaluated rigorously according to appropriate transdisciplinary criteria. As a result, we are not learning as much as we could from institutional experiments, weakening arguments for institutional reform. Those seeking to support transdisciplinary research often have difficulty in identifying researchers with the skills needed to develop, run or assess transdisciplinary projects. Improving the ways in which transdisciplinary researchers participate within existing institutional structures needs to be examined alongside questions of institutional change. We turn to this in our concluding section.

## 27.5 Conclusion: Advancing Transdisciplinary Research Integration

Our chapter leads to three conclusions:

1. Integration is the core methodology underpinning the transdisciplinary research process;
2. Insufficient attention has been paid to learning from new institutional arrangements designed to promote transdisciplinarity or to exploiting the advantages of existing institutions and
3. For transdisciplinary research to progress, advances must be made in strengthening both integration methodology and institutional support. One cannot effectively move forward without the other.

Here we build on the last of these points. For transdisciplinary research to progress, three essential elements must be addressed: funding, capacity and demonstrated success. Funding, an institutional factor, is relatively straightforward, as funding bodies are increasingly recognising the need for, as well as giving priority to, transdisciplinary research, although a specific funding procedure for the phase of problem identification and structuring should be developed (Pohl and Hirsch Hadorn, 2006). However, the capacity to undertake and to evaluate transdisciplinary research is much weaker. Each of these relies on a combination of institutional and methodological factors. Thus capacity requires institutionalised courses plus coherent course content, while evaluation requires accepted processes plus agreed criteria. Despite growing experience with integration in transdisciplinary research, the integrative core is still poorly understood, making it hard to train researchers in integration and to evaluate its effectiveness. Further, researchers with skills in transdisciplinary integration are not organised into an effective college of peers that would (a) foster the development of integration methodology, and (b) critically evaluate each others' work. The limited understanding about integration, combined with the lack of a strong peer group make demonstrating success difficult. Further, it means that the available funding is not exploited to the maximum. Indeed it is possible to think of a self-reinforcing cycle, demonstrated in Fig. 27.2.

The interaction of these three elements can form either a vicious or a virtuous cycle, with the interactions able to go in either direction around the circle. In a vicious cycle, lack of capacity to do quality transdisciplinary research feeds funder disenchantment, which in turn reduces the resources available to develop or refine methodologies, reinforcing the sense that transdisciplinary research is inefficient

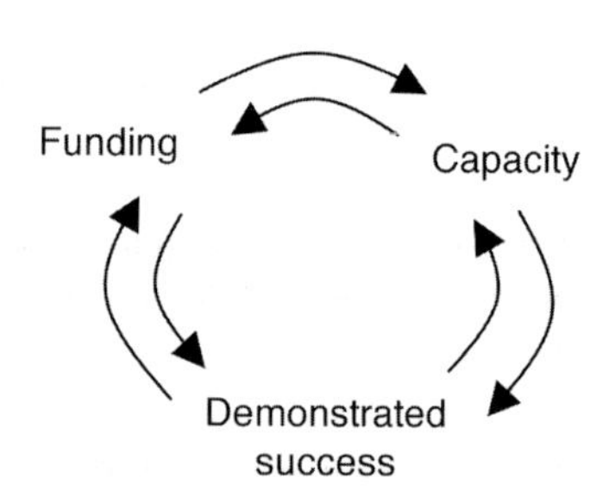


**Fig. 27.2** Self-reinforcing relationships of funding, capacity and the ability to demonstrate success

and a poor investment. Alternatively, where clear criteria for success are established through rigorous methodological advances, funding becomes easier to justify, and more recruits are attracted to a new, well-resourced field. We suggest that transdisciplinary research is at a critical point with interest and support from funding groups that is yet to be translated into real methodological or capacity development.

This Handbook reflects the current state of transdisciplinary research and provides a solid foundation to build on. We suggest that to promote a virtuous cycle the essential next steps are to develop integrative methodology and to form the college of peers. Developing integrative methodology requires an agreed way of describing it, which we deal with next.

Each discipline has standard ways of describing its methods, which highlight critical aspects of what was undertaken, as well as providing accepted shorthand for a range of more straightforward facets. At this stage there are no agreed guidelines for describing integration in transdisciplinary research, making it difficult to assess and compare studies.

A standard framework covering six questions, developed in an Australian natural resource management integration symposium (Bammer and LWA Integration Symposium Participants, 2005) could be adopted to fill this role. Descriptions of integration in transdisciplinary research would address:

1. What was the transdisciplinary integration aiming to achieve and who was intended to benefit?
2. What were the elements (e.g. discipline and practice perspectives) that were being integrated in a transdisciplinary manner?
3. Who was doing the transdisciplinary integration?
4. How was the transdisciplinary integration being undertaken?
5. What was the context for this transdisciplinary work, which might have affected any of the other elements (e.g. the aims, methods, impact)?
6. How was impact measured?

Deliberate attention to such questions not only makes it clear how the transdisciplinary research was conducted, but can also help the researchers reflect on what they are doing and pinpoint areas for improvement. Earlier in this chapter we started the process of developing categories for describing the two central elements: forms of collaboration (question 3) and means of integration (question 4). We quickly realised that further progress in understanding these key facets is hindered because different studies, for example those described in this volume, lack a consistent descriptive method.

Finally, as we outlined earlier, another existing institutional arrangement that transdisciplinary research could copy is to develop a college of peers, expert in integration methodologies and experienced in applying them to transdisciplinary projects. Essentially this is how existing disciplines work and the college of peers provides the core mechanism for evaluation, improving quality, development and rapid dissemination of new theory and methods, as well as building capacity and developing the careers of new researchers. These may then provide the triggers for more radical institutional experimentation, evaluation and, potentially, change (Kuhn, 1962; Rosenblum, 1997).

As this Handbook demonstrates, there is no shortage of people with the experience and ability to make up such a college, but they are dispersed across different sectors and contexts. We suggest that organised networks are an effective way to begin to build such a college and have developed two networks with this end in mind.

The first, td-net, the network for transdisciplinarity in sciences and humanities (www.transdisciplinarity.ch), is sponsored by the Swiss Academies of Arts and Sciences. This network aims to advance transdisciplinary research in a variety of areas, starting with experiences gained in the environmental and sustainability sciences. It has a comprehensive bibliography, is sponsoring the publication of this Handbook, and supports a range of projects, including the development of guidelines for transdisciplinary research.

The second, complementary, network, is the Integration and Implementation Sciences Network (www.anu.edu.au/iisn), hosted by The Australian National University. It aims to develop a new specialisation for researchers focusing on integrative skills and provides a list of researchers, organisations, professional associations and journals, as well as sponsoring projects.

We invite you to join us!

## References


Bammer, G.: 2005, Integration and Implementations Sciences: Building a New Specialization, *Ecol Soc* 10 (2): 6 [online].

Bammer, G. and LWA Integration Symposium Participants.: 2005, Guiding Principles for Integration in Natural Resource Management (NRM) as a Contibution to Sustainability. In: G. Bammer, A. Curtis, C. Mobbs, R. Lane and S. Dovers (eds), Australian Case Studies of Integration in Natural Resource Management (NRM), *Supplementary Issue of the Australasian Journal of Environmental Management* 12, 5–7.

Bechtel, W. (ed.).: 1986, *Integrating Scientific Disciplines*, Martinus Nijhoff Publishers, Dordrecht.

Chubin, D.E., Rossini, F.A., Porter, A.L. and Mitroff, I.I.: 1979, Experimental Technology Assessment: Explorations in Process of Interdisciplinary Team Research, *Technol Forecast Soc Change* 15, 87–94.

Clark W. and Holliday L.: 2006, *Linking Knowledge with Action for Sustainable Development: The Role of Program Management – Summary of a Workshop*, The National Academies Press, Washington.

Directorate General for Research.: 2002, *Provisions for Implementing Integrated Projects: Background Document*, European Commission.

Giri, A.K.: 2002, The Calling of a Creative Transdisciplinarity, *Futures* 34, 103–115.
Guston, D.H.: 2001, Boundary Organizations in Environmental Policy and Science: An Introduction, *Sci Tech Hum Val* 26(4), 399–408.
Klein, J.T.: 1990, *Interdisciplinarity. History, Theory & Practice*, Wayne State University Press, Detroit.
Kuhn, T.S.: 1962, *The Structure of Scientific Revolutions*, University of Chicago Press, Chicago.
Loibl, M.C.: 2006, Integrating Perspectives in the Practice of Transdisciplinary Research. In: J.-P. Voß, D. Bauknecht and R. Kemp (eds), *Reflexive Governance for Sustainable Development*, Edward Elgar, Cheltenham, pp. 294–309.
Lüdeke, M.K.B., Petschel-Held, G. and Schellnhuber, H.-J.: 2004, Syndroms of Global Change. The First Panoramic View, *GAIA* 13(1), 42–49.
National Academies (eds).: 2005, *Facilitating Interdisciplinary Research*, The National Academies Press, Washington.
Nicolini, M.: 2001, *Sprache – Wissenschaft – Wirklichkeit. Zum Sprachgebrauch in inter- und transdisziplinärer Forschung*, Bundesministerium für Bildung, Wissenschaft und Kultur, Wien.
Pohl, C. and Hirsch Hadorn, G.: 2006, Die Gestaltungsprinzipien für transdisziplinäre Forschung des td-net und ihre Bedeutung für die Evaluation. In: S. Stoll-Kleemann and C. Pohl (eds), *Evaluation inter- und transdisziplinärer Forschung: Humanökologie und Nachhaltigkeitsforschung auf dem Prüfstand*, oekom, München, pp. 165–138.
Pohl, C. and Hirsch Hadorn, G.: 2007, *Principles for Designing Transdisciplinary Research – Proposed by the Swiss Academies of Arts and Sciences*, oekom, München.
Rosenblum, D.: 1997, In the Absence of a Paradigm: The Construction of Interdisciplinary Research, *Issues Integr Stud* 15, 113–123.
Rossini, F.A. and Porter, A.L.: 1979, Frameworks for Integrating Disciplinary Research, *Res Pol* 8, 70–79.
Star, S.L. and Griesemer, J.R.: 1989, Institutional Ecology, 'Translations' and Boundary Objects: Amateurs and Professionals in Berkeley's Museum of Vertebrate Zoology, 1907–39, *Soc Stud Sci* 19, 387–420.
Stichweh, R.: 2003, Differentiation of Scientific Disciplines: Causes and Consequences. In: G. Hirsch Hadorn (ed.) Unity of Knowledge (in Transdisciplinary Research for Sustainability), *Encyclopedia of Life Support Systems (EOLSS)*, Developed under the Auspices of the UNESCO, Eolss Publishers, Oxford, UK, Retrieved July 31, 2006, from http://www.eolss.net.
UNEP.: 2006, *Towards a UNEP Environment Watch System*, Retrieved June 24, 2006, from http://science.unep.org/Environment_Watch_Documents/ConceptPaper_en.pdf
van Kerkhoff, L.: 2005a, Integrated Research: Concepts of Connection in Environmental Science and Policy, *Environ Sci Pol* 8, 452–463.
van Kerkhoff, L.: 2005b, Strategic Integration: The Practical Politics of Integrated Research in Context, *J Res Pract* 1(2), Article M5.
Weingart, P. and Stehr, N. (eds).: 2000, *Practising Interdisciplinarity*, University of Toronto, Toronto.
Wilson, E.O.: 1998, *Consilience. The Unity of Knowledge*, Abacus, London.

# Part VI
# Summary and Outlook

# Chapter 28
# Core Terms in Transdisciplinary Research

**Christian Pohl and Gertrude Hirsch Hadorn**

**Abstract** The following explanations of core terms in transdisciplinary research are meant to guide readers who are not familiar with transdisciplinary research or who are confused by the variety of meanings given to terms. It is important to keep in mind that the explanations refer to the use of terms in the context of transdisciplinary research – they do not provide general definitions. For instance the meaning of 'actor' as described below may not hold from a sociological or psychological perspective and is wrong in the context of a theatre. The explanations are taken from the 'Principles for Designing Transdisciplinary Research' (Pohl and Hirsch Hadorn, 2007) and from the Handbook (Chapter 2). Authors of the Handbook were invited to refer to a preliminary and shorter version of term descriptions when writing their contributions, but they were free to use the terms in their own way.

**Keywords:** Definition of core terms

*Actors*: Persons and institutions in public agencies, the private sector and civil society who are involved in one way or another in a problem field (see *problem field*). Their relation to the problem field is the reason why transdisciplinary researchers work with them. [. . .] Participatory research (see *participatory research*) goes beyond doing research on actors, and implies that actors can help shape the research process (see *research process*). (Pohl and Hirsch Hadorn, 2007)

*Applied research*: Problems (see *problems*) for applied research arise from difficulties in describing and explaining the variability of specific processes in a certain type of problem fields (see *problem field*), and from difficulties in developing specific measures helping an actor (see *actors*) to better achieve his/her goals. A discipline (see *scientific disciplines*) or an integrated grouping of disciplines which specialize in a certain problem field as applied research are building the scientific knowledge base for dealing with the diversity (see *diversity*) and complexity

✉ C. Pohl
td-net, Swiss Academies of Arts and Sciences, Berne, Switzerland; Department of Environmental Sciences, ETH Zurich, Zurich, Switzerland
e-mail: pohl@scnat.ch

(see *complexity*) of the selected processes in the problem field. The experience of the actor who requires systems (sec *systems knowledge*) as well as transformation knowledge (see *transformation knowledge*) to improve practice is providing the knowledge base in the life-world. (Chapter 2)

*Basic research*: Problems (see *problems*) for basic research arise from difficulties in describing and explaining a subject by general methods and models. Basic research aims at advancing the state of the art within a discipline (see *scientific disciplines*), which is the only knowledge base to identify problems and structure research questions (see *problem identification and structuring*). Basic research idealizes and reduces what is going on in real-world settings (see *life-world*) in order to formulate generally valid explanations. (Chapter 2)

*Bringing results to fruition*: According to the principle of recursiveness (see *recursiveness*), bringing results to fruition is a phase of research (see *transdisciplinary research process*) that does not occur at the end of transdisciplinary research (see *transdisciplinary research*): It takes place in the course of the research process in order to enable learning processes. Bringing results to fruition is achieved in the form of a real-world experiment (see *real-world experiment*), so that its impact can be observed and lessons can be learned for the following phase of bringing results to fruition. (Pohl and Hirsch Hadorn, 2007)

*Common good*: The common good as an ethical principle, however, refers to having the social systems, institutions, and environments on which we all depend work for the well-being of all people. (Pohl and Hirsch Hadorn, 2007) [. . .] By dealing explicitly with the question of whether proposed solutions serve the common good, transdisciplinary research (see *transdisciplinary research*) enables those involved to achieve a consensus about solutions – an important condition given the fact that actor (see *actors*) groups in the private sector, public agencies, and civil society can hold controversial positions. The question how to define the concept of the common good with regard to a specific problem field (see *problem field*) can be one of the research questions pursued by transdisciplinary research. (Pohl and Hirsch Hadorn, 2007)

*Complexity*: is used for the interrelations among heterogeneous dimensions, or plural values and norms. Thus complexity is in contrast to simplicity. (Chapter 2)

*Diversity*: means that empirical dimensions relevant to describing and analysing processes are heterogeneous in the sense that they belong to different disciplines (see *scientific disciplines*) or to the perceptions of different actors (see *actors*), and that there are plural values and norms that do not fit together in a systematic way. Diversity of and life-world (see *life-world*) perceptions involved. (Chapter 2)

*Interdisciplinary research*: refers to a form of coordinated and integration-oriented collaboration between researchers from different disciplines (see *scientific disciplines*). (Pohl and Hirsch Hadorn, 2007)

*Life-world*: refers to the human world prior to scientific knowledge. While philosophy (led by Edmund Husserl who coined the term) uses this concept within the framework of both phenomenology and constructivism as a possibility of critiquing and explaining science, Schütz's interpretive sociology links 'life-world' with the concept of the everyday world as a system of meaning: 'Life-world', for him, describes the structural properties of social reality as grasped by the agent. We use the term 'life-world' to mark the difference, within society, between the scientific and other communities (the private sector, public agencies, civil society). It was Mittelstrass (1992) who introduced the term 'life-world' into the definition of Transdisciplinarity (see *transdisciplinary research*). (Pohl and Hirsch Hadorn 2007)

*Multidisciplinary research*: approaches an issue from the perceptions of a range of disciplines (see *scientific disciplines*); but each discipline works in a self-contained manner with little cross-fertilisation among disciplines, or synergy in the outcomes. (Chapter 2)

*Participatory research*: goes beyond doing research on actors (see *actors*), and implies that actors can help shape the research process (see *transdisciplinary research process*). (Pohl and Hirsch Hadorn, 2007) [...] The aim of involving social groups is often primarily to integrate various life-world (see *life-world*) perspectives and interests into transdisciplinary research (see *transdisciplinary research*). But beyond this, participatory research is also a means of grasping the complexity (see *complexity*) of a problem (see *problems*) with the help of local knowledge, of testing the situational relevance and transferability of results, or of finding solutions for what is perceived to be the common good (see *common good*) an improving the practice-oriented effectiveness of results. (Pohl and Hirsch Hadorn, 2007)

*Problems*: are concrete, identified and structured questions within problem fields (see *problem fields*). Problems cannot be considered as given. Rather, in view of the initial random array of difficulties, it is important in the first phase of transdisciplinary research (see *transdisciplinary research process*) to determine what concrete problems there are and what they consist of. Research questions then specify these problems in such a way that they can be investigated and hopefully answered. (Pohl and Hirsch Hadorn, 2007)

*Problem analysis*: To analyse a problem (see *problems*), the problem statement is divided into sub-questions. These are dealt with and answered in relation to one another, after which the answers to the sub-questions go through a process of integration. Various forms of collaboration and modes of integration can be used to this purpose. As a principle of utmost importance, those involved must encounter one another openly before choosing the form of collaboration and means of integration. Moreover, the principle of recursiveness (see *recursiveness*) requires that decisions already taken can be reconsidered. The form of collaboration and the means of integration determine the structure and intensity of exchange between those involved. Intense exchange requires a deeper knowledge of one another's positions and a flexible attitude with regard to one's own position. (Pohl and Hirsch Hadorn, 2007)

*Problem field*: an area in which the need for knowledge related to empirical and practice-oriented questions arises within society due to an uncertain (see *uncertainties*) knowledge base and diffuse as well as controversial perceptions of problems (see *problems*). (Pohl and Hirsch Hadorn, 2007)

*Problem identification and structuring*: is the heart of transdisciplinary research (see *transdisciplinary research*). Complexity can be reduced by identifying those involved in relation to the requirements of transdisciplinary research (see *transdisciplinary research*) and by specifying the need for knowledge with respect to the three forms of knowledge. The decisions made on this basis may need to be modified in a recursive (see *recursiveness*) procedure. To create a bridge between science and practice, the problem (see *problems*) identified can be reformulated in relation to actors (see *actors*) in the life-world (see *life-world*). This is one of the areas where transdisciplinary research can develop creativity and originality, for example by finding a new way of perceiving things, which works like a bridge between fixed viewpoints. In this phase, the project must already be contextualised: first, by embedding it in science, which is achieved by referring to the state of the art in the relevant disciplines, and by learning from transdisciplinary research on similar problems; and second by embedding it in the life-world, which is achieved by considering existing needs, interests, technologies, regulations, practices and power relations that transdisciplinary research will have to take into account. (Pohl and Hirsch Hadorn, 2007)

*Real-world experiment*: is a recursive (see *recursiveness*) application of bringing results to fruition (see *bringing results to fruition*). . . . The effects of a project are observed, with a view to finding surprises (unexpected impacts). As a result, the assumptions, models and explanations developed in the project are revised in such a way that they can explain these surprises (increase of knowledge). New instances of bringing results to fruition are then planned and conducted etc. (Pohl and Hirsch Hadorn, 2007)

*Recursiveness*: is a general principle of transdisciplinary research (see *transdisciplinary research*). It points to the iterative procedures that characterises both the entire research process (see *transdisciplinary research process*) and its individual phases. This implies that the research process has to be shaped in such a way that concepts and methods can be repeatedly tested (see *real-world experiment*), and that underlying assumptions can be modified if they are found to be inadequate. (Pohl and Hirsch Hadorn, 2007) 'Recursiveness' means about the same as 'iterativeness': however, 'iterative' is defined by the *Oxford English Dictionary* as 'characterized by repeating or being repeated', while its denotation for 'recursive' – listed as first occurring in English in 1904 – is: 'involving or being a repeated procedure such that the required result at each step except the last is given in terms of the result(s) of the next step, until after a finite number of steps a terminus is reached with an outright evaluation of the result.' The difference is a fine one but it is relevant in the context of the present publication, where 'recursiveness' is used in the sense of this second OED definition. (Pohl and Hirsch Hadorn, 2007)

*Scientific disciplines*: shape scientific research by forming the primary institutional and cognitive units in academia, on which the internal differentiation of science into specialised curricula, professions and research, is based. Members of a discipline are specialists who build a scientific community. Members communicate within their community, share basic assumptions and examples about meaningful problems, standards for reliable and valid methods, as well as what is considered a good solution to a problem. What modern science gains and preserves is based to a large extend on disciplinary structures. However, boundaries between disciplines are changing: by increasing specialisation through internal differentiation within the disciplines, and by the integration of disciplines. (Chapter 2)

*Sustainable development*: is a global socio-political model for changing practices and institutions in order to achieve more equitable opportunities within and between generations while taking into account limitations imposed by the state of technology and social organization on the environment's ability to meet present and future needs. Promoting sustainable development therefore necessitates overcoming narrow preoccupations and compartmentalised concerns by involving people from civil society, the private sector and public agencies as actors in participatory deliberation and decision making. Thus sustainable development is a way to conceive the common good as the basic principle of public legislation in a complex world. Agenda 21, a program of the UN, is a comprehensive blueprint of action to be taken globally, nationally and locally. (Chapter 2)

*Systems knowledge*: addresses questions about the genesis and possible further development of a problem (see *problem*), and about interpretations of the problem in the life-world (see *life-world*). [. . .] Systems knowledge confronts the difficulty of how to deal with uncertainties (see *uncertainties*). These uncertainties are the result, on the one hand, of transferring abstract insights from a laboratory, a model or a theory to a concrete case underlying specific conditions. Furthermore, empirical or theoretical knowledge about a problem may be lacking, and depending on the interpretation of a problem, these uncertainties may be assigned different degrees of importance, which leads to diverging assessments of the need for action and of target knowledge (see *target knowledge*) and transformation knowledge (see *transformation knowledge*). (Pohl and Hirsch Hadorn, 2007)

*Target knowledge*: addresses questions related to determining and explaining the need for change, desired goals and better practices. [. . .] In the case of target knowledge, the question is what *the multiplicity of social goals* means for research, for society's practice-related problems, and for transdisciplinary collaboration between science and actors in the life-world (see *life-world*). Transdisciplinary research (see *transdisciplinary research*) faces the challenge of clarifying a variety of positions and prioritising them in the research process (see *transdisciplinary research process*) according to their significance for developing knowledge and practices that promote what is perceived to be the common good (see *common good*). This is necessary not only when the need for action has to be identified and objectives have to be determined, but also when describing the systems to which they refer and the possibilities of inducing change. (Pohl and Hirsch Hadorn, 2007)

*Transdisciplinary research process*: consists of three phases: (1) Problem identification and structuring (see *problem identification and structuring*), (2) Problem analysis (see *problem analysis*), and (3) Bringing results to fruition (see *bringing results to fruition*). The importance of each of the three phases must be taken into account when allocating time, finances and personnel. Transdisciplinary research (see *transdisciplinary research*) does not necessarily progress through the phases in the order mentioned above. (Pohl and Hirsch Hadorn, 2007)

*Transdisciplinary research*: is needed when knowledge about a societally relevant problem field is uncertain (see *uncertainties*), when the concrete nature of problems is disputed, and when there is a great deal at stake for those concerned by problems and involved in dealing with them. Transdisciplinary research deals with problem fields (see *problem field*) in such a way that it can: (a) grasp the complexity (see *complexity*) of problems, (b) take into account the diversity (see *diversity*) of life-world (see *life-world*) and scientific perceptions of problems, (c) link abstract and case-specific knowledge, and (d) develop knowledge and practices that promote what is perceived to be the common good (see *common good*). (Pohl and Hirsch Hadorn, 2007)

*Transformation knowledge*: addresses questions about technical, social, legal, cultural and other possible means of acting that aim to transform existing practices and introduce desired ones. [. . .] In the case of transformation knowledge (see *transformation knowledge*), *established technologies, regulations, practices and power relations* must be taken into account. This is the mere consequence of pragmatism, since options for change have to rely on existing infrastructure, on current laws, and to a certain degree on current power relations and cultural preferences, in order to have any chance at all of being effective. When these social, cultural and technological givens are not considered, this leads to the often criticised discrepancy between knowledge and practice. For transdisciplinary research (see *transdisciplinary research*), the challenge here is to learn how to make what is established more 'flexible'. (Pohl and Hirsch Hadorn, 2007)

*Uncertainties of knowledge*: Because of empirical diversity (see *diversity*) and complexity (see *complexity*), there is debate about which parameters are relevant, how they are connected in concrete processes, and what disciplines (see *scientific disciplines*) need to be involved. [. . .] Uncertainties exist regarding the description and explanation of the genesis and possible further development of such problem fields (see *problem field*). Disputes inevitably exist in the life-world (see *life-world*) regarding whether and how certain actors' (see *actors*) practices need to be changed, because the groups directly or indirectly involved have a variety of interests, most of which are often incompatible. (Pohl and Hirsch Hadorn, 2007)

## References

Pohl, C. and Hirsch Hadorn, G.: 2007, *Principles for Designing Transdisciplinary Research. Proposed by the Swiss Academies of Arts and Sciences*, oekom Verlag, München, 124pp.

# Chapter 29
# Enhancing Transdisciplinary Research: A Synthesis in Fifteen Propositions

**Urs Wiesmann, Susette Biber-Klemm, Walter Grossenbacher-Mansuy, Gertrude Hirsch Hadorn, Holger Hoffmann-Riem, Dominique Joye, Christian Pohl and Elisabeth Zemp**

**Abstract** The debate on transdisciplinarity is still fairly young and the process of transdisciplinary research is still being developed. This final chapter is an attempt to stimulate the debate on, and the development of, transdisciplinary research. With the 15 propositions, the editors of the Handbook take a position on the definition, scope and process of transdisciplinary research; then we give hints on how to deal with some of the most persistent stumbling blocks in transdisciplinary practice; and finally, we highlight the corner stones needed to face the scientific, the institutional and the societal challenge.

**Keywords:** Propositions · Key-features · Research practice · Stumbling blocks · Capacity building

## 29.1 Introduction

The case studies and discourses presented in this handbook illustrate the broad-ranging relevance and multi-facetted challenges of transdisciplinary research and provide a wealth of concrete hints and recommendations for transdisciplinary research.

In this chapter the Handbook's editorial committee intends to summarise key-features and means for enhancing transdisciplinary research and to point to some of the challenges ahead. We have chosen the format of propositions to emphasise that the views presented are those of the editors. Although based on the contents of the chapters in the Handbook (Hirsch Hadorn et al., 2008), the 15 propositions might not be shared by all authors in all aspects. In order to keep statements and

✉ U. Wiesmann
Centre for Development and Environment (CDE), Institute of Geography, University of Berne, Berne, Switzerland
e-mail: urs.wiesmann@cde.unibe.ch

G. Hirsch Hadorn et al. (eds.), *Handbook of Transdisciplinary Research*,

arguments brief, we have abstained from positioning them in the broad range of relevant discourses and therefore references are not included.

At the same time our propositions highlight the fact that the debate on transdisciplinarity is fairly young and that the process of transdisciplinary research is still being developed. Whether it will ever reach the format of an established discipline with sound paradigmatic foundations is questionable, since transdisciplinarity crosses boundaries between forms of knowledge; taking into account the diversity, complexity, uncertainty and values of issues. However, developing means to deal with the resulting challenges in research is what drives a growing community of transdisciplinary researchers. In this sense our contribution aims at further stimulating the debate on, and development of, transdisciplinarity research.

Our 15 propositions are arranged in three sections (Fig. 29.1). First we take a position on the definition, scope, process and outcome of transdisciplinary research;

| **Scope, Process and Outcomes of Transdisciplinary Research** | |
|---|---|
| Proposition 1 | Definition |
| Proposition 2 | Scope and relevance |
| Proposition 3 | Recursive processes |
| Proposition 4 | Knowledge forms |
| Proposition 5 | Contextuality and generality |
| Proposition 6 | Specialisation and innovation |
| **Dealing with Stumbling Blocks in Transdisciplinary Practice** | |
| Proposition 7 | Participation and mutual learning |
| Proposition 8 | Integration and collaboration |
| Proposition 9 | Values and uncertainties |
| Proposition 10 | Management and leadership |
| Proposition 11 | Education and career building |
| Proposition 12 | Evaluation and quality control |
| **Corner Stones for Enhancing Transdisciplinary Research** | |
| Proposition 13 | Facing the scientific challenge |
| Proposition 14 | Facing the institutional challenge |
| Proposition 15 | Facing the societal challenge |

**Fig. 29.1** Propositions for enhancing transdisciplinary research

then we give hints on how to deal with some of the most persistent stumbling blocks in transdisciplinary practice; and finally, we highlight some challenges ahead and a possible way forward.

## 29.2 Scope, Process and Outcomes of Transdisciplinary Research

In the first section the propositions highlight our understanding of transdisciplinary research. In the following sections, this understanding forms the basis on which we address the question of how to enhance this mode of research.

**Definition:** Transdisciplinarity is, on the one hand, rooted in the rise of the so-called knowledge society, which refers to the growing importance of scientific knowledge in all societal fields. On the other hand, it acknowledges that knowledge also exists and is produced in societal fields other than science. The difference is that systematisation leading to specialisation is more pronounced in science than in other societal fields. Transdisciplinary research focuses on the links between different sciences and between science and other parts of society. This leads to the following general definition.

**Proposition 1:** Transdisciplinary research is research that includes cooperation within the scientific community and a debate between research and the society at large. Transdisciplinary research therefore transgresses boundaries between scientific disciplines and between science and other societal fields and includes deliberation about facts, practices and values.

**Scope and relevance:** Transdisciplinary research has arisen from a growing number of complex problems in the life-world for which knowledge-based solutions are sought but for which knowledge of a single scientific discipline or societal field is insufficient. This leads to the following proposition on the scope, definition and relevance.

**Proposition 2:** Transdisciplinary research is an appropriate form of research when searching for science based solutions to problems in the life-world with a high degree of complexity in terms of factual uncertainties, value loads and societal stakes. Through bridging different scientific and social knowledge components it can significantly improve the quality, acceptance and sustainability of such solutions. However, deliberation about facts, practices and values are ongoing when bringing results to fruition in the life-world as well as in scientific communities.

**Recursive processes:** In a transdisciplinary research process, determining the problems involves making fundamental decisions about what aspects are seen as important and what constitutes disputed ground. Furthermore, decisions must reflect the uncertainties in the knowledge surrounding the problems. These challenges can be addressed by restructuring problems and correcting assumptions in the course of research. Therefore transdisciplinary research requires a research design that is basically recursive. The following proposition deals with this key-characteristic of transdisciplinary research processes.

**Proposition 3:** Transdisciplinarity implies that the precise nature of a problem to be addressed and solved is not predetermined and needs to be defined cooperatively by actors from science and the life-world. To enable the refining of problem definition as well as the joint commitment in solving or mitigating problems, transdisciplinary research connects problem identification and structuring, searching for solutions, and bringing results to fruition in a recursive research and negotiation process. Transdisciplinarity thus dismantles the traditional sequence leading from scientific insight to action.

**Knowledge forms:** Transdisciplinary research analyses complex empirical questions (systems knowledge), it aims at determining goals for better dealing with problems (target knowledge) and investigates how existing practices can be changed (transformation knowledge). Transdisciplinary research takes interrelations between knowledge forms into account and includes heterogeneous sources of knowledge by iteratively integrating knowledge components and forms. The following proposition deals with this further key-characteristic of transdisciplinary research processes.

**Proposition 4:** In relation to the nature of problems that are addressed in transdisciplinary research, the canon of participating disciplines and competences from the natural, technical and social sciences, and the humanities, as well as from the life-world cannot be pre-defined. It is to be determined during the research process which bodies of knowledge have to be integrated to take into account, produce and integrate systems knowledge, target knowledge and transformation knowledge.

**Contextuality and generality:** Turning to the outcomes of transdisciplinary research, we see that the nature of the problems addressed has far reaching consequences. In terms of concrete results and problem-solving contributions, these must be contextualised and go beyond counselling by developing transmissible knowledge. This tension is addressed in the following proposition.

**Proposition 5:** Transdisciplinary research is by necessity shaped by concrete problem contexts and related societal settings and its results are basically valid for these contexts. However, taking into account the prerequisite of contextualisation, transdisciplinary research also aims at generality by providing insights, models and approaches that can be transferred to other contextual settings after careful validation and adaptation.

**Specialisation and innovation:** Transdisciplinary research is committed to developing state-of-the-art scientific innovation at the interplay between transdisciplinary and disciplinary research. It is driven by the tension between specialisation in transdisciplinary methods and the triggering of transformation of disciplines. Therefore transdisciplinary researchers form a mixed college of peers. This tension leads to the next proposition.

**Proposition 6:** The quality of transdisciplinary research is bound by sound conceptions of integration and thus requires development of an own form of specialisation. However, transdisciplinary research is not meaningful without sound disciplinary contributions and it has the potential to stimulate innovation in participating disciplines. Bringing this potential to fruition requires an emerging college of peers able to bridge disciplinary and transdisciplinary specialisation.

The six propositions on the definition and scope (1 & 2), the process (3 & 4) and the outcomes (5 & 6) of transdisciplinary research form the basis and reference on which we deal with stumbling blocks and the way forward in the following sections.

## 29.3 Dealing with Stumbling Blocks in Transdisciplinary Practice

Given the general characteristics outlined in the above propositions and the experiences presented in the case studies of this handbook, some stumbling blocks of transdisciplinary practice seem to be common and persistent. Therefore, we briefly qualify these stumbling blocks and postulate proposals on how to deal with them.

**Participation and mutual learning:** Collaboration between science and society in transdisciplinary research implies participatory processes. At the same time participation is one of the major stumbling blocks in transdisciplinary practice. Neglecting the diversity of goals, values, expectations and related power constellations in both society and science exposes the danger of purely symbolic participation, which results in solidifying roles and positions with low innovative potential of transdisciplinarity. This neglect can lead into diffuse 'all-inclusive' processes in which positions, roles and contributions lose credibility to the extent that all major stakeholders in society and science begin to doubt the relevance of participatory processes and lose interest. These considerations lead to the following proposition:

> **Proposition 7:** Participatory processes in transdisciplinary practice require carefully structured, sequenced and selected negotiations and interactions. The different resources, goals and values at stake and their social representation in society and science need to be considered. Building on approaches of mutual learning that bridge roles and positions without dissolving them is a promising entry point to goal oriented participation.

**Integration and collaboration:** Closely related to participation is another core feature of transdisciplinary research – collaboration to integrate the perspectives and knowledge of various disciplines and stakeholders. The challenge of collaborative integration that ideally starts with problem definition and continues throughout the whole research process is again bound to a wide range of difficulties in transdisciplinary practice. These may turn into stumbling blocks, in particular, when efforts are limited to communicative action only, or when so-called synthesis processes are postponed to the end of the research process. Major difficulties also arise if integration is delegated to one of the participating disciplines only, or if integrative concepts are too stringently designed and do not leave room for participating disciplines and researchers to manoeuvre. These stumbling blocks can be overcome by considering the following proposition when planning collaboration and conceptualising integration in concrete transdisciplinary endeavours.

> **Proposition 8:** Collaborative efforts of integration have necessarily to take into account the recursive nature of transdisciplinary research. Combining different means of integration, i.e. developing joint theoretical frameworks, applied models, and concrete common outputs – in an iterative or circular process – has proven to be particularly successful. At the same time, transdisciplinary work should be organised in a manner that enables a productive balance between structured collaboration and vested interests by participating partners and disciplines.

**Values and uncertainties:** Dealing with values and uncertainties is one of the core difficulties in transdisciplinary research, practice and related capacity development. In many cases this turns out to be one of the most important stumbling blocks. The differing and often conflicting values of participating researchers and stakeholders are most decisive in recursive transdisciplinary problem definition. Beyond that, they strongly influence the design and process of transdisciplinary endeavours, i.e. who is included or excluded – as well as the interpretation and application of outputs and outcomes. In addition, ontological and epistemic foundations of participating disciplines are strongly value-loaded. If these value dimensions are hidden or neglected, transdisciplinary collaboration may turn out to be largely superficial or driven by power-constellations representing underlying values. Closely related are the uncertainties that stem from the nature of the problems addressed, the respective limitations of involved system knowledge, as well as the conflicting value-loads influencing all stages of transdisciplinary processes. In this respect it also important to recall that transdisciplinary research is basically bound to socio-political contexts, giving rise to uncertainties concerning the validity of outcomes beyond these contexts. If not actively dealt with, these uncertainties may lead to very diffuse, unnamed and disputed outcomes, or – even worse – to over-interpreted and over-generalised results.

**Proposition 9:** In order to give sufficient attention to values and stakes at all stages of transdisciplinary processes, collaborations and negotiations should be dominated by a mutual learning attitude, not by positions. This is best promoted by adequate time allocation, by creating broad ownership of the problems and by building value-consciousness through reflexive processes among researchers. Reflexivity is also the core means for dealing with uncertainties and the outer boundaries of knowledge resulting from transdisciplinary endeavours.

**Management and leadership:** Project and management structures in transdisciplinary endeavours tend to become complex and overloaded as transdisciplinary research, by default, includes a range of partners and institutions. Due to the recursive nature of problem definition and research, participating institutions and disciplines may vary over time, making management and leadership even more challenging. At the same time, the management and leadership of transdisciplinary processes is often caught in a fix. On the one hand, it is a matter of dealing with the production pressure, which results from the complex and therefore costly projects forced to compete with disciplinary research on the science market. On the other hand, adequate time, space and resources must be provided for mutual learning and recursive research processes to take place. This basic conflict is heightened by the fact that the social reference and control system of participating researchers and stakeholders is anchored within their home institutions and not within the transdisciplinary team. The following proposition indicates some points that may help management of transdisciplinary research to deal with the basic squeeze between high internal and external expectations and the low formal steering powers.

**Proposition 10:** The leading of transdisciplinary projects primarily implies finding a satisfactory balance between periods of intense collaboration with clearly defined joint outputs and periods where deepened disciplinary and multi-disciplinary contributions can be elaborated. This balance of periods is best supported by management services that simultaneously ease administrative tasks for participants, provide clearly structured and timed means of communication, integration and reflexivity, and support internal and external recognition of all contributors, i.e. through providing access to extended peers.

**Education and career building:** When considering education, training and career building in and for transdisciplinary research, an initial stumbling block is the misconception of transdisciplinarity as an additional or new discipline. This position neglects the fact that transdisciplinarity is a specific form of research collaboration and of integrative efforts, and therefore is rooted in the participating disciplines. A second problem arises when training and education is designed in a way that reduces transdisciplinarity to communication and social interaction. But – although very important – communication skills and social competence alone do not make transdisciplinarity possible. Competences that relate to reflexivity on disciplinary and interdisciplinary methodologies or to conceptual and theoretical skills are equally important. However, the most common stumbling block in education, and in particular in respective career building, is related to the conflicting reference systems to which researchers are exposed in transdisciplinarity research: that of their own scientific discipline, of the interdisciplinary research context, and of the society concerned. For the individual researchers who aim at building their career, e.g. at the PhD or the post-doc level, this leads to tensions that are difficult to deal with and that may result, either in withdrawal from transdisciplinarity or in amateurish scientific transgression.

**Proposition 11:** Transdisciplinary training and education is best developed in close connection with the disciplines of origin. Besides building communication and collaboration capacities through practical exposure, emphasis should be put on reflexivity and on methodological, conceptual and theoretical skills that enable the exploration of boundaries and connections between disciplines. Related career building can be supported by careful planning and sequencing of outputs targeting the reference systems of the original discipline and the enhanced peers of transdisciplinarity.

**Evaluation and quality control:** When all the above points are taken into account it is obvious that external evaluation and internal quality control of transdisciplinary research are great challenges and may turn out to be stumbling blocks. In an increasingly competitive scientific environment and in a science-critical society, independent evaluation is crucial for strengthening high quality transdisciplinary research. If respective assessments only refer to the frontier of knowledge in one or several disciplines and do not respect the integrative and context-specific achievements of such research, transdisciplinarity will be discredited by default. Transdisciplinarity may also be discredited by poor outputs and outcomes stemming from a lack of internal quality control. It is important to note that discourses and procedures related to quality control are often hindered or even refused consideration so as not to transgress into other partners' fields of competence and assigned roles.

**Proposition 12:** Evaluation of transdisciplinary research has to go beyond traditional reference systems. It should include qualifying integration and collaboration of disciplines and stakeholders, the recursive design of the research process, and the way the project is based on, and can provide input to, scientific knowledge and societal problem handling. In order to strengthen internal quality control efforts, researchers should concentrate on finding the delicate balance between respecting specific competence and transgressing them in constructive and critical dialogue within transdisciplinary teams.

## 29.4 Corner Stones for Enhancing Transdisciplinary Research

In the previous section we mentioned some important difficulties in transdisciplinary practice and we gave some hints on how to deal with them. We have pointed to the most important corner stone for further enhancing transdisciplinary research, namely improving its practice and thereby enhancing its credibility in science and society. However, this will not be sufficient, as additional challenges have to be faced at a minimum of three more levels, which we will briefly discuss: the scientific, the institutional and the societal challenge.

**Facing the scientific challenge:** A large part of the scientific community still perceives transdisciplinary research, at worst as a semi-scientific application of several disciplines or, at best as a promising new meta-discipline. These positions overlook the fact that transdisciplinarity is a mode of research that is geared to the science and society interface, aims at knowledge based contributions to life-world problems and is rooted in and builds on the participating disciplines. The effect of the predominant image of transdisciplinarity is at least threefold. Firstly, it hinders conceptual and methodological development and innovation at the scientific and science–society interfaces. Secondly, it prevents the development of acknowledgement and reference systems between disciplines and transdisciplinary endeavours. Thirdly, it promotes fraudulent labelling against the background of the novelty image of transdisciplinarity.

**Proposition 13:** Good and concrete transdisciplinary practice must be supplemented by efforts at the levels of its scientific foundations and its scientific recognition. Such efforts must go beyond systematising transdisciplinary research procedures and aim at theoretical, methodological and topical development and innovation at the interface with participating disciplines – to the benefit of both sides. Facing these challenges requires development of extended peer networks and other collaborative networks that bridge transdisciplinary and disciplinary reference and quality control systems.

**Facing the institutional challenge:** Related to its image in the scientific and science policy community, transdisciplinary research occupies a peripheral institutional position in academia. It is often associated with institutions of applied research and demand driven consultancy, or it is packaged in temporarily limited projects or programmes. This peripheral position has the advantage of having a more immediate connection to the science–society interface. However, it has the great disadvantage that it does not promote theoretical and conceptual development and reflexivity, and even worse, that it is unlikely to stimulate innovative synergies in disciplinary research and curricula. One probable reason is that although many researchers and disciplines participate in transdisciplinary endeavours, the core of their reference system perceives this participation as a scientific service rather than as a genuine interest of the respective discipline. Thus, the weak institutional position hinders or may even prevent transdisciplinarity from realising its full potential.

**Proposition 14:** In order to enhance transdisciplinary research, its scientific foundations and its innovative potential for participating disciplines, the institutional position in science and academia has to be strengthened. This means incorporating aspects of transdisciplinarity into research, curricula and career building within established disciplinary institutions, and may include promoting specialist transdisciplinary institutions. The growing network of peers will need to play a key role, allowing transdisciplinary practice to be promoted more pro-actively by the scientific community.

**Facing the societal challenge:** A key task of transdisciplinarity is to address the science and society interface, implying that the role and image of science in society matter, as does the conception of society in science. However, persisting conventions of these roles, images and conceptions conflict with the goal of transdisciplinary research to address life-world problems characterised by a high degree of complexity in terms of factual uncertainties, value loads and societal stakes. If left undebated, these conventions often lead into dead-locks and false expectations in transdisciplinary practice.

**Proposition 15:** Efforts to enhance transdisciplinarity should be accompanied by and embedded into a societal debate on the role of science in society, particularly when dealing with factual uncertainties. At the same time, the scientific community is urged to constantly renew the debate on the role of values and stakes in research. Contributing to solving life-world problems through transdisciplinary research requires science to be conscious and explicit in terms of values and in terms of the boundaries of knowledge and findings – and it requires a corresponding image of science in society.

We, the editors of this handbook, are convinced that transdisciplinary research forms a major avenue for enhancing science based contributions to solve complex problems in the life-world. At the same time we are convinced that transdisciplinarity holds the potential to stimulate innovation in a broad range of disciplines. Enhancing transdisciplinary practice and its scientific foundations is therefore a rewarding challenge. We hope that our 15 propositions, and this handbook as a whole, will encourage many scientists to face this challenge and enter into transdisciplinary practice and debate.

## References

Hirsch Hadorn, G., Hoffmann-Riem, H., Biber-Klemm, S., Grossenbacher-Mansuy, W., Joye, D., Pohl, C., Wiesmann, U. and Zemp, E. (eds): 2008, *Handbook of Transdisciplinary Research. Proposed by the Swiss Academies of Arts and Sciences*, Springer, Heidelberg.

# Index